E. Stahl · K.-W. Quirin · D. Gerard

Dense Gases for Extraction and Refining

Translation from the German Edition
by M. R. F. Ashworth

With 107 Figures and 46 Tables

Springer-Verlag Berlin Heidelberg NewYork
London Paris Tokyo

Prof. Dr. Egon Stahl †
D-6930 Eberbach, Friedrich-Ebert-Str. 29

Dr. Karl-Werner Quirin
D-6638 Dillingen, Dr. Prior-Str. 48

Dr. Dieter Gerard
D-6639 Rehlingen, Lothringer Str. 12

Translator:
Professor Dr. M. R. F. Ashworth
Universität des Saarlandes
Organische und Instrumentelle Analytik
D-6600 Saarbrücken

ISBN-13:978-3-642-72894-5 e-ISBN-13:978-3-642-72892-1
DOI: 10.1007/978-3-642-72892-1

Library of Congress Cataloging-in-Publication Data
Stahl, E. (Egon)
[Verdichtete Gase zur Extraktion und Raffination. English] Dense gases for extraction and refining / E. Stahl, K. W. Quirin, D. Gerard; translated from the German edition by M. R. F. Ashworth. p. cm. Translation of: Verdichtete Gase zur Extraktion und Raffination. Bibliography: p. Includes index.
ISBN-13:978-3-642-72894-5 (U.S.)
1. Gas extraction. I. Quirin, K. W. (Karl-Werner). 1952– II. Gerard, D. (Dieter), 1954–
III. Title.
TP156.E8S7313 1987
665.7'3--dc 19

Softcover reprint of the hardcover 1st edition 1988

2152/3020-543210

Foreword

Dense gases possess dissolving properties for lipophilic compounds which can be utilised for substance extraction and fractionation. This has been known for a long time but it is only recently that it has begun to arouse general interest. There are many reasons for this. First, it is fascinating that simply altering the pressure and/or temperature can alter the dissolving power of the gas. Secondly, carbon dioxide, for example, is available in large quantities and is cheap, non-flammable and neither subject to crises (e.g. political) nor hostile to the environment.

Anyone wishing to learn something about the present state of knowledge of this subject is obliged to study the patent literature (fairly extensive), the widely scattered original publications, the few review articles (some out of date) or heterogeneous symposium reports. There is no compilation devoted entirely to the present state of extraction and fractionation with dense gases. We therefore decided, responding to a request of the Springer Publishing Co., to prepare such a work. We were able to profit from over ten years of practical experience of our research team. The main emphasis has been laid on the possibilities of applying dense gases for obtaining and refining naturally occurring materials. The physico-chemical principles have been treated only in so far as they contribute to a better understanding of the high pressure extraction. Detailed considerations of phase equilibria and calculations have been omitted; authoritative compilations of these are available. It seemed important to us to refer in a special chapter to further, non-extractive possibilities of use of dense gases, e.g. in combating pests or in fluid chromatography.

We have attempted to provide as full a picture as possible but are quite certain that some things have escaped our attention. Accordingly we ask for any suggestions which could be incorporated into a second edition. This book will have fulfilled its purpose if it gives those interested in the use of dense gases a picture of the subject and useful stimulation.

We thank the Friends of the University of the Saar and the "Fonds der Chemischen Industrie" for their support; the publishers for their helpfulness; Mrs. M. C. Atherton for typing; and colleagues of the research group for valuable discussions and suggestions and for providing additional material, some of which had not yet been published.

Saarbrücken, March 1986

Egon Stahl
Karl-Werner Quirin
Dieter Gerard

Contents

Symbols and Abbreviations

Symbols and abbreviations used in the book are listed here. The final disposition of all the symbols is given in the text

A	aroma fraction
AV	acid value
a, b	parameters for intermolecular attraction and repulsion, resp.
bp	boiling point
C	component
CP	critical point
D	diffusion coefficient (cm^2/s); logarithm of the reduction of bacterial count
d	density
E	extract; extract flow; enhancement factor; energy
$E^{1\%}_{1cm}$	extinction of a 1 % solution of thickness 1 cm
F	degree of freedom
f	supercritical fluid
f_i^0	fugacity of the component i in the reference state
g	gaseous; gram
h	hour
hRf	measure of the migrated distance in thin-layer chromatography
IV	iodine value
K	partition coefficient
k	number of the components of a system; capacity ratio
k_w	dissolving coefficient (min^{-1})
L	lower critical end point
LCST	lower critical solution temperature
l	liquid
l_{ij}	correction parameter for molecular interactions
M	mortality percent
min	minute
m_{rel}	relative mass of solvent flow (g solvent per g sample)
n	refractive index
p	pressure (bar)
p_c, p_r	critical pressure; reduced pressure
ppb	parts per billion
ppm	parts per million
P	phase
PV	peroxide value
Q	quadruple point

r	correlation coefficient
R	universal gas constant; residue; reflux
S	stabiliser, solubility (mg substance per g solvent)
SM	solid matter
s	solid; entropy (kJ/kg K)
SV	saponification value
T	absolute temperature (K)
T_c, T_r	critical temperature; reduced temperature
TP	triple point
t	temperature (°C); tonne
t_r	time of retention
U	upper critical endpoint
UCST	upper critical solution temperature
V	isochore; molar volume
vol	volume
wt	weight
x	mole fraction; concentration of a component in the condensed phase
x_t, x_b	concentration of a substance in the top fraction, bottom fraction, resp.
x_f	feed fraction
y	mole fraction; concentration of a component in the gas phase, in the solvent
Y	yield
Z^0	compressibility factor at the critical point
α	separation factor
α_D	specific rotation
η	dynamic viscosity
ϱ_c	critical density
γ	activity coefficient
φ	fugacity coefficient

Abbreviations of Parts of the Installations

B	buffer volume
BP	back pressure regulator
C	compressor, condenser
D	safety disc, jet capillary
E	extractor, equilibrium cell
ES	extract separator
F	filter
FC	fractionation column
FM	flow meter
G	gas supply
H	heat exchanger
I	integrator
L	level indicator
M	manometer, mixing device

MP	metering pump
MV	metering valve
P	pump
R	recorder, switching relay
RC	regeneration column
RH	resistance heating
RS	raffinate separator
RV	reducing valve
S	separator
SC	separation column
T	T-piece
V	valve
VP	vacuum pump

Other Abbreviations

CSD	critical solvent de-ashing
DAB	Deutsches Arzneibuch (= German pharmacopoeia)
DDT	dichlorodiphenyltrichloroethane
DGC	dense gas chromatography
DGE	dense gas extraction
DSSE	donor solvent supercritical extraction
DTC	dithiocarbamate
ECD	electron capture detector
E-TG	epoxy-triglycerides
FDA	Food and Drug Administration
FE-TLC	fluid extraction-thin layer chromatography
FFA	free fatty acids
FID	flame ionisation detector
FS	free sterols
GRAS	generally recognised as safe
HC	hydrocarbons
HCB	hexachlorobenzene
HCH	hexachlorocyclohexane
HPLC	high pressure liquid chromatography
H-TG	hydroxy-triglycerides
LSE	liquid solvent extraction
MEK	methyl-ethyl-ketone
MPI	Max-Planck-Institute
NCB	National Coal Board
NRTL	non-random two liquid
NSI	nitrogen solubility index
PAA	polyacrylamide
PCB	polychlorinated biphenyls
PE	polyethylene
PMMA	polymethylmethacrylate
PP	polypropylene

PS	polystyrene
PTFE	polytetrafluoroethylene (teflon)
PUFA	polyunsaturated fatty acids
PVC	polyvinyl chloride
REM	raster electron microscope
ROSE	residual oil supercritical extraction
RST	regular solvent theory
S	sterol
SE	sterol esters
SFC	supercritical fluid chromatography
SGE	supercritical gas extraction
TG	triglyceride
TLC	thin layer chromatography
UNIQUAC	universal quasi-chemical theory
USDA	U.S. Department of Agriculture
USP	U.S. Pharmacopoeia
WAF	water- and ash-free
WE	wax ester

I. General Picture of Separation Procedures

I.1. Principles of Substance Separation

The basic idea of all procedures of separation of a mixture of different substances is to cause it to yield at least two substances of differing properties by adding an additional substance or supplying energy to it in a suitable apparatus. The choice of separation procedure is primarily determined by whether the mixture is heterogeneous or homogeneous. If it is heterogeneous, mechanical procedures can separate it into its various phases, e.g. filtration, centrifuging or pressing. To separate homogeneous mixtures, one must profit from the differing physico-chemical properties of the individual components. A separation of phases is obtained in the apparatus by adding an auxiliary material or energy. The physico-chemical procedures of separation are generally divided into those determined by equilibria and those by speed. No further consideration is given here to the latter, such as molecular distillation, ultrafiltration, electrodialysis or gas diffusion. Reviews of these procedures are to be found elsewhere [1, 2]. In the physico-chemical separation, procedures determined by equilibria, a state of equilibrium is set up between two immiscible phases, the compositions of which differ from each other. A selection of such procedures of separation is given in Table 1.

In some of the procedures in Table 1 the components to be separated from the mixture are taken up with an auxiliary substance and thereby removed from the mixture. These can be regarded as extraction procedures. The differences are to be sought chiefly in the thermodynamic state of the auxiliary material used. Fig. 1 gives a simple schematic idea of this by portraying the p(T) diagram of an auxiliary.

In adsorption, a solid material is used as "extraction agent"; in absorption, a stationary liquid auxiliary is used to separate gas mixtures. In classical solvent extraction a mobile liquid phase is introduced which is not miscible with the material to be extracted, whether solid or liquid. In distillation with a carrier gas ("stripping") the auxiliary material is gaseous. The procedures of steam distillation or azeotropic or extractive distillation work on the liquid-vapour phase boundary of the auxiliary.

Extraction with dense gases is an extension of the domain of classical extraction. The thermodynamic state of the auxiliary material is then characterised by the pressure and temperature values in the region of the critical point. Separation with dense gases has some of the characteristics of distillation but more of classical extraction. Zosel [4] thus suggested the term "destraction" but this was unable to gain a foothold because an auxiliary material is needed for the separation and it thus counts rather more as an extraction procedure. Extraction using dense gases is based mainly on the different

Table 1. Physico-chemical methods of separation determined by equilibria

Procedure	Mixture to be Separated	Separation Medium	Principle of Separation	Example
Distillation	Liquid	Heat energy	Differences in vapour pressure	Ethanol-water
Extractive and azeotropic dist.	Liquid	Heat energy + auxiliary (liquid)	Differences in vapour pressure	Obtaining anhydrous acetic acid with using dichloroethene
Steam distillation	Solid, liquid	Heat energy + auxiliary (water)	Differences in vapour pressure	Obtaining essential oils
Stripping (distillation with carrier gas)	Liquid	Gas	Differences in volatility	Removal of volatile matter from fatty oils
Extraction:				
a) solid-liquid	Solid	Solvent	Different solubilities in the solvent	Extraction of oil seeds with hexane
b) liquid-liquid	Liquid	Solvent (not miscible with sample)	Different solubilities in the liquid phases	Extraction of acetic acid from aqueous solution with ethyl acetate
Extraction with dense gases	Solid, liquid	Dense gas	Different solubilities in the solvent, differences in volatility	Preparation of extracts of spices
Absorption	Gas	Liquid	Different solubilities in the liquid	Scrubbing of CO_2, H_2S from natural gas using ethanolamines
Adsorption	Gas + liquid or 2 liquids	Solid	Different tendencies to layer formation on boundary surfaces	Drying of gases or liquids with blue silica gel
Chromatography	Solution of various substances	Solid or liquid (stat. phase) + liquid (mobile phase)	Multiplicative adsorption/partition	Thin-layer chromatography
Crystallisation	Liquid	Removal or supply of heat energy	Differences in solubility, differences in melting point	Obtaining sugars from solutions
Drying/freeze drying	Moist solid/ice and solid	Heat energy	Differences in volatility of solid and water; evaporation/sublimation of water	Dehydration of foods

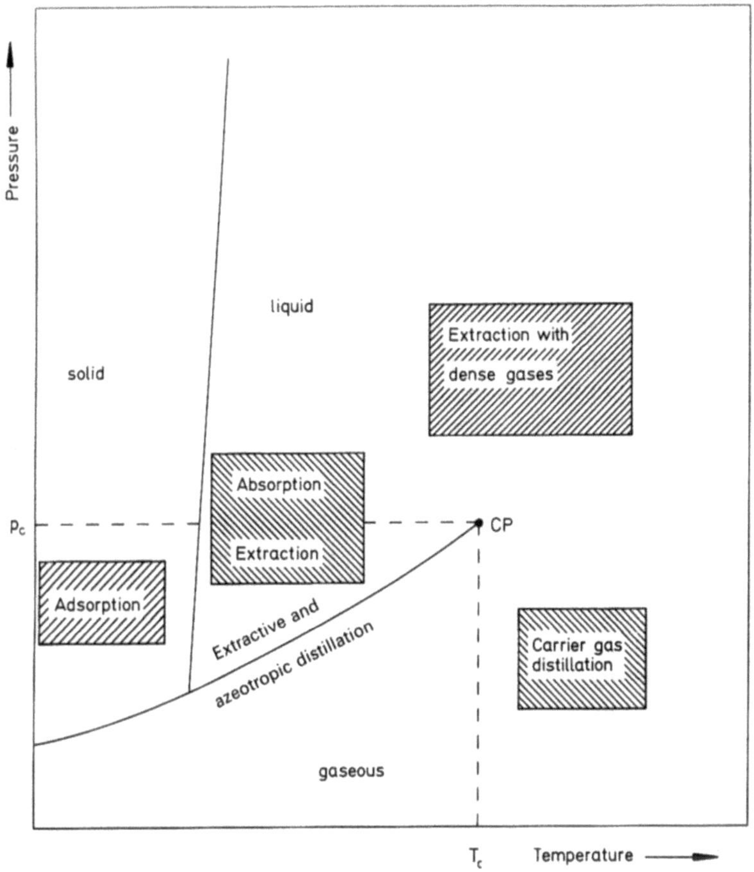

Fig. 1. The thermodynamic state of an auxiliary material in separation procedures [3]

solubilities in the extraction agent of the components to be separated, even though the vapour pressures of the compounds have a distinct influence on the separation, especially in the isolation and refining of easily volatile essential oils and aroma substances. The "volatility" of the components to be separated, i.e. the transition into the dense gas phase, is thus a function of pressure and temperature, whereas in distillation it depends largely on the temperature.

I.2. Conventional Methods of Extraction

The quality of the product isolated is the principal measure of the value and efficiency of an extraction procedure; naturally the native chemical composition of the components should remain unchanged. Preliminary concentration often has to precede the real extraction procedure with fresh plant material which contains from 60 to 90% water. Such a preliminary procedure should be only non-drastic drying in order to avoid chemical changes of the components. The isolation of components from natural products is a problem that has not yet been entirely satisfactorily solved, especially with sensitive natural materials and highly volatile oils. The influence of temperature, oxygen, length of treatment etc. in thermal procedures generally leads to more or less marked changes of the substance mixture obtained.

Steam Distillation

Procedures purely of distillation have found little application for isolation of natural products and are not discussed further here. In contrast, distillation with the help of an auxiliary material is rated highly. Steam distillation is the oldest and, up to the present day, still the most important method for obtaining essential oils, i.e. the characteristically smelling volatile oils contained in plant material. These relatively volatile oils with boiling points in the range 150 to 250 °C are separated from other material with the help of the carrier, steam. The oil/water ratio in the distillate, which determines the duration of the distillation, is fixed through the vapour pressure of the essential oil components at the boiling point of the mixture of auxiliary material and sample.

In steam distillation the starting material is subjected to a temperature of ca. 100 °C; this can lead to artefacts of the essential oil constituents which are often thermolabile. In addition, the water can exert a hydrolytic influence, bringing about chemical changes of the oils. Many steam distillations are accompanied by formation of decomposition products, such as hydrogen sulphide, ammonia, acetaldehyde or acetic acid; these are mostly water soluble but may partly enter the oil and render subsequent refining necessary.

Steam distillation under reduced pressure has the advantage of a lower temperature and thus yields a higher oil quality. A disadvantage is the unfavourable oil/water ratio in the distillate, especially with less volatile oils, which can lead to long distillation times. Vacuum steam distillation is used more for refining crude oils than for obtaining essential oils. Distillation at pressures above atmospheric, using superheated steam, is sometimes applied to plant material with oils which are difficult to distil. The oil/water ratio in the distillate is then more favourable than under normal pressure. However, the higher temperature leads to larger amounts of decomposition products.

Classical Solvent Extraction

Extraction with classical organic solvents is an important procedure for obtaining lipophilic plant components. Extracts prepared from dry material using this technique are usually termed oleoresins or resinoids; those from fresh plant products, of special importance in the perfume industry, are named "essences concretes". Classical solvents display poor selectivity so that the desired components often contain interfering side products. An undesirable dark colour impairs the possibility of use of many oleoresins. Solvent extracts have often a viscous to wax-like consistency. A refining stage is therefore necessary before application.

As far as its properties are concerned, the choice of a solvent for extracting natural materials demands a compromise; all the technical steps of extraction, separation and recovery of the solvent from the solution of extract (miscella) and from the extraction residue, have to be considered.

The *selectivity*, i.e. the ability to dissolve the desired product to a greater extent than the other constituents of the mixture, can be varied to a certain extent through the nature of the solvent or solvent mixture. A suitable solvent must also possess a high *capacity*, that is, a marked ability to take up the desired substances. These two properties are, however, usually mutually exclusive.

A further, very important property is the chemical *non-reactivity* and *stability*. An ideal solvent should be inert towards the extract and residue and stable towards influences of the environment, such as heat, light, air and water. It should also have no corrosive action on the installation.

The influence of the factors of *density*, *viscosity* and *surface tension* must be considered individually. In solid-liquid extraction, low viscosity and surface tension increase the speed of percolation and the yield of extract since the passage of the solvent through narrow capillaries is thus facilitated and the extraction accelerated. For liquid-liquid extraction two immiscible solvents must be found, each of which can dissolve the desired components. The difference in density between the phases and the surface tension values of both should be as large as possible to give a fast phase separation. If the surface tension is low and the density difference small, stable emulsions can be formed which are difficult to break. Solvent viscosity should also be low, just as in solid-liquid extraction, so that mixing and mass transport are faster. Both types of procedure have in common that low density and viscosity keep the pressure drop in flowing systems at a low value and thus reduce the energy required to transport the solvent.

The *boiling point* should be as low as possible so that extract and residue are not subjected to excessive heat during recovery of the solvent. This separation is often the most expensive part of the whole procedure. The *specific heat* and the *latent heat of evaporation* should not be too high so as to save energy during heating and evaporation of the solvent.

Ideally the solvent should not be flammable or at least have as high a *flash point* as possible and the narrowest possible *explosion range* of mixtures with air. The *MAK* value (maximal allowable concentration at the place of work) [5] of the extraction agent should be as high as possible; the health of those working in the establishment is then not endangered and the environment is not impaired by solvent vapours.

An important procedure for exhaustive extraction of plant material is percolation; dense gas extraction of natural substances is carried out according to this principle. The extraction process is then determined by the properties of both the sample to be extracted and the solvent. The properties of the ideal solvent have been already discussed above. Since the relationships are complex, the features of the drug material which play a part are discussed phenomenologically in the main.

The *degree of comminution* of a sample has a considerable influence on the extraction. The constituents of broken up plant cells are readily accessible to the extraction agent and need only to be washed out; those of intact plant cells, however, are dissolved out only by diffusion. Washing out is a faster process than diffusion, so that, after a certain time, pure diffusion prevails [6]. If the extraction yield in percolation is plotted against the course of the procedure, i.e., against the amount of solvent passing through, it is seen that the extraction process can be divided into three phases (Fig. 2).

At first the extraction curve climbs almost linearly; the concentration in the solvent stays roughly constant in the washing out phase (I). After this stage the concentration of extract in the solvent falls rapidly (II), which indicates the end of the washing out process. In phase III of the extraction curve the extracted amount increases only

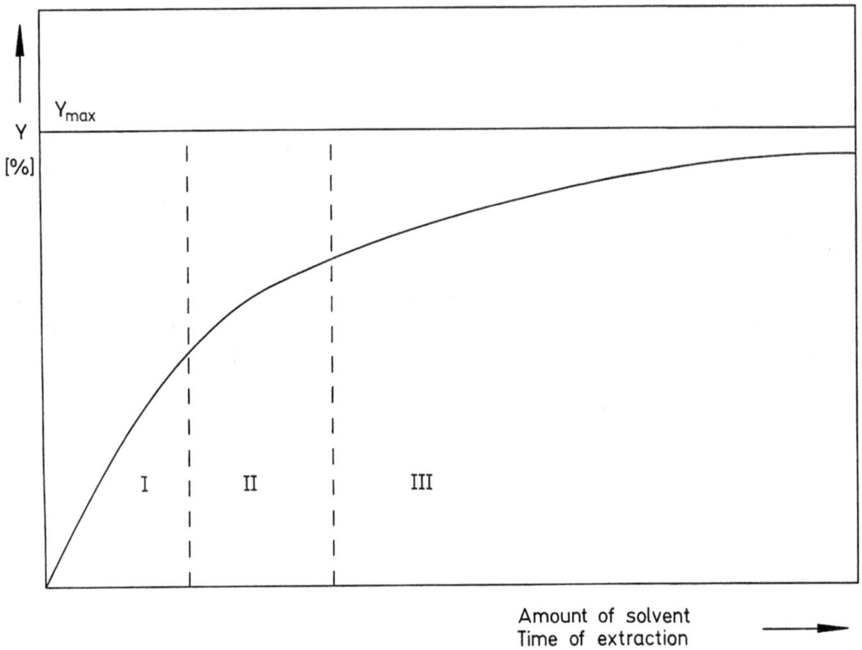

Fig. 2. Course of extraction of natural plant material; y = yield in %; y_{max} = maximum possible yield; I = washing out phase; II = transition from washing out to diffusion; III ≙ diffusion phase

slightly with time, i.e. amount of solvent. In this phase the extraction process is due to pure diffusion, which is then the rate-determining stage of this procedure. Melichar [7] stated that the degree of comminution itself does not influence the concentration ratio of miscella and the drug sample. But it is important for the course of the extraction, determining the point of transition from washing out to diffusion and hence the duration of extraction. If the plant material is compact, the yield-determining factor is the grain size, when it is not possible to tolerate especially long extraction times as a result of slow diffusion processes.

The *swelling* of plant material in the solvent is highly significant for the extraction because it widens the cell capillaries (increase of porosity) and thereby augments diffusion. This reduces the duration of extraction or may even make an extraction possible. The *water content* of the sample can be important in this connection. An impressive example is the removal of caffeine from crude coffee [8] using supercritical carbon dioxide. The extraction of caffeine from whole raw beans succeeds in a reasonable time only when the coffee material is moist. Another example is the removal of nicotine from crude tobacco with supercritical gases; it depends markedly on the moisture content of the tobacco [9]. These examples show that a certain moisture content exerts a positive influence on the extraction. However, the water content of the sample should not have a negative influence on the wetting properties of the solvent.

If the substances to be extracted are chemically bound, the sample may have to be subjected to *preliminary treatment*. Thus alkaloids often occur as salts and are difficult to extract with lipophilic solvents. Treatment of the drug with alkali liberates the free bases which are than easily extractable.

An additional factor of influence is *adsorption* of substances, already dissolved, on the skeletal matter of the sample [7]. This prevents the maximum possible yield from being attained.

A general evaluation of these parameters mentioned is not possible because the behaviour of plant material for extraction is from the very beginning so varied and, in addition, can be influenced by the different extraction agents.

References

1. Ullmanns Enzyklopädie der technischen Chemie, Verlag Chemie, Weinheim 1972
2. Brunner, G.: in Enzyklopädie Naturwissenschaften und Technik, Verlag Moderne Industrie, Landsberg a. Lech 1981
3. Brunner, G., Peter, S.: Chem. Ing. Tech. *53*, 529 (1981)
4. Zosel, K.: Angew. Chem. *90*, 748 (1978)
5. Henschler, D.: Gesundheitsschädliche Arbeitsstoffe, Toxikologisch-arbeitsmedizinische Begründung von MAK-Werten, Verlag Chemie, Weinheim
6. Schultz, O. E., Klotz, J.: Arzneim. Forsch. *3*, 471 (1953)
7. Melichar, M.: Pharmazie *17*, 290 (1962)
8. Zosel, K.: Germ. Pat. 2005293 (Studiengesellschaft Kohle mbH) filed 1970, granted 1974
9. Roselius, W., Vitzthum, O., Hubert, P.: Germ. Pat. 2043537 (Studiengesellschaft Kohle mbH), filed 1970, granted 1975

II. Basic Principles of Extraction with Dense Gases

II.1. Historical Review

The first observations of the occurrence of supercritical phases were made as early as the beginning of the nineteenth century by Cagniard de la Tour [1]. He noted that the gas-liquid phase boundary disappeared when a certain temperature was exceeded. Andrews [2] carried out further studies of dense carbon dioxide in 1869; he was the first to work with binary mixtures of carbon dioxide and nitrogen. The first to demonstrate the solubility power of supercritical gases for solids were Hannay and Hogarth [3]. Their work, published in 1879 and 1880, showed that the solubility of inorganic salts in ethanol displayed no discontinuity when the critical point was passed. The concentration of cobalt chloride, for example, in the supercritical phase exceeded markedly the value to be expected from the vapour pressure according to the Poynting correction [4]. Van der Waals [5] carried out pioneer work in the development of a thermodynamic conception of the supercritical state of one- and two-component systems.

The considerable experimental problems associated with investigations in the region of the critical state at first hindered further intensive research in this domain. The subsequent numerous studies were restricted mainly to setting up phase diagrams, for example of binary mixtures of hydrocarbons [6] in the years after 1930; the practical side which the solubility power of dense gases offered especially for extraction and separation, aroused little interest. On the other hand it was recognised early that critical phenomena played a significant role in the formation of ore deposits and minerals under hydrothermal conditions [7], in the migration of petroleum deposits in the earth's crust under the influence of compressed natural gas [8] and in the formation of deposits of quartz on turbine blades in working with supercritical water vapour [9] which contained dissolved mineral matter. It was also recognised very early that the dissolving capacity of dense gases was determined by the pressure or, more directly, by the density [10, 11]. Booth and Bidwell [12] gave a detailed account of the work on solubility studies in the critical region up to the year 1948.

After vigorous development during the last 30 years the practical side today embraces growing artificial crystals [13], obtaining the viscous, so-called ternary oil from petroleum deposits [14] and also the use and testing in extraction and separation of compounds.

The first suggestions for the use of dense gases as auxiliaries for separation are to be found in the American patent literature as early as 1936. A method is given there for separating "high molecular mixtures" [15]. It is based on a volatile agent (carbon dioxide is mentioned among these) in the region of its critical temperature dissolving in the condensed phase under pressure and thus bringing about separation into two phases. Other proposed patents describe countercurrent extraction of petroleum with dense gases with a view to separating into fractions containing asphalt and free of it [16]; or aim at purifying fatty oils according to the same principle [17]. Groll suggested fractionating natural and synthetic oils by first taking up the mixture in a normal solvent and then modifying its solvent capacity by adding a gas under supercritical conditions [18].

These patents were practically ignored. Then, at the end of the fifties, Zhuze in the USSR described also a procedure for fractionating crude oil, extracting earth waxes [19] and obtaining lanolin from wool fat [20], with the aid of dense gases. The real break-through of dense gas extraction was the work of Zosel in the sixties, at the Max-Planck Institute for coal research [21, 22] which provided the incentive for extensive further research. The subsequent rapid development of this method of separation in various domains led to the first symposium in 1978 in Essen, under Stahl's stimulus, devoted entirely to the theme "Extraction with supercritical gases" [23]. In the meantime symposia have been held all over the world, almost yearly, on the physico-chemical, engineering and practical aspects of the new extraction method. Reviews of extraction with liquefied and supercritical gases with various main themes are now to be found widely in the literature, for example from: Paul and Wise [24]; Hicks and Young [25]; Irani and Funk [26]; Williams [27]; Reid [28]; Randall [29]; Paulaitis et al. [30]; and Ely and Baker [31].

The known procedures for application of dense gases for separation of substances have hitherto included use with coal and petroleum products, the vast domain of extraction of natural products and special applications, such as desalting of sea water, separation of ethanol-water mixtures and regeneration of adsorbents. The active industrial interest in the new procedure is documented by the comparatively large part of the patent literature which deals with high pressure methods in the last few years. In the field of natural materials, the central theme of this monograph, carbon dioxide has been used on the technical scale as solvent for: removing caffeine from crude coffee beans (HAG AG, Bremen) and obtaining hop extracts (HEG Hopfenextraktion GmbH, Münchsmünster)/SKW Trostberg AG; Hopfenextraktion HVG, Barth, Raiser and Co. Wolznach; Pauls and Whites International Ltd., England; Steiner Hops Ltd., England; Carlton and United Breweries Ltd., Australia all in competition with the classical methods.

At the moment energetic afforts are being made to commercialise extraction with dense gases in the realm of aromas, spices and essential oils.

II.2. Definition of Terms

As already stated above, extraction methods can be classified according to the thermodynamic state of the solvent. In extraction with dense gases an agent is used, the critical temperature (T_C) of which is in the vicinity of the operation temperature of the process. The state of the gas, established by the factors of pressure and temperature, is characterised by a density value which is of the order of magnitude of those of ordinary liquid solvents. The pressure needed for this is not the only condition and determining factor for the dissolving ability of the gas but rather the presence of a certain density. It is therefore reasonable to use the term "dense gas extraction" and this is increasingly gaining favour over other terms, such as gas extraction, destraction, fluid extraction, high pressure extraction etc.

In order to characterise the procedure it is unimportant whether the gas is in the liquefied or supercritical state because the properties of the dense gas, and hence the effects connected with these, change without discontinuity when the critical temperature is passed. The expressions "liquefied" or "supercritical" are restrictive and ought therefore to be used only when referring expressly to a particular region of state.

The state of the auxiliary in the extraction stage is shown in a p(T)-diagram in Fig. 3 with a line of dashes.

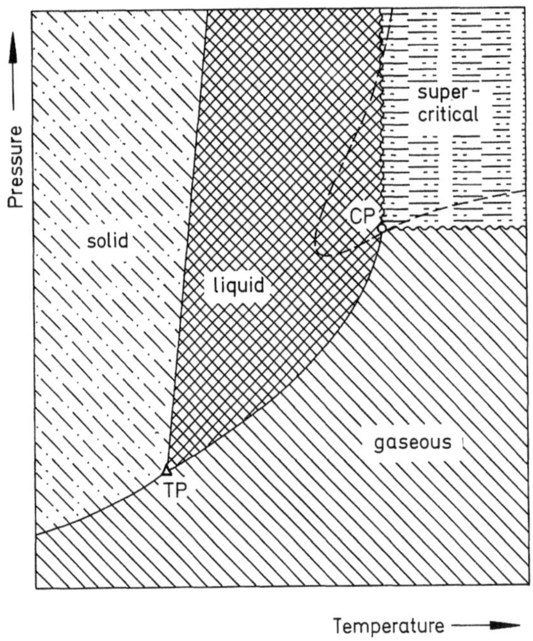

Fig. 3. p(T)-diagram of state of a pure substance (solvent); TP = triple point; CP = critical point

At a temperature below the critical value, in the region of the liquid state, it is usually unnecessary to raise the pressure appreciably above that of the critical point. In this region the increase in density of the liquid phase is only very small and the dissolving capacity barely dependent on pressure. In the same way it is generally uninteresting to use reduced temperatures (T/T_C) greater than 1.1 when working at pressures only just above the critical point; in this region the solvation power of the solvent falls markedly and characteristics of distillation are noticeable.

In accordance with this definition of the solvent, the compressed gas can be dense carbon dioxide (T_C 31 °C) at a working temperature of 20 °C (liquid) or 80 °C (supercritical); or dense toluene "vapours" (T_C = 318 °C) at 300 °C or about 350 °C (as perhaps for extraction of coal) [32, 33]. The reduced temperatures in the separating stage are generally within the range 0.9 to 1.5, whereas the reduced pressures can be between 0.8 and 5, depending on temperature. In individual cases, higher pressures can be advantageous.

II.3. Special Features of Dense Gases

There are two aspects of the use of dense gases for extraction and separation of compounds. On the one hand the aim may be to obtain as much as possible of liquid or gaseous components of a particular material, e.g. coal or wood, under drastic conditions. This procedure is a combination of extraction, pyrolysis and, to some extent, even chemical reaction of the raw material with the solvent; this is of less interest to us. On the other hand valuable chemicals or active principles may be sought from biological carrier material in as gentle and selective a way as possible, resembling an ordinary percolation with a chemically inert solvent. Here dense gases offer an alternative to the usual solvents.

Like liquid organic solvents, dense gases possess the property of dissolving involatile substances at relatively low temperatures. In addition, they display some other positive properties which render them interesting in comparison with conventional solvents.

Mass transportation is fast with dense gases. Although their density (d) is only slightly lower than that of liquids, their dynamic viscosity (η) is nearer to that of the normal gaseous state. The diffusion coefficient (D) of a dense gas is, in the vicinity of the critical point, more than ten times that of a liquid. Table 2 compares some of these important parameters for the different states.

These values are naturally dependent on pressure and temperature. In some cases gases attain densities similar to those of liquids just above their critical temperatures when reduced pressures (p/p_C) of 1–2 are used. If the temperature is clearly above the critical value, correspondingly higher pressures are required. As pressure is increased the viscosity of the compressed gas approaches the corresponding value for a liquid solvent more slowly than does the density. Increase in temperature leads to increase in viscosity of a gas but a decrease in viscosity of a dense gas or liquid, whereas in all three cases, the density falls.

Dense gases are thus in principle better able than classical solvents to penetrate into the material to be extracted and to take up and carry away the soluble components.

Table 2. Orders of magnitude of physical data of gas, dense gas and liquid [29, 34]

	d (g/cm^3)	η (g/cm s)	D* (cm^2/s)
Gas, 1 bar	$(0.6–2) \cdot 10^{-3}$	$(1–3) \cdot 10^{-4}$	0.1–0.4
room temperature			
Dense gas			
T_C, p_C	0.2–0.5	$(1–3) \cdot 10^{-4}$	$0.7 \cdot 10^{-3}$
$\sim T_C$, $4p_C$	0.4–0.9	$(3–9) \cdot 10^{-4}$	$0.2 \cdot 10^{-3}$
Liquid			
room temperature	0.6–1.6	$(0.2–3) \cdot 10^{-2}$	$(0.2–2) \cdot 10^{-5}$

* self diffusion for gas and dense gas, binary mixture for liquids

The pressure drop in the flowing system is small. Dense gases are therefore ideal for separation methods depending on equilibria, such as percolation. This opinion must be modified, however, because in cases where the solvent has only a small capacity for the compounds to be dissolved, mass transfer is usually little favoured.

Dense gases used for extraction can be completely and undrastically removed from the extract and extracted material. In the high pressure extraction the extract solution (miscella) can be caused to separate out by lowering the pressure and/or change of temperature; this is in contrast to liquid solvents which have to be removed by desolventation procedures such as (vacuum) distillation. The energy demand for regeneration is, in many cases, less for extraction with dense gases than for normal extraction with solvents. Since the excess pressure in the installation prevents the entry of oxygen during the extraction, damaging oxidation is excluded; and there is no thermal strain which might lead to rearrangement or decomposition of labile plant constituents. Examples of this are the extractions without decomposition of thermolabile proazulene from chamomile flowers [35] and from wormwood plants [36]; of valepotriates from valerian root [37]; and of unstable sesquiterpene ketones from calamus roots [38], using dense carbon dioxide.

Dense gases possess a relatively high *selectivity*. The gases used for gentle extraction of natural products are typically lipophilic solvents in the dense state, which dissolve neither polar nor polymeric substances. In addition they are weak solvents which rarely attain concentrations of extracted material of more than a few % by weight in ordinary percolation of natural substances. This means that generally more solvent is required for extraction with dense gases than with organic solvents. On the other hand, skilful treatment with dense gases can yield extracts of high content of active principles, e.g. from pyrethrum flowers [39] or valerian roots [37]; or give high-grade products without having to carry out many of the refining stages necessary after classical extraction. The dissolving properties of dense gases can be adapted to the separation problem by variation of the parameters of pressure and temperature in both the dissolving and recovering stages; they are then equally suitable for selective extraction or refining crude extracts.

Addition of a compound of volatility intermediate between that of the gas and that of the condensed phase, a so-called entrainer, to the dense gas can augment appreciably its solvent power; in certain cases the selectivity remains unimpaired or may even increase [40, 41]. By profiting here from a "miscibility gap", there is

Table 3. Physico-chemical data of some gases suitable for extraction of natural products

Gas	T_C (°C)	p_C (bar)	ϱ_C (g/cm³)	b.p. (°C)
Ethene	9.9	50.5	0.20	−103.7
Trifluoromethane	25.9	46.9	0.52	− 82.2
Carbon dioxide	31.0	72.9	0.47	− 78.5 (subl.)
Ethane	32.2	48.2	0.20	− 88.6
Dinitrogen monoxide	36.5	71.7	0.46	− 88.5
Sulphur hexafluoride	45.6	37.7	0.73	− 63.8 (subl.)
Propene	91.9	45.4	0.22	− 47.4
Propane	96.8	42.4	0.22	− 47.1

then a good chance of recovering the extraction agent merely by change of temperature.

Process temperatures in the 0–100 °C range are needed for non-aggressive extraction of natural materials. The gases used as solvent must have critical temperatures within this range. A selection of possible gases is given in Table 3, in increasing order of critical temperature.

Some gases in the table are less suitable for application in larger scale practice. Trifluoromethane and some other possibly suitable fluoro-(halo) hydrocarbons are too expensive or are suspected of destroying the ozone belt of the earth's atmosphere. Dinitrogen monoxide (nitrous oxide) must be used with care because it supports combustion and tends to spontaneous decomposition under certain conditions [42, 43]. The pure hydrocarbons, especially ethane and propane, are suitable for some problems although they are flammable and form explosive mixtures with air. The most important is, however, carbon dioxide; it is interesting that it is isosteric with carbon disulphide, the oldest lipophilic solvent used for extraction.

The use of carbon dioxide as solvent has the advantages listed below:

- it is without risk physiologically. It occurs in sparkling beverages and is an end-product of human metabolism. The GRAS-status (generally recognised as safe) for carbon dioxide in extraction of foodstuffs has been applied for [44].
- it is sterile and bacteriostatic.
- it is neither combustible nor explosive; no protective equipment against explosion or fire is needed.
- it is harmless to the environment and creates no waste products such as gas, water and unused solvent; considerable expense is thus avoided.
- it is available in large amounts under favourable conditions and largely independent of the petrochemical industry; large reserves in liquid form are technically available.

Despite these considerable advantages the industry shows reserve at present towards extraction with dense gases. Objections to their use, especially where know-how is lacking, are the risk of mastering the new technology and considerable investment costs required for working under high pressure; and technological problems, as yet unsolved, in connection with the continuous operation of the procedure. These arguments will lose significance, however, as development progresses and energy and conventional solvents become scarcer and more expensive.

The attractiveness, already existing, of extraction with dense gases, particularly carbon dioxide, is shown by the attempts at industrial application; and interest in the new procedure has been demonstrated by the large number of patent applications in recent years. For obvious reasons an insight into the state of research on the industrial side is only incomplete.

II.4. Thermodynamic Fundamentals

II.4.a Description of the Solvent Circulation

To facilitate an understanding of the extraction procedure using dense gases, the changes of state of the solvent circulated with compressor or pump are illustrated below by means of t,s-diagrams (s = entropy). This representation is important

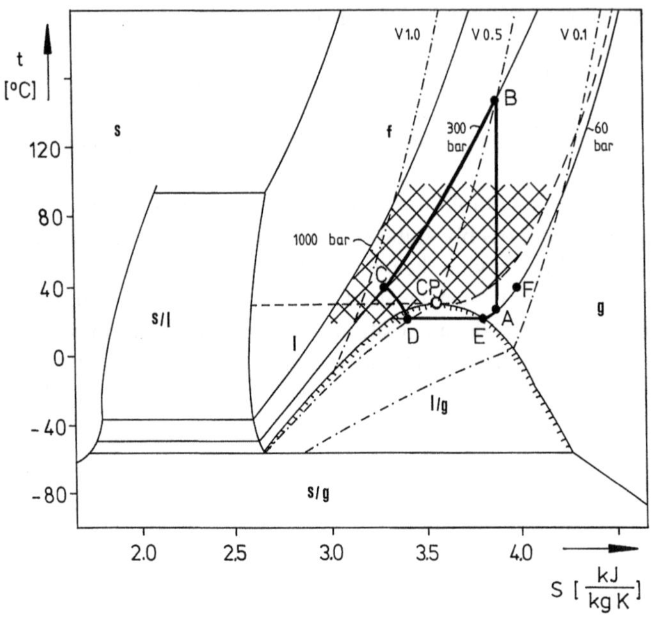

Fig. 4. Extraction circuit (A–F), using a compressor in the t,s-diagram of carbon dioxide; extraction at 300 bar/40 °C, separation at 60 bar/40 °C. AB isentropic compression; BC isobaric cooling; CD adiabatic expansion; DE isobaric, isothermal evaporation; EF isobaric warming; FA isobaric coooling; V isochores with densities in g/cm^3; data from [45–47]

also in order to obtain thermodynamic data, e.g. of the thermal balance of individual steps of the procedure and hence for the design of constructional parts of the plant. Carbon dioxide is taken as the solvent example; it has been studied comprehensively and is particularly relevant.

A larger part of the phase diagram of carbon dioxide is shown in Fig. 4 for orientation. The part of the diagram suitable for extraction of natural products is shaded. It covers the temperature range from 20 to 100 °C and the pressure range from $p_C = 73$ to 1000 bar.

The areas are drawn in on the phase diagram in which the carbon dioxide exists as homogeneously solid (s), liquid (l), gaseous (g) or supercritical phase (f) as well as the two-phase regions s/l, s/g and l/g. The area of the supercritical state is denoted by the line of dashes (which does not represent a phase boundary), distinguishing it from the liquid and gaseous regions. The part of the line to the right of the critical point (CP) agrees with the critical isobar (73 bar). Three additional isobars, corresponding to pressures of 60, 300 and 1000 bar, are drawn in alongside. The dot-dash lines which unite state of equal density (quoted in g/cm^3) are isochores (V). The shaded area to the left of the isochore V 0.5 denotes the region of high solvent power, suitable for rapid separation of lipophilic constituents. To the right of this isochore the solubility power diminishes rapidly as the gas density falls. Fractionation can be achieved in this region by profiting from differences in solubility in the extraction or separation step.

An extraction circulation using a compressor is plotted with the thick line through the points A to F in Fig. 4. From point A, gaseous carbon dioxide at 60 bar and 25 °C, as can be obtained from commercially available steel cylinders, an ideal compression, isentropic and without loss, leads to point B, at 300 bar, for instance. The gas, heated through the compression, is cooled isobarically to the extraction temperature e.g. 40 °C (point C) before being introduced into the extraction vessel. After extraction the gas, laden with extracted material, is allowed to expand adiabatically to a pressure below the critical value, e.g. 60 bar (point D). It thereby cools to the liquefaction temperature and is present as wet vapour. It yields a liquid phase rich in extracted material and a gas phase free of extract. In order to bring about complete separation of the extract, the liquefied part is first subjected to isobaric and isothermal evaporation (point E) and the gas then warmed isobarically to the separation temperature of 40 °C for example (point F). After passing through the separator the gas cools to the ambient temperature (point A) and is then drawn again into the compressor. During this circulation process, differences in density of the carbon dioxide of about a factor of ten arise, especially between the extraction state (point C) and the separation state (point F); they can be estimated with the help of the plotted isochores (V).

Fig. 5 shows the solvent circulation with the same extraction and separation states but using a pump which draws in the liquefied carbon dioxide. An isobaric thermostatisation to the extraction temperature follows the isentropic rise in pressure from A to B or B′. Depending on the increase in pressure and the desired temperature, it may be necessary to cool or warm the dense gas (B — C or B′ — C′, respectively). The transition from the extraction state (C or C′) to the separation state (F) is inde-

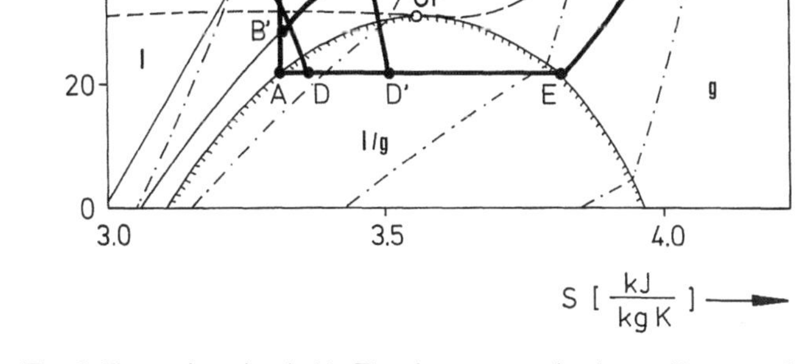

Fig. 5. Extraction circuit (A–F) using a pump in the t,s-diagram of carbon dioxide; extraction at 300 bar/40 °C or 100 bar/40 °C, separation at 60 bar/40 °C. AB isentropic increase of pressure; BC isobaric cooling or warming; CD isenthalpic expansion; DE isobaric, isothermal evaporation; EF isobaric warming; FE isobaric cooling; EA isobaric, isothermal condensation; V isochores with densities in g/cm³

pendent of the pressure generation and traverses the same changes of state as when using a compressor. After the extract has separated (F) the regenerated gas is first cooled to the temperature of liquefaction corresponding to the pressure for separation (E), then condensed in an isobaric and isothermal stage (A) and again introduced into the pump.

The type of pressure generation chosen has a marked influence on the design of the remainder of the installation. Using a compressor, a large fraction of the energy needed for pressure generation is converted into heat which has to be removed in order to attain the extraction temperature. It is possible to reduce the power requirement for the compression and hence the overall warming of the gas by applying pressure in two stages with intermediate cooling. If the temperature of the precompressed gas is brought to a value just above the critical temperature, considerable energy for the compression to the extraction pressure can be saved.

Using a pump, the major part of the energy provided serves for increasing the pressure. This rise in temperature through compressibility of the liquid phase is small in comparison with that when a compressor is used; the amount of heat to be supplied or removed to attain the extraction temperature of, say, 40 °C, is thus also small. However, a considerable amount of heat energy must be removed from the gas in a condenser in order to be able to liquefy the gas which is at the separation pressure (= suction pressure of the pump). Further, it is sensible to supercool the gas at the phase boundary l/g, i.e. in the boiling state. When the pump then applies suction, its function will not be impaired as a result of evaporation through pressure loss (cavitation). Since the density of the drawn-in solvent, using a pump, is about ten times that when using a compressor, the conveying capacity of a pump is correspondingly more effective or its size can be much less for the same performance.

The presentation of the course of the process using a pump or a compressor is simply one example and refers to carbon dioxide as extracting gas. Different aspects and other courses of the process may be encountered, depending on whether other gases are used, the separation of the extract is in a single stage or multistage through pressure reduction, temperature increase or lowering, or a combination of changes of temperature and pressure.

The ideal procedure is also found only to a limited extent in practice. The individual steps are not reversible; the performance of the constructional items responsible for the changes of state (compressor, pump, heat exchanger with suitable units for supply) are accompanied by heat losses and have a definite efficiency which cannot be taken into account here. Further, the physical data for pure carbon dioxide have to be partly corrected in consideration of the charged states in reality.

II.4.b Considerations of Phase Equilibria

Extraction with dense gases is a procedure which for separation of materials takes advantage of phase equilibria in a system under definite conditions in the same way as distillation or classical extraction with solvents. As a rule the system consists of at least three components, the dense gas (which performs the function of an auxiliary) and the two substances to be separated. The separation procedure is based on a (quasi-)binary equilibrium in cases where the material to be separated is a pure substance or can be regarded as a pseudocomponent by reason of its homogeneity, and the component remaining, for example plant carrier material, is completely insoluble.

Binary mixtures

Simple binary systems of the dense gas and a substance to be dissolved are the most suitable for illustrating the behaviour of the phases. They can be completely described by three independent variables e.g. pressure (p), temperature (T) and mole fraction (x) which fix the composition and can be represented in three-dimensional pTx-diagrams. Useful also are two-dimensional sections through such diagrams at constant temperature or pressure, or the projection in one of the three planes, for instance in the pT-plane.

The GIBBS phase rule

$$F = C + 2 - P \tag{1}$$

describes the relation, in thermal equilibrium, between the number of independent variables (degrees of freedom) F, the number of components, C and the number of phases, P. Accordingly, in a binary system a homogeneous phase is characterised by a volume (three independent variables), the co-existence of two phases is represented by an area and that of three phases by a line with only one degree of freedom.

Figure 6a shows the simplest case of a pTx-diagram in which there are only gas-liquid equilibria and the two liquid phases are miscible in all proportions. Component 1 is the more volatile solvent, component 2 a less volatile, lipophilic substance. The lines of dashes A_1—CP_1 and A_2—CP_2 in the two pT-planes are the vapour pressure curves of the pure substances, each of which ends at a critical point. The line CP_1—CP_2 represents the critical curve of binary mixtures of 1 and 2; in this simple case it runs continuously between the critical points of the pure substances. This behaviour is encountered only in systems of two components which have similar molecular masses, polarities and critical data (e.g. methane/propane or carbon dioxide/n-hexane). The two-phase region is enclosed by two convex surfaces which meet in the lines A_1—CP_1, CP_1—CP_2 and CP_2 A_2. The uppermost surface thereby forms the phase boundary to the gaseous region, the lower surface the boundary to the homogeneous liquid phase. The form and course of these two phase boundaries are of vital interest for the high pressure extraction. Above the critical curve there is a homogeneous mixture for every composition, i.e., substance 2 can be taken up indefinitely by the supercritical extraction agent 1 beyond the critical curve.

Fig. 6b shows the vapour pressure curves of the two pure substances and the critical curve of the mixture corresponding to the gas-liquid equilibrium in the p(T)-projection. The hatching of the phase boundary indicates the heterogeneous region of state. The critical line does not always have to describe a convex curve (Type 1) as in Fig. 6a. Depending on the system, it can also run uniformly between the critical points of the pure substances (Type 2) or show a temperature minimum (Type 3).

Fig. 6c shows a section in the pT-plane at constant composition x_{2a} of the binary mixture. For the pure substances ($x_2 = 0$ or $x_2 = 1$), the vapour pressure curves are in the pT-plane; they end at the critical point, i.e., at the point where it is still just possible at the highest temperature and pressure to distinguish the liquid and gas phases which are in equilibrium. On the other hand, with the binary mixture, a loop

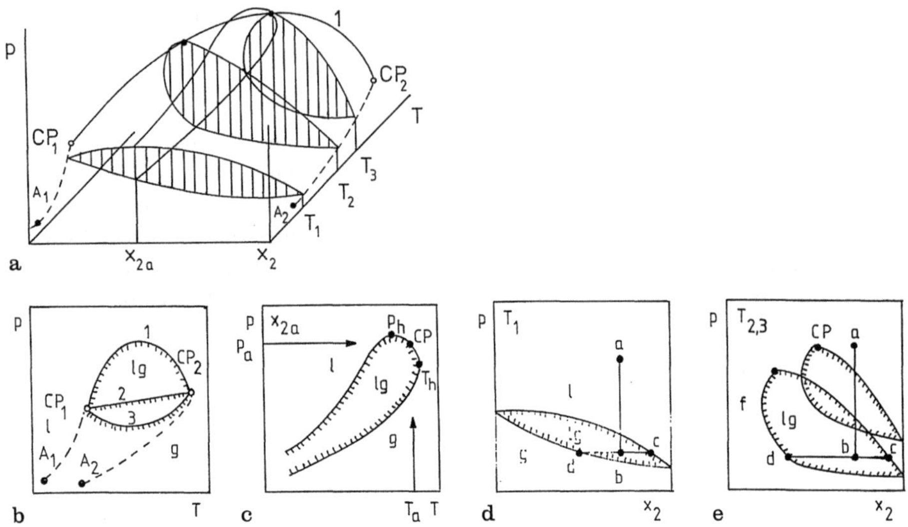

Fig. 6. pTx diagram of a binary mixture (**a**) with p(T)-projection (**b**); p(T) section at constant composition x_{2a} (**c**) and some isotherms in the $p(x_2)$ diagram (**d, e**)

of bubble point and of dew point loci is obtained in the pT-plane; the critical point, CP, at which the co-existing phases gain identical major properties and are thus indistinguishable, coincides generally with neither the point of highest temperature, T_h, nor that of the highest pressure, p_h; these three points can occur in any sequence on the phase boundary line.

Particular evaporation and condensation phenomena can be observed with mixtures near the critical point which are not observed with pure substances. This is due to the isopleth course (p(T)-curve for x = constant). If the temperature is increased at constant pressure, P_a, in Fig. 6c, the mixture, still liquid, traverses the bubble point curve and enters the two-phase region. When the temperature is raised still further, the volume of the gas phase yielded passes through a maximum and then diminishes until finally there is again a domain of homogeneous state. A condensation effect corresponding to this evaporation effect can occur when the pressure is increased isothermally. If the gaseous mixture is compressed at the temperature T_a, a liquid phase is formed intermediately when the two-phase region is traversed. Its amount first increases, then decreases and finally it disappears. The domain of homogeneous state which is attained each time after traversing the two-phase region could be termed liquid in the first case, gaseous in the second. This distinction thus becomes meaningless in the critical region. Systematic investigations, especially of Schneider, have shown that, in the critical region, all three possible two-phase equilibrium types (liquid/liquid, liquid/gas and gas/gas) display continuous transitions [48, 49].

The planes defined through a constant temperature, T_1, T_2 or T_3, cut the two surfaces corresponding to the gas-liquid equilibrium in the way shown. The resulting isothermal $p(x_2)$-sections are given in Figs. 6d and 6e. At a temperature of the compressed gas, T_1, below the critical value, a spindle-shaped figure is obtained, like that of a distillation diagram. If a homogeneous mixture of composition (a) is decompressed to the value at point (b), a separation of phases takes place into liquid of the composition (c) and vapour of composition (d), in equilibrium with each other. An increase of temperature to above the critical value of the solvent leads to characteristic loops which correspond to the $p(x_2)$-isotherms, T_2 and T_3. No gas-liquid transition can now be observed on the left side. At the maximum the horizontally running tie-lines which connect the phases in equilibrium at constant temperature and pressure shrink to a single point, the critical point, CP, at the corresponding temperature. Pressure release into the heterogeneous region often brings about an effective separation on account of the asymmetric form of the isotherms.

There are many behaviour patterns of binary mixtures other than the simple phase behaviour shown in Fig. 6. At temperatures below the critical value of the more volatile component, 1, liquid-liquid separations and three-phase lines liquid-liquid-gas can occur. The critical curves of the liquid-liquid and gas-liquid equilibria are usually not strictly distinguishable. They can have different forms and, depending on the composition of the binary mixture, can show a continuous or discontinuous course. Studies based on the Van der Waals equation have shown that it is possible to subdivide binary systems into various types of phase diagram according to the course of the critical lines and especially the way in which they are related to the critical points of the pure substances and the three-phase lines [50–52].

According to van Konyenberg and Scott [52], 6 principal classes of binary equilibria can occur on considering only gas and liquid phases. The classification is most easily recognised in the pT-projection of the relevant pTx-diagrams. In addition to the simplest case (class 1) shown in Fig. 6, Fig. 7 contains the p(T)-diagrams of classes 2, 3 and 4 and also a type of equilibrium related to class 3 but including solid phases. Two characteristic p(x)-isotherms are given in each p(T)-diagram. The classes are numbered starting from largely symmetrical binary mixtures and continuing via increasing differences in the molecular size, volatility and polarity of both components which determine the phase behaviour. The critical lines (continuous lines), three phase lines (dot/dash) and the vapour pressure curves of the pure substances (dashes) in the p(T)-projection mark the boundaries of surfaces which enclose two-phase regions in the three-dimensional pTx-diagram. The isotherms are characteristic sections of these surfaces, the two-phase regions being indicated by hatching.

In the phase diagram of class 2 the critical curve of the gas-liquid equilibrium runs continuously between the critical points, CP_1 and CP_2, of the pure substances, like with class 1. In addition, a liquid-liquid mixed phase appears at low temperatures; it has a critical line which begins at the upper critical end point, U, the point of intersection with the three phase line, llg, and runs almost parallel to the p-axis. Correspondingly the critical solution temperatures (UCST) for liquid-liquid equilibria, show only a very weakly positive (type 1) or negative (type 2) dependence on temperature. Examples of phase diagrams of class 2 are mixtures of carbon dioxide with octane, decane [49] or 2-hexanol [53]. The three-phase lines occurring can have an upper and/or a lower critical end point, at which two of three co-existing phases become identical. Each critical end-point is denoted with a triangle.

In the single-phase domain where the less volatile component, 2, is taken up completely by the dense gas, 1, possibilities for extraction are denoted by points (L) in the p(T)-diagrams of Fig. 7. Starting from these points a change of conditions is necessary to bring about a transition to the two-phase region and consequently

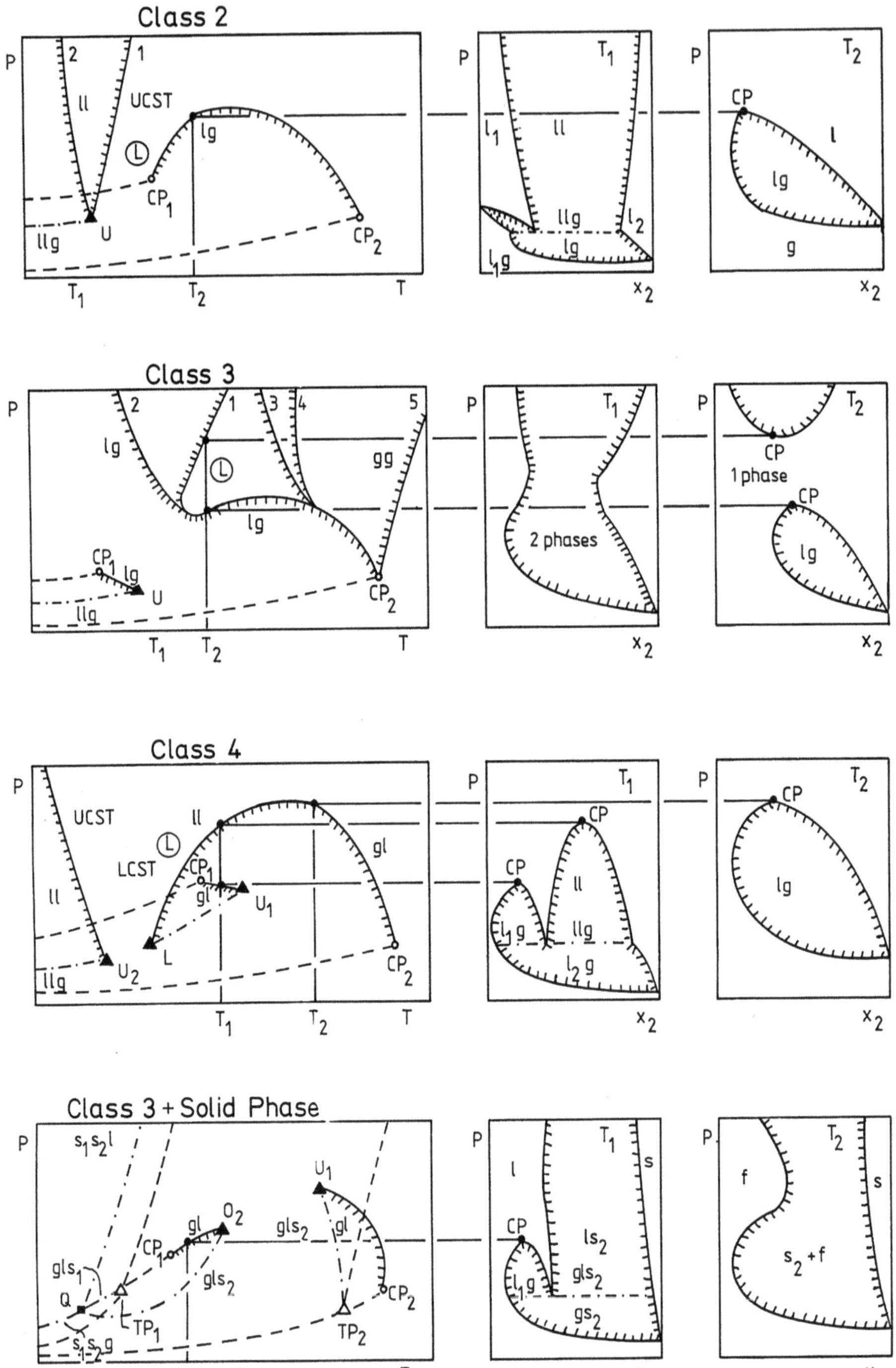

Fig. 7. Representation of different classes of binary system in the p(T)-projection with two selected p(x)-isotherms in each case

separation of the dissolved substance. This is possible in almost all cases by lowering the pressure. Sometimes, e.g. class 3, type 1, a powerful increase in pressure can lead to phase separation. Lowering of temperature and, in certain states, depending on the pressure, increase in temperature can also accomplish separation.

If the mutual solubility of both components of a class system 2 decreases further, the critical curve of the liquid-liquid equilibrium is displaced towards higher temperatures until it finally enters the gas-liquid region and passes continuously into the critical curve, lg. In the class 3 diagram now present, this curve no longer runs continuously between the critical points, CP_1 and CP_2. It consists of two branches. The one is linked to the critical point of the less volatile component, CP_2 and rises to very high pressures when the temperature falls. In some cases pressure maxima and minima may be traversed before the steep, weakly positive (type 1) or negative (type 2) gradient is attained. The pressure extremes disappear when the miscibility of the components diminishes still further. The critical curve can move to higher pressures with negative gradient (type 3), pass through a temperature minimum (type 4) or adopt immediately a positive gradient (type 5). In these last two cases so-called gas-gas equilibria are encountered at high pressures and supercritical temperatures. The second and shorter branch of the critical curve, beginning at CP_1, ends at the upper critical end point, U, where it meets the three-phase line 11g. In some class 3 types this three phase line can also be above the vapour pressure curves of the two pure substances. Examples of phase diagrams of class 3 are the systems of carbon dioxide with water [54, 55], hexadecane [56] and squalane [57].

In phase diagrams of class 4 the critical line for the gas-liquid equilibrium beginning at point CP_2, attains a maximum when the temperature is lowered and then changes continuously into a critical line corresponding to a liquid-liquid equilibrium; this line is to be assigned to the lower critical separation temperatures (LCST); it meets the three-phase line llg at the lower critical end point, L. At lower temperatures an additional liquid-liquid separation occurs. As in diagrams of class 2, its critical line gives the pressure dependence of the upper critical separation temperatures (UCST) and begins or ends at the upper critical end point, U_2. Examples of class 4 diagrams are binary mixtures of carbon dioxide with nitrobenzene, or of ethane with squalane.

Class 5 systems differ from class 4 systems only in that the liquid phases are completely miscible at temperatures below the lower critical end point, L, and that a dissociation into two liquids occurs in only a small domain between the critical temperatures of the pure substances. Unlike the previous five classes, diagrams of class 6 cannot be derived from the Van der Waals equation. They occur in some aqueous systems and display, like classes 1 and 2, an unbroken critical curve, gl, between the critical points of the two components. In addition, they possess a region with liquid-liquid separation, the critical curve of which is bound at low pressures by an upper and a lower critical end point.

Some characteristic p(x)-isotherms are shown in Fig. 7. If the temperature is below the upper critical end point such that the three-phase line, llg, is cut (e.g. in the diagram of class 4 at temperature T_1 with $T(CP_1) < T_1 < T(U_1)$), three different two-phase regions appear. Liquid-gas equilibria, l_2g, occur in the region below the three-phase line marked with a dot/dash line. The right hand branch of the curve, at high x_2 proportions, corresponds to the solubility of the supercritical compound 1 in the liquid compound 2; the left hand branch of the curve depicts the solubility of the liquid compound 2 in the supercritical compound 1. At high concentrations of component 1, a small two-phase domain, l_1g, appears above the pressure of the three-phase line; this domain disappears at a pressure above the binary critical point CP on the critical line CP_1—U_1. As the temperature T_1 is increased this two-phase domain becomes smaller and smaller until it finally disappears at the upper critical end point U_1. Further, there is a larger two-phase region, ll, corresponding to a liquid-liquid separation, that vanishes above another critical point of the mixture. This point lies on that part of the critical line L—CP_2 which is ascribed to the liquid-liquid equilibria. If the temperature is so high, like T_2 in class 4 or T_1 and T_2 in class 3, that the three-phase line is no longer cut, one or two separate two-phase regions appear; in class 4 in the transition from T_1 to T_2 the upper critical point changes continuously from liquid-liquid to liquid-gas equilibria.

Solid phases have been included in the lowest row of diagrams in Fig. 7. The line of dashes is now the vapour pressure curves, lg, the sublimation curves, sg, and the melting point pressure curves, sl, of the pure substances. Point Q is a quadruple point, where the four phases, s_1, s_2, l and g, exist in equilibrium; four different three-phase lines start from it, giving the possible combinations of any three of these four phases. The triple point temperature of the less volatile component, 2, is above the critical temperature of component 1, so that there is no temperature interval in which both components occur together in the liquid state. In the diagram the three-phase curve of the type gls_2 reaches such high pressures that the critical curve is intersected, yielding both critical end points, U_1 and U_2. In the vicinity of these end points the solubility of the condensed phase is markedly dependent on pressure and temperature.

The p(x)-isotherm, T_1, of this system is situated above the critical temperature of the pure component, 1, but below the temperature of the critical end point, U_2. At pressures above the three-phase curve, gls_2, the dissolving capacity of phase 1 for the solid substance, 2, is given by the phase boundary which runs almost parallel to the p-axis at small x_2-values. At temperatures at which the three-phase line, gls_2, is not cut, the diagram shown for the isotherm, T_2, is yielded, in which a liquid phase is no longer present. The curvature of the isotherms to the right, towards high x_2-values, at above a certain pressure in the system is due to the increased solubility of the solid substance, 2, in the dense gas, 1, as a result of solvation, observed in dense gas extraction. At very high pressures, however, the isotherm curves to the left and the solubility decreases.

Streett has published a detailed treatment and discussion of types of binary equilibrium under high pressure [58]; information is to be found also in the work of Schneider and of Rowlinson, already quoted.

Ternary Mixtures

Four variables in all are required to describe a system with three components. In order to use a pictorial representation in two dimensions the variables of pressure and temperature are kept constant so that the composition is given by the other two. The third coordinate of composition is then automatically fixed. Such a phase diagram is then usually given in an equilateral triangle (according to Gibbs) in which each of the three coordinates ends at 100% of one of the components. The points for the binary mixtures are then on the sides of the triangle and each point within the triangle

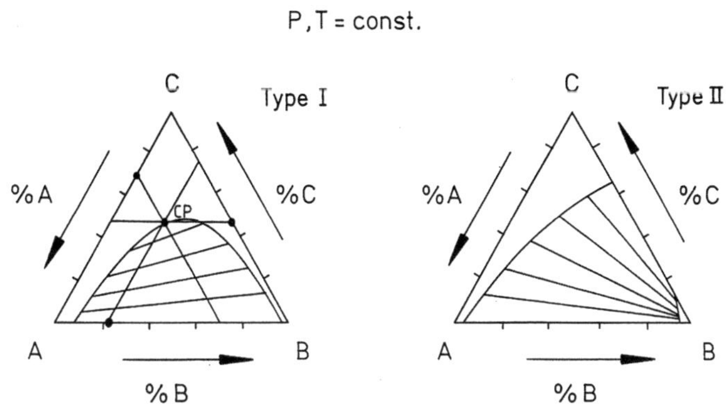

Fig. 8. Phases of a simple ternary system (see text)

gives a definite composition of the ternary mixture. The composition of a point, e.g. CP in Fig. 8, is then derived by drawing through it parallels to the sides of the triangle. The amount is then obtained of that component which is opposite to the side of the triangle. For instance, the parallel line to side AB yields the relative amount of component C. If it is desired to show also the temperature dependence, for example a temperature axis is drawn perpendicularly to the surface of the triangle.

In regions where the three components are not miscible with one another, two-phase areas appear in the triangle, analogous to those in the binary systems; these are surrounded by the boundary or binodal curve. Diagrams of two types are distinguished through the course of the binodal curve: type I (according to Treybal) with a "miscibility gap" in only one of the three binary systems and a closed phase boundary line; and type II with "miscibility gaps" on two of the three coordinate axes. In thermal equilibrium the tie-lines give the composition of the two separated phases. However, they do not run horizontally, in contrast to the two component systems. The place at which the tie-lines shrink to a single point corresponds to the critical, isothermal solution or separation point. Application of the Gibbs phase rule to the two-phase region gives three degrees of freedom, i.e. pressure, temperature and a composition can be changed independently of one another.

The triangular diagrams in Fig. 8 are suitable for depicting the separation of a binary mixture with the help of a dense gas as agent. A and C, which are miscible in all proportions in the diagrams, are taken as the liquid components to be separated; B is the volatile auxiliary serving as solvent. Diagram I describes the conditions in the dissolving stage and diagram II those in the separating out stage. In the first case the dense gas, B, and component C of the mixture form a homogeneous, supercritical phase, whereas in the binary system, AB, there is a miscibility gap. As the binodal curve shows, gas B is, under the conditions of the extraction, capable of dissolving an appreciable amount of the less volatile components; in accordance with the course of the tie-lines, larger amounts of C than of A can be taken up in the gas phase. This preferred extraction of C leads finally to the separation of the binary mixture. The conditions can then be adjusted by, for example, lowering pressure, so that in the separation stage a second miscibility gap in the system BC occurs. The extraction phase separates into a liquid and an almost pure gas phase, B, which contains only traces of the less volatile components. The boundary curve, which gives the solubility of the gas in the liquid phase, shows a relatively high solubility of gaseous B in liquid phases rich in C and, in contrast, a low solubility in liquid phases rich in A.

Just as with binary systems, ternary phase diagrams can be more complex than the examples shown in Fig. 8. For instance, in the vicinity of the critical point of one of the pure components, ternary systems may have three-phase domains, liquid-liquid-gas, as in binary diagrams. Examples of ternary systems which have been investigated thoroughly at high pressure are: acetone-water-ethylene and butanone-water-ethylene [59]; carbon dioxide-tridecane-hexadecane and carbon dioxide-tridecane-1-hexadecanol [60]; and hydrocarbon-water-carbon dioxide mixtures [61].

Ternary systems play a part also in the separation of aqueous alcohol solutions with dense gases [62–64] and in the solution of poorly volatile substances in a dense gas using an added material of intermediate volatility (entrainer, modifier) (see Chapt. III.3.). Brunner and Peter [65] have described ternary and quaternary (concentration tetrahedron) systems.

II.4.c Calculations of Phase Equilibria

The choice of suitable operating conditions in the extraction with dense gases demands knowledge of the phase equilibria of the substances to be dissolved in particular regions of temperature and pressure. The experimental determination of the equilibrium concentrations of the substances to be separated in a dense gas is highly demanding. One attempts therefore to reduce the number of experiments required by correlating and extrapolating, using appropriate calculation methods.

The considerations of this theme below are not intended to provide complete calculation procedures. These do exist in large numbers, in many cases more or less empirically derived and usually not universally suitable but restricted by certain conditions (state, molecular size, polarity). Reference must be made here to the original literature.

The macroscopic conditions of phase equilibrium of classical thermodynamics state that pressure and temperature of both phases I and II and the chemical potentials or fugacities of all components, i, in both phases must be the same [66]. Since pressure and temperature are normally pre-chosen variables, the relation holds:

$$f_i^{I} = f_i^{II}; \qquad i = 1 \dots k \tag{2}$$

where f_i is the fugacity of component i in a mixture of k components. Fugacities can be calculated with the help of an equation of state but it must be valid for both phases I and II over the whole region of interest. As a rule this is the case only with similar and relatively non-polar mixture components.

No phase equation describing both the liquid and the gas phase exists for highly unsymmetrical mixtures of substances which differ considerably in molecular mass, structure or polarity. The two phases have to be considered and described separately, the gas phase with the aid of a fugacity coefficient, φ, and the condensed phase by means of an activity coefficient, γ, which take into account the fact that the mixture is not ideal.

The gas phase equation is given by:

$$f_i = y_i \varphi p \tag{3}$$

where y_i is the mole fraction of component i in the gas phase and p the total pressure.

The fugacity coefficient, φ, which describes the deviation from the behaviour of an ideal gas, can be calculated for the gas phase with a thermodynamic equation of state [67].

The activity coefficient, γ, of the condensed phase is defined through

$$f_i = x_i \gamma_i f_i^0 \tag{4}$$

where x_i is the mole fraction of component i in the liquid phase and f_i^0 its fugacity in an arbitrary reference state. The activity coefficient characterises the deviation of the behaviour of component i from this reference state or from Raoult's law when the fugacity of the pure liquid under the phase conditions of the system is taken as reference standard.

In the region of high pressure the supercritical gas phase resembles more and more the liquid state as far as its density is concerned; equation (3) for the gas phase then becomes less valid. In this case a qualitative forecast of the equilibrium is possible if equation (4) only is used to calculate the fugacities; then

$$x_i^I \gamma_i^I = x_i^{II} \gamma_i^{II} \tag{5}$$

must apply (x_i^I is the mole fraction of component i in the extraction phase and x_i^{II} that in the condensed phase). For a system of k components there are k such equations, with the added condition that the sum of the mole fractions in both phases must always be unity. For binary mixtures the problem is reduced to solving the two equations

$$x_1^I \gamma_1^I = x_1^{II} \gamma_1^{II} \tag{6}$$

and

$$(1 - x_1^I)\, \gamma_2^I = (1 - x_1^{II})\, \gamma_1^{II} \tag{7}$$

where the index 1 refers to the solvent and x_1^I and x_1^{II} give the mole fractions in the phases containing more and less solvent, respectively.

However, the evaluation of activity coefficients as functions of mole fractions is more problematic in this system, the behaviour of which deviates so markedly from the ideal, than in cases where the fugacity coefficients can be calculated from (3) with the help of a simple cubic equation of state which requires relatively little information about the components of the mixture.

When the solubility of solid components is investigated, i.e. a solid-liquid or solid-gas equilibrium is considered, the situation is simplified in so far as the miscibility of the solvent with the solid phase can normally be neglected and this phase regarded as a pure substance. As a result, one has to consider only interactions in the gas phase and the influence of pressure on the fugacity of the solid substance; interactions in the condensed phase can be neglected.

Two principal procedures come into question for calculating the phase equilibria; they consider the problem situation from different sides: either on the basis of equations of state and the principle of corresponding states; or on the basic of a quasi-lattice model and microscopic statistical considerations.

The use of equations of state permits, with a simple algebraic expression and the help of only a few individual constants, a relation between pressure, temperature and volume to be established. The equations are based on principles of the Van der Waals theory and possess usually a term for attraction and one for repulsion. It is a semi-empirical form, so that a whole series of modifications of the original Van der Waals equation (8) exist; examples are those of: Redlich-Kwong [68] (9); Bak and Bjerre [69]; Soave-Redlich-Kwong [70] (10) and Peng-Robinson [71] (11); they employ differing terms for the attraction but retain the same repulsion term:

$$\text{VdW:}\quad p = RT/(V - b) - a/V^2 \tag{8}$$

$$\text{RK:}\quad p = RT/(V - b) - a/T^{0.5} \cdot (V + b)\, V \tag{9}$$

$$\text{SRK:}\quad p = RT/(V - b) - a(T)/(V + b)\, V \tag{10}$$

$$\text{PR:}\quad p = RT/(V - b) - a(T)/(V^2 - 2bV - b^2) \tag{11}$$

R = universal gas constant; T = absolute temperature; V = mole volume; p = pressure, a, b = parameters for intermolecular attraction and repulsion, respectively.

Mention may be made also of the equations of Dieterici [72], Kay-Redlich-Ngo [73] with modified attraction and repulsion terms, or equations built up in a complex way, e.g. of Behar [74], Deiters [75] etc. There are also purely empirical equations of state, e.g. of Kumar and Starling [76] which reproduce the experimental data satisfactorily but with terms lacking any physico-chemical significance. The operation and adaptation of the individual parameters to changed situations is correspondingly difficult.

The equations (8 to 11) contain two parameters, of which b is independent of temperature and a in (10) and (11) depends on temperature. These parameters, or the temperature dependence of a, are determined in various ways, with the help of critical data or other experimental values as described in the original literature. Thermodynamic properties of a dense gas as pure substance, such as vapour pressure curves, enthalpies and the pressure and temperature dependence of densities, can then be calculated using these adapted equations of state. The accuracy of the data thus obtained is, however, basically unsatisfactory in the critical region. Density values in the critical and liquid regions are too low from equations (9) and (10) but satisfactory from (11). The compressibility factor Z^C at the critical point is given for example as 0.375 by [8], as 0.33 by (10) and 0.307 by (11), independently of the type of gas, whereas the experimental values range from 0.27 to 0.29. Claims to accuracy in the critical region are therefore renounced and the parameters usually chosen so that the best possible description of liquid densities and of vapour pressure curves results.

The individual interactions between the molecules and their influence on the free energy of mixing are unknown for mixtures of substances. With the help of a model a mean is obtained of the individual interactions, by regarding the mixture as a hypothetically pure substance (one fluid model). The quantities a and b of this imaginary substance, needed for the equations of state, are derived with the assistance of so-called mixing rules; according to van der Waals [77] these are of simple quadratic form.

For binary mixtures,

$$a_m = (1 - x)^2 a_{11} + 2x(1 - x) a_{12} + x^2 a_{22} \tag{12}$$

$$b_m = (1 - x)^2 b_{11} + 2x(1 - x) b_{12} + x^2 b_{22} \tag{13}$$

or, generally,

$$a_m = \sum_i^k \sum_j^k x_i x_j a_{ij} \quad \text{with} \quad a_{ij} = (a_{ii} a_{jj})^{0.5} (1 - l_{aij}) \tag{14}$$

$$b_m = \sum_i^k \sum_j^k x_i x_j b_{ij} \quad \text{with} \quad b_{ij} = 0.5(b_{ii} + b_{jj}) (1 - l_{bij}) \tag{15}$$

where x_i and x_j are the mole fractions of components i and j, respectively, and k the number of components of the system. a_{ii}, a_{jj} and b_{ii}, b_{jj} are parameters of the pure substances i and j, resp.; a_{ij} and b_{ij} characterise the interactions between different molecules and, as said, can be obtained from the parameters of the pure substances and a correction parameter, l_{aij} or l_{bij}, normally independent of pressure, temperature and composition (two parameter mixing rules).

The simple form

$$b_m = \sum_i^k x_i b_i \tag{16}$$

can often be used when i is the volatile solvent, for example, and j the less volatile component [78].

Along with the general equilibrium conditions and an equation of state, the mixing rules permit the composition of the phases of substance mixtures to be calculated. For a binary system:

$$x_1^I, x_1^{II} = f(T, p, a_{11}, a_{22}, b_{11}, b_{22}, l_{a12}, l_{b12}) \tag{17}$$

For non-polar systems, good forecasts of the gas-liquid equilibrium are possible by suitably adapting the interaction parameter l in combination with the simple cubic two-parameter equations of state (9–11).

Fig. 9 shows the course of the 60 °C phase equilibrium isotherm in the pressure-concentration diagram of carbon dioxide-n-hexadecane, compared with the calculated curve. It can be seen that the solubility of the carbon dioxide in the liquid hydrocarbon at low pressures has been as correctly described with the help of equations (9), (14) and (16) as has the critical point of the liquid-gas equilibrium. The system is, however, only inadequately described at lower temperatures where an additional liquid-liquid equilibrium appears.

Since the classical mixing rules depend on random mixing of the molecules, they cannot take account of the intermolecular interactions of solvent and substrate. At

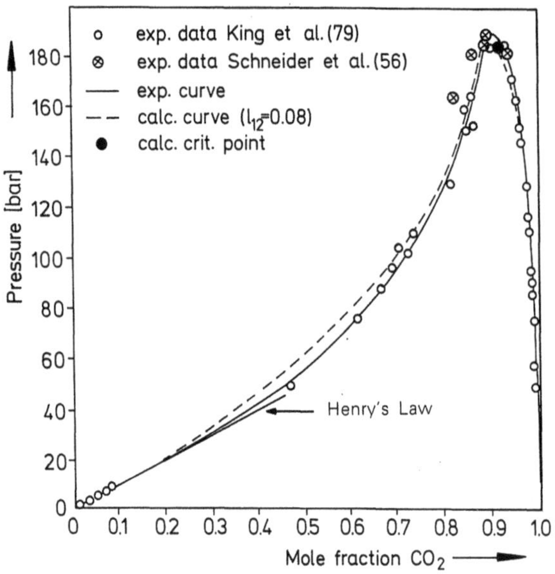

Fig. 9. p,x-diagram of the system carbon dioxide-n-hexadecane at 60 °C; experimental values and curve calculated from the RK-equation (q) [79]

best, they apply to non-polar, similar molecules. Both the equations of state and the mixing rules have to be modified to be able to describe mixtures of components showing great differences in molecular mass and/or polarity. It must be noted that in these unsymmetrical systems the neighbourhood of one sort of molecule differs from that of the other sort (local composition concept).

Vidal [80] and Heyden [81], among others, gave modified mixing rules, non-quadratic and independent of density, for the parameters a and b of the cubic equations of state; these enable good predictions to be made in regions of low compressibility (liquid or high density gas). Modified rules, dependent on density, which can be used in domains of high or low pressure, were proposed by Whiting and Prausnitz [82], Mollerup [83], Won [84] and others.

Activity coefficients in the liquid phase are determined with the help of excess values which give the divergence of the properties from those of an ideal mixture. A series of models exists for determining the free excess enthalpy of unsymmetrical systems, e.g. NRTL (non-random two liquid) [85], UNIQUAC [86] etc. These are derived from lattice models and are based on the conception of local composition. The universal quasi-chemical model, UNIQUAC, takes also the size and shape of the molecules into consideration by dividing each molecule into segments and according it a definite external surface area. A further interesting development is the UNIFAC model [87] which considers characteristic groups, such as $-CH_3$, $=CO$, $-OH$, from which the whole molecule is built up, instead of the molecules themselves. When once the additive parameters of interaction of individual molecular groups are known, this procedure of group contribution, together with parameters for molecular size and form, enables forecasts to be made about liquid-liquid equilibria of a large number of different mixtures. The UNIFAC method requires for this only little information in addition to the structural formulae of the components and tabulated interaction parameters of the groups; in contrast, other model schemes of calculation need critical data, molar volumes or vapour pressure values of the pure substances or experimental data of the mixture for matching of parameters.

The methods based on lattice models were applied first to equilibria in the low pressure region and to liquid mixtures (e.g. destillative separations). This was because a definite place relative to its immediate neighbours can be accorded to each molecule as a result of the low compressibility (identity of the partial molar volumes in all liquid phases). Advanced lattice models can be used also for equilibria with dense gases where the phases may show large differences in density. This feature is accommodated by having empty spaces arbitrarily distributed over the lattice [88]. A pure substance is regarded as a binary system of occupied and empty lattice places, a binary mixture as a ternary system of two different sorts of occupied lattice place and one sort of empty place. Changes in density due to alterations of pressure and temperature are accounted for by the concentration of the empty places but the size of the lattice places remains unchanged. Modern lattice theories based on these principles permit a good description of the phase behaviour of substances of various polarities and molecular masses, even in the critical region [89].

The progress hitherto achieved using models and methods of calculation has been invaluable in helping understand phase equilibria and correlating measured values. It must be remembered, however, that the state of the components of the system can vary in the region from below the triple point up to above the critical point and, depending on their combination, lead to an enormous number of possible systems. To describe all these quantitatively naturally poses problems and cannot be accomplished with a universal prescription. For example, consideration has to be given to the large isothermal compressibility of a dense gas; to special interections which can lead to liquid-liquid separations in the critical region, or to large negative solution volumes of solids; or to possible mutual influences of dissolved substances in the gas phase in systems of more components. In this connection, experimental data assist in adaptation of parameters for calculation; and they are in the final analysis the yardstick for assessing the quality of the models and the accuracy of the calculations.

References

1. Cagniard de la Tour, C.: Ann. chim. phys. (2) *21*, 127, 178 (1822)
2. Andrews, T.: Trans. Roy. Soc. (London) *A 159*, 547 (1869)
3. Hannay, J. B., Hogarth, J.: Proc. Roy. Soc. (London) *29*, 324 (1879) and ibid. *30*, 178 (1880)
4. Poynting, J. H.: Phil. Mag. *12*, 32 (1881)
5. Van der Waals, J. D.: Die Continuität des gasförmigen und flüssigen Zustandes, Johann Ambrosius Barth, Leipzig (1881)
6. Sage, B. H., Webster, D. C., Lacey, W. N.: Ind. Engng. Chem. *28*, 1045 (1936)
7. Niggli, P.: Z. Anorg. Allgem. Chem. *75*, 161 (1912)
8. Katz, D. L., Kurata, F.: Ind. Engng. Chem. *32*, 817 (1940)
9. Kennedy, G. C.: Econ. Geol. *45*, 629 (1950)
10. Tyrer, D.: J. Chem. Soc. *97*, 621 (1910)
11. Morey, G. W.: Proc. Am. Soc. Testing Materials *42*, 98–1022 (1942)
12. Booth, H. S., Bidwell, R. M.: Chem. Rev. *44*, 477 (1949)
13. Laudise, R. A., Sullivan, R. A.: Chem. Engng. Prog. *55*, 55 (1959)
14. Holm, L. W., O'Brien, L. J.: J. Pet. Technol. April *1971*, 431

15. Pilat, S., Godlewicz, M.: U.S. Pat 2188012 and 013 (1936)
16. Messmore, H. E.: U.S. Pat. 2420185 (1943)
17. Palmer, G. H., Fanwood, N. J.: U.S. Pat. 2658907 (1950)
18. Groll, H. P. A.: Germ. Pat. Appl. 1079636 (1953)
19. Zhuze, T. P.: Petroleum (London) *23*, 298 (1960)
20. Zhuze, T. P., Jushkevic, G. N. und Gekker, J. E.: Maslob. Zhir. Prom. *24*, 34 (1958)
21. Zosel, K.: Germ. Pat. 1493190 (1964)
22. Zosel, K.: Angew. Chem. *90*, 748 (1978)
23. Schneider, G. M., Stahl, E., Wilke, G.: Extraction with Supercritical Gases, Verlag Chemie, Weinheim, 1980
24. Paul, P. F. M., Wise, W. S.: The Principles of Gas Extraction, Mills and Boon, London, 1971
25. Hicks, C. P., Young, C. L.: Chem. Rev. *75*, 119 (1975)
26. Irani, C. A., Funk, E. W.: Recent Developments in Separation Science. Vol. III, part A, p. 171, CRC Press, West Palm Beach, Florida, 1977
27. Williams, D. F.: Chem. Eng. Sci. *36*, 1769 (1981)
28. Reid, R. C.: Supercritical Fluid Extraction, a Perspective, Hougen Lecture Series, 1981
29. Randall, L. G.: Sep. Sci. Technol. *17*, 1 (1982)
30. Paulaitis, M. E., Krukonis, V. J., Kurnik, R. T., Reid, R. C.: Reviews in Chemical Engineering, Vol. I, No. 2, p. 179 (1983)
31. Ely, J. F., Baker, J. K.: NBS Tech. Note (US) 1070 (1983)
32. Whitehead, J. C., Williams, D. F.: J. Inst. Fuel *48*, 182 (1975)
33. Kershaw, J. R., Jezko, J.: Sep. Sci. Technol. *17*, 151 (1982)
34. Gouw, T. H., Jentoft, R. E.: J. Chromatogr. *68*, 303 (1972)
35. Stahl, E., Schütz, E.: Arch. Pharm. *311*, 992 (1978)
36. Stahl, E., Gerard, D.: Parfuem. Kosmet. *64*, 237 (1983)
37. Stahl, E., Schütz, E.: Planta Med. *40*, 262 (1980)
38. Stahl, E., Keller, K.: Planta Med. *47*, 75 (1983)
39. Stahl, E., Schütz, E.: Planta Med. *40*, 12 (1980)
40. Peter, S., Brunner, G., Riha, R.: Fette, Seifen, Anstrichm. *78*, 45 (1976)
41. Panzner, F., Ellis, S. R. M., Bott, T. R.: Ind. Eng. Chem. Symp. Ser. No. 54, p. 165 (1978)
42. Brandt, B. B., Rozlovskii, A. J.: Doklady Akademii Nauk SSSR, *132*, 1129 (1960)
43. Ribovich, J., Murphy, J., Watson, R.: J. Hazard. Mat. *1*, 275 (1975)
44. Anon.: J. Am. Oil Chem. Soc. *60*, 520 (1983)
45. BUSE Broschüre: Daten und neue Zustandsdiagramme von CO_2, Bad Hönningen, 1981
46. Landolt-Börnstein Tabellen, 6. ed. Vol. II, Springer Verlag, Berlin, 1971
47. IUPAC Commission on Thermodynamics and Thermochemistry, Carbon Dioxide International Thermodynamic Tables of the Fluid State. Eds. Agnus, S., Armstrong, B., de Reuck, K. M., Pergamon Press, London, 1976
48. Schneider, G. M.: Adv. Chem. Phys. *17*, 1 (1970)
49. Schneider, G. M. in Chemical Thermodynamics, Vol. 2, Specialist Periodical Reports, Chemical Society, Kap. 4, London 1978
50. Schneider, G. M.: Angew. Chem. Int. Ed. Engl. *17*, 716 (1978)
51. Rowlinson, J. S.: Liquids and Liquid Mixtures, 2. ed. Butterworth, London 1969
52. Van Konyenburg, P. H., Scott, R. L.: Phil. Trans. Roy. Soc. (London) *298*, 495 (1980)
53. Alwani, Z., Schneider, G. M.: Ber. Bunsenges. Phys. Chem. *80*, 1310 (1976)
54. Takenouchi, S., Kennedy, G. C.: Amer. J. Sci. *262*, 1055 (1964)
55. Tödheide, K., Franck, E. U.: Z. Physikal. Chem. N.F. *37*, 387 (1963)
56. Schneider, G. M., Alwani, Z., Heim, W., Horvath, E., Franck, E. U.: Chem. Ing. Tech. *39*, 649 (1967)
57. Liphard, K. G., Schneider, G. M.: J. Chem. Thermodyn. *7*, 805 (1975)
58. Streett, W. B.: in Chemical Engineering at Supercritical Fluid Conditions, Eds. Paulaitis, M. E., Penninger, J. H. L., Gray, R. D., Davidson, P., Ann Arbor Science, 1983, p. 3
59. Elgin, J. C., Weinstock, J. J.: J. Chem. Eng. Data *4*, 3 (1959)
60. Konrad, R., Swaid, J. Schneider, G. M.: Fluid Phase Equil. *10*, 307 (1983)
61. Snedeker, R. A.: Ph. D. Thesis, Princeton University, Princeton, NJ, 1955

62. Paulaitis, M. E., Gilbert, M. L., Nash, C. A., paper presented at the 2nd World Congress of Chemical Engineering, Montreal, Canada, October 4–9, 1981
63. McHugh, M. A., Mallett, M. W., Kohn, J. P., paper presented at the Annual Meeting of the AIChE, New Orleans, November 9–12, 1981
64. Brunner, G., Kreim, K.: Chem.-Ing.-Tech. *56*, 550 (1985)
65. Brunner, G., Peter, S.: Ber. Bunsenges. Phys. Chem. *83*, 1137 (1979)
66. Haase, R.: Thermodynamik der Mischphasen, Springer, Berlin 1956
67. Prausnitz, J. M.: Molecular Thermodynamics of Fluid Phase Equilibria, Prentice - Hall, Englewood Cliffs, NJ. 1969
68. Redlich, O., Kwong, J. N. S.: Chem. Rev. *44*, 233 (1949)
69. Bjerre, A., Bak, T. A.: Acta Chem. Scand. *23*, 1733 (1969)
70. Soave, G.: Chem. Eng. Sci. *27*, 1197 (1972)
71. Peng, D. Y., Robinson, D. B.: Ind. Eng. Chem. Fundam. *15*, 59 (1976)
72. Partington, J.: An Advanced Treatise on Physical Chemistry, Vol. 1, Longmans, London (1951)
73. Redlich, O., Ngo, V. B. T.: Ind. Eng. Chem. Fundam. *9*, 287 (1970)
74. Behar, E., Jain, C.: Rev. IFP *36*, 173 (1981)
75. Deiters, U.: Chem. Eng. Sci. *36*, 1139, 1147 (1981); and ibid., *37*, 855 (1982)
76. Kumar, K. H., Starling, K. E.: Ind. Eng. Chem. Fundam. *21*, 255 (1982)
77. Van der Waals, J. D.: Die Kontinuität des gasförmigen und flüssigen Zustandes, Vol. 2, Barth, Leipzig, 1900
78. King, M. B. et al.: in Chemical Engineering at Supercritical Fluid Conditions, Eds. Paulaitis, M. E., Penninger, J. M. L., Gray, R. D., Davidson, P., Ann Arbor Science (1983), p. 31
79. King, M. B., Bott, T. R., Sheldon, J. R., Mahmud, R. S.: Ber. Bunsenges. Phys. Chem. *88*, 812 (1984)
80. Vidal, J.: Chem. Eng. Sci. *33*, 787 (1978)
81. Heyden, G.: Proceedings, 2nd World Congress of Chemical Engineering, Montreal, Canada, Vol. 5, 41 (1981)
82. Whiting, W. B., Prausnitz, J. M.: Fluid Phase Equilibria *9*, 119 (1982)
83. Mollerup, J.: Fluid Phase Equil. *7*, 121 (1981)
84. Won, K. W.: Fluid Phase Equil. *10*, 191 (1983)
85. Renon, H., Prausnitz, J. M.: AIChE J. *14*, 135 (1968)
86. Abrams, D. S., Prausnitz, J. M.: AIChE J. *21*, 116 (1975)
87. Fredenslund, Aa., Jones, R. L., Prausnitz, J. M.: AIChE J. *21*, 1086 (1975)
88. Mermin, N. D.: Phys. Rev. Letters *26*, 957 (1971)
89. Kleintjens, L. A.: Fluid Phase Equilibria *10*, 183 (1983)

III. Methods, Apparatus and Plants

III.1. Procedures for Determining Phase Equilibria

The phase behaviour of a substance or a mixture in contact with a dense gas must be known in order to be able to design a proedure of separation or extraction. This phase behaviour, of poorly volatile substances in highly compressed gases, is generally ascertained experimentally. However, it is an advantage to have at the same time a suitable method of mathematical calculation which is able to correlate the experimental values and, if it is dependable enough, even partly to take the place of the experiments.

The aim of the experimental work is to obtain the three values characterising the state of a system, namely, pressure, temperature and composition keeping to the equilibrium conditions. Several possible methods are available for this; these are considered below but without discussion of the constructional details of the apparatus used, (material, joints, optical windows, electrical circuits etc.) and their manipulation. Reference is made in this connection to a review of Hála [1] of investigations in regions of intermediate pressure (from a few tens to a few hundred bar) in which gas-liquid equilibria play the principal role. Schneider [2] and Tsiklis [3, 4] have reviewed and discussed apparatus for measurements in regions of high pressure (from a few hundred to a few thousand bar) in which mainly liquid-liquid equilibria are observed at low temperature and gas-gas equilibria at high temperature. Young [5] also has comprehensively reviewed experimental methods for studying the phase behaviour of binary mixtures under high pressure. Possibilities especially for determining the critical points of gas-liquid equilibria of binary mixtures have been discussed by Hicks and Young [6].

III.1.a The Synthetic Method

In the synthetic method accurately known amounts of substrate and solvent are placed in a suitable autoclave. This mixture is homogeneous at first. The conditions are then altered so that the transition from one phase to two phases is just recognisable. This may be achieved by a combined change of pressure and temperature, although this route to attaining the phase boundary implies a limited versatility; or, in a suitable apparatus, by change of temperature at constant pressure or of pressure at constant temperature. If, for example, the temperature is varied and the experiment repeated at various constant pressures, one autoclave content yields a series of points on the p(T)-isopleth. If the relative proportions of solvent and substrate are varied during each of these experimental series, the part of the solubility surface of a (pseudo-) binary mixture which is of interest can be determined. However, exact p(T)-values can be obtained only in those phase regions where the change of pressure with temperature is neither too small (in the method of temperature variation) nor too large (in the method of pressure variation). It must be mentioned in addition that the synthetic method is not very suitable when the solubility of the condensed phase is only slightly dependent on pressure or temperature; this applies in general to phase regions which are further away from the critical point.

The transition from the single to the two-phase system is best recorded in the synthetic method through direct visual observation of the appearance of cloudiness or a meniscus. A condition is that the refractive indices of the co-existing phases differ sufficiently from each other. Further, to observe the phase transition enough substance must dissolve in the dense gas; this latter condition applies also when the phase change is observed through alteration of a physical property of the mixture and not visually. An example is with the help of the values of pressure, volume and temperature, where one is kept constant, the second varied and the third recorded. If, for example, the phase boundary is sought at constant temperature by altering the pressure, the compressibility of the mixture can be followed by recording V(p)-curves and this shows a marked change at the phase transition. There are no limitations with regard to the density of the coexisting phases, however. At high pressures and solubilities a point can be reached at which the density of the compressible gas phase equals or exceeds that of the condensed phase. Investigations in this barotropic region are just as possible with the synthetic method as are measurements in the critical region. pVt-data can be obtained along with the solubility values. With mixtures of more than two components it is not possible under the given conditions to obtain data in addition to the observation whether the system has one phase or two phases; this is because the composition of the phases cannot be ascertained. The high pressure cell of the synthetic method can be made small and simple since no samples are taken from it. The apparatus is therefore cheap and easy to handle rapidly.

Numerous authors have described apparatus which they have employed for studies of phase equilibria using the synthetic method. E.g. Schneider [7], Alwani and Schneider [8], Lentz [9], Steiner and Schadow [10], Buback and Franck [11] and Schröder and Arndt [12]. The firm of Nova Swiss offers a complete apparatus with a simple observation cell with fixed volume.

III.1.b The Analytical Method

Like the synthetic method, the analytical method is based on a static measuring principle. The heart of a suitable apparatus is, just as in the synthetic method, a pressure-tight cell, in which the equilibrium can be more quickly attained by vigorous stirring of the contents. Associated equipment is fitted to the cell here also for generating and regulating pressure, for thermostatisation and for measuring and recording the phase parameters. An experiment is carried out by placing the substance concerned in contact with the dense gas in the measuring cell and adjusting pressure and temperature so that the mixture is in a heterogeneous state. The mixture is stirred energetically at constant pressure and temperature. After the equilibrium is reached the two phases are allowed to separate and the composition of each phase is determined. In some cases this is performed by measuring physico-chemical properties of the phases in the pressure vessel itself. More generally, however, a sample is taken of each phase and investigated outside the pressure vessel; more analytical possibilities are then available.

It is not necessary to know the exact amounts of substrate and gas which are used in this analytical method. The composition of the co-existing phases adjusts itself to the given conditions and is determined by a suitable analytical procedure immediately after sampling. Systems with more than two components can thus be studied. It is also possible to determine isotherms and isobars in a large phase region using a single cell filling.

The analytical method is unsuitable for measurements in the critical region and with barotopic systems on account of the imperfect phase separation. The pVT-data of the compressible gas phase, needed to calculate the substrate concentrations, cannot be obtained using the analytical method. The sampling technique is often complicated and difficult. The volume of the cell should be as large as possible but that of the sample as small as possible, so that diminution of pressure and disturbance of the equilibrium are largely avoided when a sample is taken. However large cell volumes mean expensive autoclave constructions, especially for high pressures, and uniform temperature distribution over an extensive sample space is sometimes difficult to attain at high temperatures. Further small sample amounts mean great demands on the analysis, especially with small amounts of substance in the gas phase.

Apparatus for measurements of phase equilibria using the analytical method have been developed by numerous authors. Differences in construction are encountered, depending on the equilibrium to be measured and the pressure and temperature ranges. Among those using the method may be mentioned: Tsiklis and Maslennikova [13], Takenouchi and Kennedy [14], Tödheide and Franck [15] and Streett and Hill [16]. The analytical method demands great care during sampling, usually performed with the help of fine capillaries, because the reduction of the sample to atmospheric pressure brings the danger of premature condensation of less volatile components, thereby leading to falsified results. Some authors, e.g.

Table 4. Application possibilities of the analytical and the synthetic methods for determining phase behaviour of mixtures [19] (+ easily practicable; ○ possible; — not possible)

	Static methods	
	analytical	synthetic
Phase equilibria of binary systems, no poorly volatile components	+	+
Binary systems with poorly volatile components	+	○
Equilibria with more than two components	+	—
Determination of the two-phase region	○	+
Ascertaining if several phases are present	—	+
Measurements in the critical region	—	+
Solubility of poorly volatile components in compressed gases	+	—

Gerritsen and Hartmann [17] and Maier and Stephan [18], linked the cell directly to a gas chromatograph for analysing the sample taken. The apparatus of the last named authors is combined simply, releasing pressure of the sample directly in the injection block of the gas chromatograph.

Comparison of the analytical and synthetic methods shows that the two methods are complementary and that, depending on the problem, the one or the other will be preferable. Brunner [19] published a discussion of the possibilities of application of both methods for determining the phase behaviour of mixtures; the results are given in Table 4.

III.1.c The Dynamic Method

The dynamic method for determining phase equilibria has many features in common with the analytical static method and can thus be termed analytical dynamic. The circulation apparatus for determining liquid-liquid or gas-liquid equilibria, used by several investigators, for example by Tsang and Streett [20], Kuk and Montagna [21] and Radosz [22], could be regarded as an intermediate between these two methods because the equilibrium is established in a dynamic system but the samples are taken under conditions of the analytical method (directly coupled with a gas chromatograph).

In both the analytical and dynamic methods a sufficiently large amount of substance (which need not be known exactly) is brought into contact with the more volatile component in a pressure vessel under conditions which lead to formation of two phases. The pressure and temperature values can be independently fixed. The composition of the phases is a function of these values and is investigated analytically outside the vessel. The composition of the gas phase is of particular interest for the dense gas extraction; it expresses the capacity of the dense gas as solvent for a solid or liquid. In the analytical dynamic method the amount of gas sample taken is not replaced by a sealing liquid such as mercury (or, with very small samples, not replaced at all) as is the case with the analytical static method. Instead, the vessel is refilled with fresh, pure gas under the same conditions of state. On passage through the sample-containing vessel, this gas again takes up substrate, which is present in excess, to an extent determined by pressure and temperature; dissolved substances can be collected at the other end after expansion. The dynamic method is thus based on the same flow-through principle as a drug extraction or percolation.

It is possible in this way to take appreciably larger samples than in the static method, if desired. The available range of measurement is not determined alone by the limits of the analytical procedure but can be extended towards lower values through size of sample. This is highly advantageous, especially in investigations of substances of low volatility and solubility. On the other hand, solubilities exceeding 10% cannot generally be measured accurately using the dynamic method. The reason for this is that the sample is able to occupy only part of the available volume of the pressure vessel and the flow of a certain amount of the more volatile component is needed to dissolve and transport it. Hence it is not possible to distinguish the one-phase character of the system, i.e. complete miscibility of a substrate with the dense gas, from a high solubility value.

The gas sample removed in the experiment is decompressed in a suitable vessel so that the phases separate. The amount of substance recovered is determined analytically, e.g. by weighing, photometry or gas chromatography. The amount of regenerated gas can be ascertained easily by, for example, passing through a calibrated flowmeter. The molal concentration or the concentration in mass percent can thus be directly calculated without knowledge of the pVT-data when the molecular mass is known (in the case of pure substances) or the mean molecular mass (for mixtures). In contrast, the values obtained by the analytical static method cannot be converted to suitable concentrations unless the pVT-data of the gas are known; these data have to be determined separately or taken from the literature.

Special attention has to be devoted to the problem of attainment of equilibrium in the dynamic method. The substance under investigation must be filled into the sample autoclave in a suitably available form with as large a surface as possible, if necessary on a carrier. The mass flow of the dense gas phase has to be sufficiently slow so that, corresponding to the volume of the sample vessel, the time of contact with the condensed phase is adequate. Saturation is then achieved and maintained until the major part of the sample has been extracted. Experience gained so far shows that contact times of 10 to 20 min are normally sufficient; the faster mass transport behaviour of dense gases compared to that of classical solvents plays a significant part here. No change in equilibrium concentration was noted even when the amount and flow of solvent and the position of the substrate in the extraction vessel were varied under identical conditions of pressure and temperature; this demonstrates that equilibrium can be attained rapidly [23]. In addition good agreement has been found between the values obtained using the dynamic method and the static procedure [24].

It is more difficult and demanding in the dynamic system than in the static procedures to attain and keep a constant pressure and temperature over a longer period. Care must be taken with liquid substrates to exclude effects of entrainment and floating in the solvent passage. It is also important to separate the dissolved substances completely from the flowing gas. No difficulties arise with poorly volatile materials, such as fatty oils, but special measures must be adopted with substances of intermediate volatility, e.g. essential oils. These include pressure reduction to the value of the solubility minimum [25] or to atmospheric pressure, collecting the substances in cooling traps; or by adsorption on active carrier materials. The dynamic method is thus less suitable for investigations of very volatile substances, e.g. the organic solvents. It encounters problems also with substance mixtures when the individual components exhibit different solution properties; the composition of the substrate or extract can then change markedly during the investigation. Information can be furnished only when enough substance is available from all components to guarantee during the course of the experiment a constant charge of gas corresponding to the conditions.

The dynamic procedure has been used by many authors. For example, Michels, Dumoulin and van Dijk [26] described an apparatus for determining gas-liquid equilibria; it enabled samples to be taken from both phases. Numerous articles have been published in recent literature concerned with quantitative studies of solubility as a condition for devising separation procedures based on dense gas extraction; in this work the dynamic method was employed to determine gas-solid or gas-liquid equilibria, especially on the gas side. Using a type of high pressure gas chromatograph Czubryt, Myers and Giddings [27] investigated the solubility of oligomeric alcohols at pressures up to almost 2000 bar. Van Leer and Paulaitis [28] and McHugh and Paulaitis [29] measured the solubilities of phenol and chlorinated phenols and of biphenyl and naphthalene derivatives. Johnson, Ziger and Eckert [30] determined the solubilities of solid hydrocarbons. Kurnik and Reid [31] investigated the phase behaviour of solid mixtures, such as naphthalene/phenanthrene, using the dynamic method. Stahl and Willing [32] studied the solubilities of solids (alkaloids) in various dense gases; they used a micro-apparatus of advantageously small dead volumes and ease of handling. Stahl and Gerard [25] determined the solubility behaviour of constituents of essential oils in supercritical carbon dioxide at pressures from 20 to 120 bar, where miscibility is not yet complete. The phase behaviour of fatty oils was studied up to a pressure of 2600 bar by Quirin [33] with the help of the dynamic method. Ohgaki, Tsukahara and Katayama [34] determined the solubility of octadecane in liquefied and supercritical carbon dioxide and ethane. Kwiatkowski, Lisicki and Majewski [35] used the dynamic method to investigate the solubilities of solid organic compounds in dense gases.

III.2. Extraction of Solids

III.2.a Microextraction

Compared to classical extraction procedures, extraction with dense gases is technologically more demanding and the extraction parameters are more complex. For this reason experiments to gain an idea of the extractability and to ascertain the conditions for extraction on the laboratory or pilot plant scales are costly in time and effort and hence in money also. A procedure is highly desirable in which work is with small amounts, easily carried out, and yielding rapidly information about the qualitative extractability of substances as a function of different parameters. The information thus acquired can then serve as a spring-board for increasing to the preparative large scale in which the engineering and production details can be optimised.

An apparatus suitable for this purpose was described by Stahl and Schilz [36] as long ago as 1976. It has been modified and further developed in recent years [37]. The method depends on on-line coupling of the micro-extraction, using dense gases

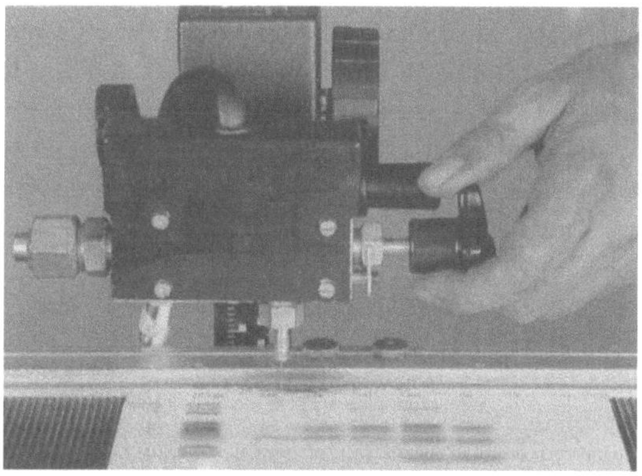

Fig. 10. Partial view of the microextraction apparatus with direct coupling to thin-layer chromatography

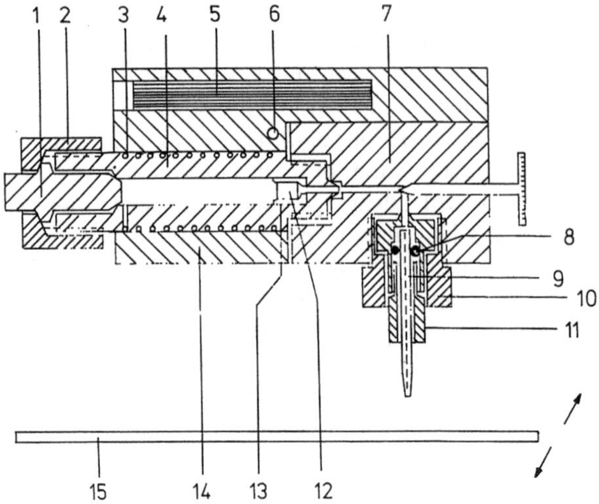

Fig. 11. Longitudinal section of the extraction and transfer parts of the microextraction apparatus.
1 conical fitting; 2 screw cap; 3 capillary for pre-warming the gas; 4 stainless steel extraction autoclave; 5 cartridge heater; 6 thermistor; 7 miniature cut-off valve; 8 O-ring; 9 glass capillary; 10 screw thread; 11 capillary holder; 12 metal sintered filter; 13 PTFE-fitting; 14 brass block; 15 TLC-plate

as a procedure of separation, with thin-layer chromatography to yield an analytical procedure of simplicity and information value. A definite amount of the dense gas – usually one to two grams – is passed through a micro-autoclave which contains the sample to be extracted (powdered plant material or a pure substance) on an inert carrier of large surface area (glass wool, kieselguhr). Depending on its state, the gas takes up a part of the sample and is then depressurised through a fine glass capillary. The solvent power depends on the density and hence the dissolved materials precipitate through this fall in pressure. They are transported by the fine stream of gas and can thus be applied directly as a start band on to the TLC-layer which is moved to and fro. Chromatography is then carried out in the usual way. Information about the high pressure extraction is thereby obtained rapidly and simply.

The microextraction apparatus (Fig. 10) and its way of functioning are explained with the help of Fig. 11. The gas supply line is omitted from the diagram. A small membrane compressor or liquefied gas pump can be employed to bring the gas to the desired pressure value for extraction. Since little gas is required, this dearer method of pressure generation can in fact be dispensed with. For example, it is possible to fill the gas, liquefied at lower temperature, into a pressure container which can be heated; this container is then warmed, pressure regulation being accomplished with a reducing valve.

The central feature of the apparatus is the extraction autoclave of stainless steel [4] with connected sample transfer to the thin-layer chromatography plate [15]. The extraction vessel has a sample capacity of 2 ml and fits into the brass block [14] which can be heated. The block and hence the extraction vessel are thermostated with the help of the cartridge heater [5]. The heating is regulated by a control device connected to the thermistor [6] which is situated in a small recess between the sample vessel and the heating element. A stainless steel capillary [3] wound round the extraction vessel, serves for pre-warming the entering gas. A metal sintered filter [12] at the autoclave exit retains fine particles of the sample. The sample vessel is closed with a small cut-off valve [7]. The pressure of the gas is reduced to atmospheric through a fine glass capillary [9], the internal diameter of which is narrowed to about 10 μm by drawing out conically from thermometer tube glass. A screw cap and O-ring hold the capillary. The gas flows at a rate between 0.1 and 1 g/min, depending on the pressure set and the diameter of the glass capillary which can be easily widened by shortening. The flow is measured before the experiment with a soap bubble flowmeter for all pressure stages.

The TLC plate [15], supported on a carriage, is situated horizontally below the vertical capillary through which the gas passes as pressure is reduced. Depending on the pressure, the layer is 1 to 5 mm from the end of the capillary. Automatic movement to and fro during the extraction yields a start band of adjustable length.

The powdered sample is packed between two glass wool plugs in a glass tube of about 1 ml content and 40 mm length, open at both ends; it fits exactly in the opening of the microextraction autoclave and is sealed to the autoclave wall with teflon tape. This guarantees clean and rapid exchange of the extraction receiver. Extraction is preferably performed in a stepwise pressure gradient. The pressure is raised in stages and the sample extracted with a constant amount of gas at each pressure stage. The TLC-plate is displaced further after each increase in pressure.

After development and visualisation of the chromatogram a diagram is obtained on which the abscissa shows, discontinuously, the extraction parameter of pressure or temperature. The ordinate direction shows the separation of the extracted substances according to their chromatographic behaviour (polarity). The presence and relative intensities of the individual zones yield information about which substance was extracted at which pressure and in which relative amounts. The behaviour of both the principal components and the accompanying substances can be seen and it is thus easy to gauge whether and which fractionated or selctive extraction is possible.

Additional information can be derived simply and quickly, e.g., about the influence of the degree of pulverisation of the sample, of any chemical pretreatment of it or of the use of an entrainer, on the extractability of individual components. A direct comparison of the extracted components from drug materials of different origin, type and variety under fixed extraction conditions is possible. For this, the extracted components of various samples must be applied as start bands along the abscissa.

The firm of NOVA SWISS, Effretikon, Switzerland, has supplied a somewhat modified apparatus for microextraction since 1984.

III.2.b High Pressure Soxhlet Extraction

A simple apparatus for extracting solids with liquid carbon dioxide is commercially available. It is performed in an ordinary glass soxhlet apparatus, situated in a pressure chamber (Fig. 12).

A suitable amount of the extraction agent, carbon dioxide, in the form of dry ice, is placed in the lower part of the pressure vessel. The vessel is then closed and put into a heated water bath. The solid carbon dioxide evaporates, increasing the gas

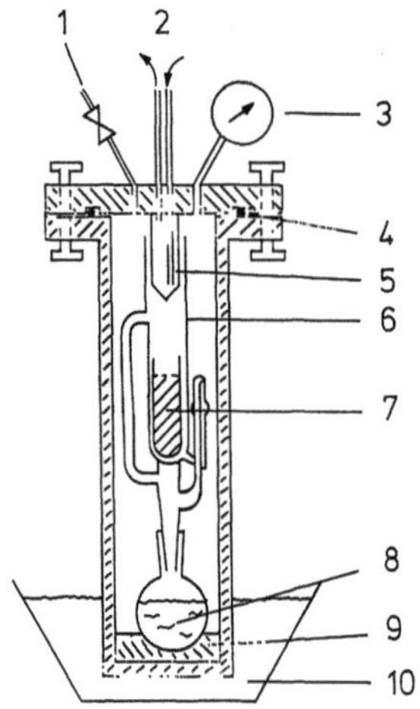

Fig. 12. High pressure soxhlet extractor (J. & W. Scientific Inc.).
1 cut-off valve; 2 cooling; 3 manometer; 4 O-ring; 5 condenser; 6 glass soxhlet apparatus; 7 sample; 8 boiling liquid carbon dioxide; 9 plate for heat transfer; 10 heated water bath

pressure in the vessel to the value for liquefying; this is determined by the temperature of the condenser in the cover of the vessel. For example, a pressure of 51 bar is attained at a condenser temperature of 15 °C. The extraction gas which has liquefied in the condenser drops into the soxhlet apparatus and extracts the sample there in a thimble. During extraction the extract becomes more concentrated in the roundbottomed flask in the liquid carbon dioxide which is maintained boiling there. After extraction the gas is released through a valve and the extract can be obtained from the round-bottomed flask after opening the pressure vessel.

This simple apparatus for extracting with liquid carbon dioxide enables small amounts of extract to be obtained for analytical purposes. Product improvement is scarcely possible because the conditions of extraction can be varied only in limited fashion by altering the temperature of the condenser. Besides CO_2 the high pressure soxhlet extraction is possible only with gases which can be liquefied in the temperature range 0 to 20 °C under a pressure of 100 bar. In this case equipment is also necessary for filling with liquefied gas.

III.2.c Preparative Extraction

Principle

The procedure of extraction with dense gases follows essentially the same course whether on the laboratory or the production scale. In contrast to microextraction, where only small amounts of gas are needed, the gas used in preparative installations must be circulated in order to keep its consumption within reasonable limits. Fig. 13 depicts the extraction circulation of a dense gas, using various methods for separating the extract.

The dense gas is led under extraction conditions through the starting material placed in a pressure vessel. This dissolves the extractable material. The miscella is allowed to pass via a throttle valve into the separating vessel (Fig. 13a) under reduction of pressure. This lowers the solvent power of the gas and the dissolved sub-

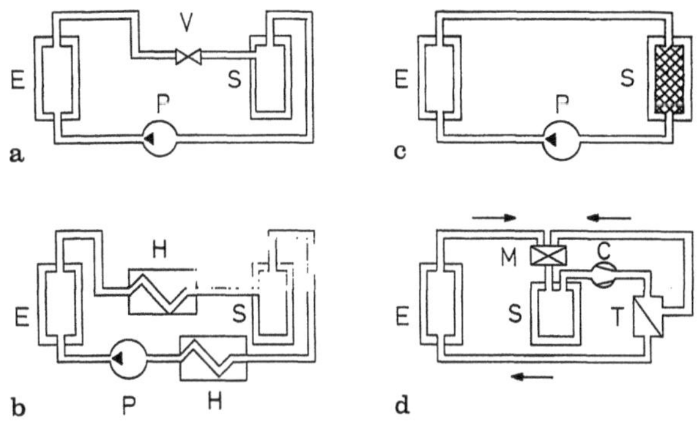

Fig. 13a–d. Extraction circulations based on various principles of extract separation.
Extract separation through: **a** change of pressure; **b** change of temperature; **c** adsorption; **d** mixing with an inert gas.
E extraction vessel; S separator; P pump; C compressor; V throttle valve; H heat exchanger; M device for gas mixing; T gas separation (membrane)

stances separate out. The recovered gas is brought again to the pressure for extraction and led anew into the sample vessel. The principle of isothermal separation does not mean at all that heat exchangers are not required in practice because every change in pressure of a gas is accompanied by a temperature change. The extract can be brought to separation also by temperature change at constant pressure (Fig. 13b). Dependent on pressure and the nature of the dissolved substance, this isobaric procedure can bring about separation by either increase or decrease of temperature. After the separation stage, the gas has merely to be brought back to the extraction temperature and transported to the extraction vessel. This separation principle alone is usually not sufficiently effective and is thus combined with other possibilities. In another procedure variant, the substances dissolved in the extracting gas are separated by adsorption on suitable materials, e.g. active charcoal (Fig. 13c). This procedure, which can be pursued isobarically and isothermally in the complete circulation, is very economic from the point of view of the operation itself since it is necessary neither to utilise high compression energies nor to transfer large amounts of heat energy. One must remember, however, that in this process extra costs are involved for regenerating the adsorbent. The technique is therefore especially interesting when only small amounts of extract or impurities are to be separated from the carrier material (decaffeination of raw coffee).

An additional method for separating the extract, which has no practical relevance at the present moment, depends on the fact that the solvent power of, for instance, carbon dioxide, is reduced by nitrogen or argon (Fig. 13d) [38]. This technique has an advantage over the customary methods for separating extract from dense gas: pure extracts are obtained directly in procedures which are conducted isothermally and sensibly isobarically. It is, however, necessary to regenerate by removing the inert gas from the extraction agent by means of a membrane procedure.

It is quite possible, to combine these individual principles for separation of the extract. It is indeed usual in practice because of advantages in certain cases. Further, a stepwise reduction of the solvent power of the dense gas can separate the extracted material into different fractions.

Constructional Parts

The high pressure extraction plants visualised vary according to the size and the nature of the material to be extracted. Nevertheless, there are principles common to all basic engineering constructional designs, and features of the required installation components which are generally valid. The function and arrangement of the individual structural units will be described here only generally, omitting usually the constructional details.

If the high pressure installation is to be used for preparing extracts of natural products destined for human consumption in some form, the choice of building material for all parts of the plant which come into contact with the extracts must respect the requirements of foodstuff regulations. This means in particular that the internal surfaces of the pressure vessels, pipelines, heat exchangers etc. must be stable against corrosion and easy to clean (e.g. stainless steel 316 Ti (AISI)). This requirement is most important since many extracts and flavouring materials are corrosive on their own or in combination with water and the dense gas. The pressurising part of the extraction plant is designed according to the particular problem. Experimental installations on the laboratory or pilot plant scale should be designed for the largest possible range of pressure, so as to enable varied tasks to be tackled. On the other hand, production plants are adapted to definite products, for which the maximum operating pressure has been established. A genuine continuous operation is usually not possible in the extraction of solids with dense gases, so that the introduction, and removal, of solids to and from the high pressure region still poses technical problems at present. Charges are made under conditions of normal pressure, so that some parts of the installation, notably the extraction vessels, are subject to the strain of surging pressure

Fig. 14. Clamp closure and pneumatic filling device of a high pressure vessel (photo from the UHDE GmbH)

changes. This can lead to material fatigue and must be allowed for when planning the pressure vessel.

On the laboratory scale the material for extraction can be handled without any problems; on the technical scale, however, solutions must be found to the questions of charging the extraction vessel. A rapid transfer of the crude material is a cost-determining factor in industrial extraction plants. The type of closure is determined by the way in which filling and emptying are performed. Granular material is relatively simple to handle; it can be easily filled or emptied through small openings in the top or bottom of the extractor. Sometimes, however, it is necessary to open the complete cross-section of the extraction container to interchange for example basket-like charges containing plant material, or to introduce or withdraw material for extraction with the aid of a pneumatic conveyer. Several means are known in practice for closing the cover of container vessels. Mention may be made of flange-, clamp-, pin-, bayonet-, sliding- and frame-closure and also of the use of screw clamps or screw caps for closing the cover of the high pressure vessel. Fig. 14 shows an elegant solution to the problem of closing, namely, a fully automatic clamping mechanism.

The closing clamp, operated hydraulically, is opened, the vessel cover is raised, likewise by hydraulic power, and pushed to the side and a pneumatic filling device is placed on the high pressure tank. Sealing elements are of especial interest in constructing high pressure installations. Metal to metal joints are suitable for pipelines and armatures, e.g. conical joints, flanged screw fittings with lens joint or cut ring joints. Pipes may be joined also by welding in large plants when the pipeline system is installed permanently.

Metal to metal joints are unsatisfactory for sealing larger profiles, such as the covers of extraction vessels, because the surfaces have then to be pressed together with very great force. It is then better to use O-ring gaskets or sealing rings derived from O-rings in form and structure, e.g. with inner springs. The non-aggressive extraction of natural products makes no severe demands on the temperature-stability of the sealing materials used. However, other requirements are made of the sealing material: It must be of foodstuff quality and possess good chemical resistance to the dense gas and the extracts. The gaskets are usually organic polymers and should not swell under high gas pressure, otherwise when the pressure is released and the sealed joint opened, a marked increase in volume of the gasket may take place; this could make it unusable.

A distinction is made between inner- and outer-O-rings, which depends on the position of the gasket (Fig. 15). Vessel covers are usually best sealed with inner-situated O-rings; the autoclave lid does not then have to be pressed on with such force and closing is possible with a clamp, for example. As pressure builds up in the vessel the cover can rise a small distance without a leak occurring. Advantages of outer-situated O-rings are that they can be exchanged faster and more easily and wear out less through friction and shearing stress. An example of their use is for closing flange joints without reduction of profile area, as for example, is desirable with pipe joints.

A particular difficulty in running a plant for extraction with dense gases is the necessity to use thermostatic control of the sample and the extraction agent in the high pressure containers. Stainless steel has a relatively low thermal conductivity and there are thus long warming and cooling times with thick-walled vessels when using external heat exchangers. If faster temperature changes are desired this can be accomplished with additional means. For instance, a heat exchanger in the form of a heating coil can be built in the interior of the high pressure vessel. However, this method is unsuitable for extraction vessels since it restricts the availability of the precious high pressure volume and makes difficult rapid charging with the sample material. It has, however, proved valuable in practice for the extract separator. Heat transfer in the extraction vessel can be improved by using an internal wall heat regulation, provided that it has been constructed according to the multi-layer system. The thermostatisation can then be achieved using electrical heating elements or a canal system through which a heat carrier flows.

A smooth flow of gas and ease of cleaning are worth attention in the design of a heat exchanger, since a change of state of the extraction agent is brought about in these parts of the installation which can lead to separation of extracted material and choking. Especially double pipe and tube bundle heat exchangers find use in high pressure extraction installations; the dense gas, with or without dissolved material, passes through the tubular part of the heat exchanger, the heating or cooling medium washes the casing. In this case, only the tubes have to be made of stainless steel, easy to clean and withstanding the high pressure.

Thermostatisation is necessary also for certain pipeline parts and armatures if there is danger of an unwanted separation of extract which can choke the system. This can be achieved with electrical strip heaters or blankets. At least, these parts of the installation should be well insulated against heat radiation losses. The temperature in the high pressure vessels and pipelines is measured with thermoelements or resistance thermometers.

Except when using the isobaric procedure, the solvent in the extraction vessel is at a higher pressure level than in the separator, in extraction with dense gases. There are several ways of overcoming this pressure difference and for transporting the dense gas, of which practically only work with a compressor or pump is relevant. In the design of this part of the plant, it must be attended to that the extraction agent does not come into contact with bearings or pistons smeared with liquid; dense gases are good solvents for lubricating oils! This would result in contamination of extraction agent and extract on the one hand, and in impairment of the function of the aggregate on the other. For this reason, almost only membrane compressors and pumps are employed in practice for compressing and transporting the extracting gas. The basic mode of operation with compressor or pump has been described in Chapter II,4.1 during discussion of the thermodynamics of the solvent circulation.

Various constructional parts are available to control the parameters of pressure and gas flow. These are set according to different principles, depending on the position

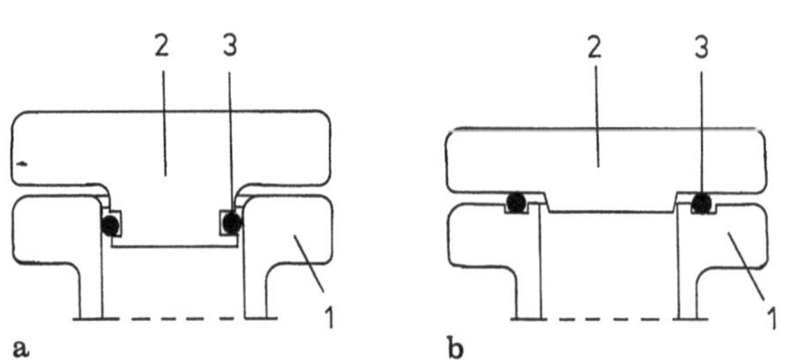

Fig. 15a, b. Sealing of pressure vessel with O-rings.
a inner-situated O-ring; **b** outer-situated O-ring; 1 pressure vessel; 2 vessel cover; 3 O-ring

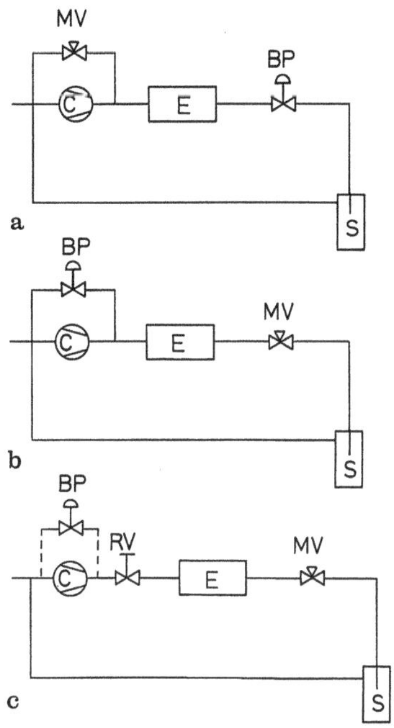

Fig. 16a–c. Ways of regulating extraction pressure an gas flow
C compressor; E extraction vessel; S separation vessel; BP back pressure regulator; RV reducing valve; MV metering valve

in the installation. Overflow or reducing valves are available as alternatives to setting the extraction pressure. The former regulate the pressure at the entry point to the valve and excess gas is allowed to flow by. The latter regulate the pressure at the exit of the valve by blocking the gas flow when the desired pressure has been attained. Special metering valves are highly suitable for regulating the gas flow. Fig. 16 shows, in a simplified extraction circulation procedure, various arrangements of these constructional parts for regulating pressure and gas flow [39].

In Fig. 16a the extraction pressure is controlled through the back pressure regulator BP, situated at the exit of the extraction vessel and in the flow of the gas stream, laden with extracted substance. The decompression of the dense gas, and hence the separation of the dissolved components, takes place at the valve. It must be noted with this arrangement that this control element can become covered with extract, even choked, which would cause a rise in pressure in the extraction part of the plant. Safety devices are consequently necessary, such as contact manometers to cut out the compressor, or safety valves on the extraction vessel which release excess gas into the atmosphere or the gas supply if the set pressure is exceeded. The flow is regulated in a bypass to the compressor via the metering valve, MV. In the design portrayed in Fig. 16b the two control elements, back pressure regulator and metering valve, are interchanged with respect to Fig. 16a. The back pressure regulator with which the extraction pressure is set, does not then come into contact with gas charged with extracted material. It fulfils at the same time the function of a safety valve. The decompression of the dense gas takes place at the metering valve, which is used to set the gas flow speed. Fig. 16c illustrates the possibility of setting the extraction pressure with the reducing valve, RV. An additional control element is necessary in this case, such as a back pressure regulator, to ensure that the pressure in front of the reducing valve does not rise indefinitely when little gas is needed in the extraction circulation.

A design such as that in Fig. 16a comes especially into consideration for a production plant; however, the control valve MV is not required because, in a large scale plant, the conveying capacity of the compressor is adjusted to the extraction volume and the sample material and is fully exploited in the extraction process. More is demanded of the control elements of a laboratory or pilot plant because it is used for experimental purposes and work is conducted with varying parameters. It is especially necessary that pressure can be regulated over a wide range and that various definite flow speeds of the dense gas, independent of the pressure, can be set. All the techniques depicted in Fig. 16 are suitable for the largest possible variation in regulating these parameters in a laboratory installation; that in Fig. 16b is especially advisable on account of the less problematic pressure setting. The Fig. 16c variant is of interest only when small amounts of gas are required for the experiments, as, for example, in microextraction. Expenses can be lowered in this case by replacing the compressor by a gas storage vessel in which the necessary pressures can be attained by appropriate temperature changes.

The gas flow is very conveniently measured using mass flowmeters which record specifically according to the gas and largely independently of the conditions of state. When measurement is made according to the principle of volume estimation (suspended body, spring-supported float), the value depends on the density of the gas and the pVT-data must be taken into account to determine the amount of gas. The flowmeter should be in the gas stream which contains no extract if the results are to be accurate and free of interference. This is possible in two places, in front of the extraction vessel or after the separator. The latter is mostly preferred in practice because the flowmeter does not then have to be designed for high pressures.

Examples of Installations

Constructional components, and devices for control and measurement have now been described and in the following a description is given of some installations of various sizes, suitable for obtaining substances from solid sample material on a preparative scale. Plants for extracting liquid starting material using dense gases are mentioned in Chapter III.4 under the heading of refining procedures.

Laboratory or pilot plant installations perform valuable service in establishing the procedure parameters when the engineering design of large scale plants is contemplated. An installation on the laboratory scale must be designed so that the most widely differing problems can be studied. The following points must then be specially considered:

- covering the largest possible range of procedures for extraction with dense gases (concerning the factors of pressure, temperature and gas flow)
- flexibility of scale of extraction
- possibility of extract fractionation by multiple stage separation techniques
- continuous control and recording of the parameters of the process

The flow scheme in Fig. 17 shows the construction and function of an extraction installation on the preparative laboratory scale, as built by Stahl and co-workers [42, 43].

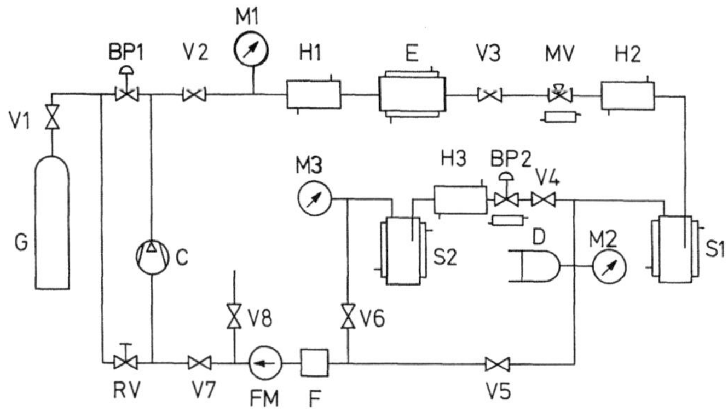

Fig. 17. Flow plan of the extraction installation.
G gas supply; C membrane compressor; E extraction vessel; S1, S2 separation vessels; F active charcoal-blue silica gel-filter; FM mass flowmeter; M1–M3 strain gauge pressure transducers; H1–H3 heat exchangers; RV reducing valve; BP1, BP2 back pressure regulators; MV metering valve; V1–V8 cut-off valves; D safety diaphragm

The gaseous extraction agent, e.g. carbon dioxide, is taken from the supply reservoir, G, an ordinary commercial steel cylinder, via the cut-off valve, V1, and led into the installation through the reducing valve, RV. The pressure reducer determines the suction pressure of the compressor and the separation pressure in separation vessel, S1, for single-step separation or in vessel S2, for two-stage separation. The pressure set up is controlled and recorded by an indicator gauge via the pressure transducer, M2 or M3. Compression of the gas is carried out with the membrane compressor, C. The extraction pressure is shown on the manometer, M1, and is adjusted with the help of the back pressure regulator, BP1, which allows gas in excess of that for the desired pressure to flow back into the storage vessel, G. The compressed gas is brought isobarically to the desired temperature using the heat exchanger, H1, and flows through the extraction vessel, E, likewise thermostated, where it encounters, and takes up extractable material from, the sample. The flow speed desired is set with the heated fine metering valve, MV. Here, the gas plus dissolved extract suffers loss of pressure from that for extraction to that for separation out (single-stage separation in S1, valves V4 and V6 closed, V5 open). After the isobaric evaporation of the extraction agent, partially liquefied through the decompression in heat exchanger H2, the extracted material is collected in separation vessel S1.

Another variant involves fractionated separation by stepwise pressure reduction in the vessels S1 and S2, in series next to each other (valve V5 closed, valves V4 and V6 open). The separation pressure in S1 is set with the heated back pressure regulator, BP2, which functions simultaneously as a further valve for depressurising. The pressure in S2 corresponds to the suction pressure of the compressor (set on the reducing valve, RV). In addition to this separation through pressure stages, a stepwise separation of the extract can be achieved by varying the temperatures of the process or by combined pressure and temperature changes. The separation part of the installation is secured against an uncontrolled rise in pressure by a safety diaphragm.

After having left the separator the gas, now containing no extract, reaches the mass flowmeter via the adsorption filter, F. The meter is coupled with a recorder which

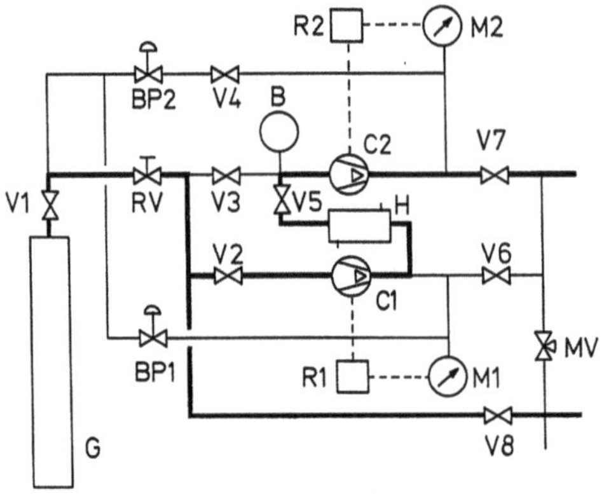

Fig. 18. Pressure generation and regulation in a laboratory installation for pressures up to 3000 bar.
G gas supply; RV reducing valve; BP1, BP2 back pressure regulators; MV fine metering value; V1–V8 out-off valves; B buffer volume; C1 membrane compressor to 1000 bar; C2 membrane compressor to 3000 bar; R1, R2 switching relays; M1, M2 contact manometers; H heat exchanger

registers the amount of gas used. Since the sample is often changed in a laboratory apparatus, a built-in blue silica gel-active charcoal filter facilitates necessary cleaning of the equipment, prevents interferences with the gas flow measurement and with the compressor as a result of separation of extract, and excludes contamination of the gas supply with extract. A two-stage compression is advantageous for the gas supply of a laboratory scale extraction installation in the pressure domain above 350 bar [33]. Fig. 18 portrays suitable equipment for generating and regulating pressure for this. The path taken by the gas in the high pressure region is accented by the solid drawn line.

The gas for extraction is taken from the storage vessel via the reducing valve, RV, and fed to the compressor, C1 (V2 open, V3 closed). This supplies compressor C2 (V5 open, V6 closed), with pre-compressed gas at 90 to 150 bar. An intermediate cooling of this gas to temperatures just above T_c (31 °C for carbon dioxide) is important for the second compression stage to be effective. The gas from compressor C2, now brought to the desired final pressure (shown on manometer M2) attains the extraction part of the system via valve V7. Manometers M1 and M2 are designed as contact manometers which, when a definite pressure is exceeded, operate a safety switch to stop the electrically driven compressors.

It is an advantage to have a buffer volume, B, between the two compressors of a two-stage compression. This ensures an adequate, pressure-constant supply of gas to the compressor C2. Without this, fluctuations occur, depending on whether the suction strokes of the two compressors coincide or the compression stroke of C1 coincides with the suction stroke of C2. The pressure regulation can be performed up to 1000 bar with the aid of the back pressure regulator, BP2. Regulation is more of a problem in the higher region and has to be done via a bypass which is opened by the fine metering valve, MV. The part of the highly compressed gas from C2 which is not needed in the extraction circulation returns thereby directly into the suction pipe of C1. The setting of the fine metering valve, MV, permits a rough setting of the extraction pressure; the fine regulation can be undertaken at the back pressure regulator, BP1, by changing the preliminary pressure of compressor C2 and hence its conveying capacity.

Fig. 19 shows the advantages of a two-stage compression. The energy requirement for adiabatic-reversible compression of CO_2 to 600 bar is depicted. The starting

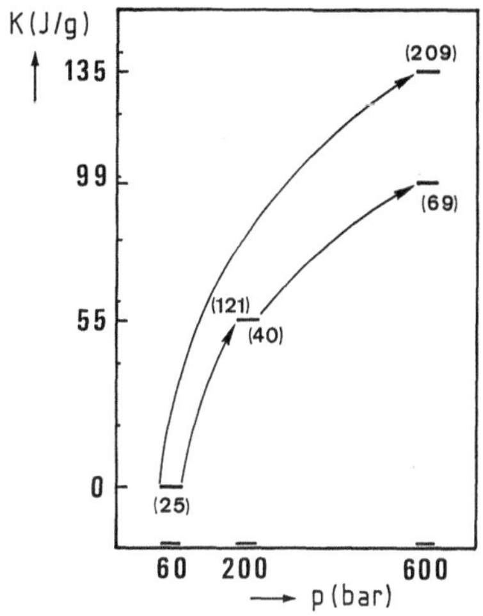

Fig. 19. Scheme illustrating the energy requirement for isentropic compression of carbon dioxide with and without intermediate cooling.
(figures in parantheses = temperatures in °C)

material is gaseous CO_2 (60 bar, 25 °C), as taken from the usual commercially available steel cylinders. Isentropic compression is always accompanied by rise in temperature; temperature values are thus given in Fig. 19, the figure at the beginning of the arrow denoting the temperature before, and that at the end of the arrow, the temperature after compression. The two-stage increase in pressure from 60 to 600 bar with intermediate cooling, for example to 40 °C, can be seen to save more than a quarter of the energy of compression which would be needed in a single-stage procedure. It is also evident that the pressure increase from 60 to 600 bar, carried out in this way, requires 99 kJ/kg CO_2 which is not even twice that amount of energy needed from compression from 60 to 200 bar, namely, 55 kJ/kg CO_2.

The laboratory installation described consists of three separate units: The pressure generation, i.e., the permanently set-up compressor station with the gas supply, is connected to the extraction part itself (Fig. 20) via a pipe system. The transportable stand contains the pressure vessels, the control arrangements, such as armatures, thermostats and heat exchangers, and the sensors for taking measurements.
The control group (on the left in the figure) forms the third unit and contains the source of electric power, indicating instruments and recorders of the parameters of the process. The laboratory installation permits preparative extraction of natural products in amounts up to 5.7 litres at 350 bar and 300 ml at pressures of 2000 bar.

In order to optimise the process of extraction with dense gases, experiments on a scale exceeding that of the laboratory are advisable, e.g. in a pilot plant with 50 litres extraction volume. The data obtained in this way can be more reliably extrapolated to the production scale.

Fig. 21 shows a pilot plant of the firm NOVA SWISS Effretikon, Switzerland which uses a 50 l extraction vessel. The flow sheet of this plant is shown in Fig. 22 and serves to describe it [44].

In contrast to the laboratory installation described above, the compression and conveying of the extraction agent are carried out with the help of a liquid pump. The gas for extraction is taken in the liquid state from the bottom of the supply vessel, G, in which it, as a liquid, is on the verge of boiling, and then cooled through

Fig. 20. Frontal view of the laboratory installation for extraction

Fig. 21. Pilot extraction plant with 50 l extraction volume in the pressure region up to 320 bar (firm of NOVA, SWISS, Effretikon, Switzerland)

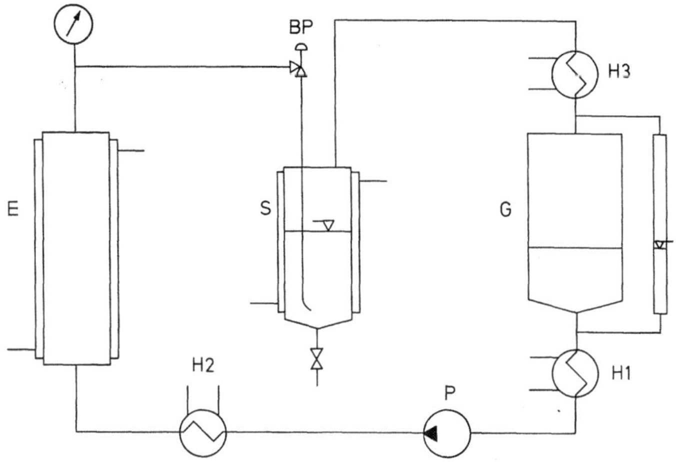

Fig. 22. Simplified working scheme of the 50 l plant.
E extraction vessel; S separator vessel; G gas supply; L level indicator; BP back pressure regulator; M manometer; P liquid pump; H1 to H3 heat exchangers

the heat exchanger, H1, so as to prevent cavitation of the solvent as a result of pressure fall in the suction line of the pump. It is then brought to the extraction pressure with the aid of the liquid pump, P. The conveying capacity of the pump can be varied via the piston stroke. The pressure is set at the back pressure regulator, BP, which is built into the main circuit so that the whole gas amount conveyed is exploited for extraction. The gas is thermostated isobarically to extraction conditions in the heat exchanger, H2, and then traverses the extraction vessel, E, from below upwards. The gas, laden with extract, suffers pressure reduction at the valve BP to that for separation out. In a departure from the principle of the laboratory installation, the extraction gas, now partly liquefied, is transferred together with the extract, directly into the separator, S, where the extraction agent is in the wet vapour region. To prevent the liquefied gas from overflowing out of the separator, which would lead to loss of extract, the vessel is fitted with a level indicator which controls the supply of heat for evaporating the liquefied gas. The level is shown with the help of sensors built into the separation vessel at various heights; these function either according to the capacity principle or measure the temperature, in which case the state of the extraction gas is ascertained with the help of the separation pressure derived from the phase diagram of the gas. The gaseous solvent, free of extract, is finally liquefied in the condenser H3 and flows back into the storage vessel, thereby completing the process cycle.

Since the liquid pump must always be supplied with liquefied gas through the suction line, the level of liquid extraction agent in vessel G must be controllable. This can be done for example by using a bypass to G in which a magnetic float shows the level. The pressure for separation out and hence the suction pressure of the pump, is determined in this type of process course by the temperature in the gas storage vessel. The gas for the extraction is in gas-liquid equilibrium in it, so that, at a given constant volume, a change of temperature is always accompanied by a change of pressure.

Aspects of planning and layout of high pressure extraction plants for natural materials have been discussed also by Lack [44a] who described a 10 l installation.

It was for pressures up to 325 bar. If desired, the extractor could be equipped with a stirring device. The maximum throughput of solvent was 60 kg/h. The work contained also a number of calculated ideal cases of cycles to ascertain the energy demand for various extraction and separation states, making use of T,s-diagrams of carbon dioxide.

No detailed information about engineering features of production plants can be given here since the know-how of the users is, understandably, not revealed. Some aspects of general interest should, however, be touched on. Production plants vary considerably in size as a result of the multiplicity of the products which can be obtained by dense gas extraction. For preparing smaller amounts of product, e.g. of high grade active principles for the pharmaceutical industry, of precious essential oils or other specialities, installations with extraction volumes of the order of 100 l can be economically worked. Plants must have a capacity of several cubic metres, however, to permit economical extraction of mass products, such as coffee, hops or oil seeds. Fig. 23 shows part of an extraction plant with three extractors designed for quasi-continuous counter-current operation, having a capacity of up to 1.5 t of starting material per day.

Large plants of this sort with extraction volumes of several thousand litres demand a high throughput of dense gas, up to 70 tonnes carbon dioxide per h for example. An exchange surface of 250 m^2 is needed for thermostating an amount of this size [45]. The gas in such a plant can no longer be supplied with membrane pumps or compressors. Water-lubricated plunger pumps, for example, are used for compression and conveying in such cases.

Fig. 23. Extraction plant of content 500 l and designed for 385 bar pressure (works photo FLAVEX GmbH)

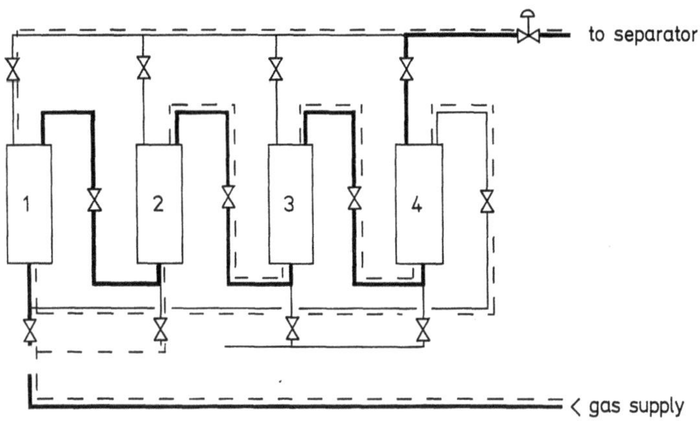

Fig. 24. Scheme of a multiple-vessel counter-current extraction

At the present time it is not possible to carry out continuous extraction, i.e. to convey solids into and out of the high pressure vessels under the conditions of extraction. Instead, the extraction volume is advantageously distributed over several pressure vessels, whereby the batch process begins to resemble a genuine counter-current extraction. This is shown schematically in Fig. 24.

Four extraction vessels are joined in series in this arrangement (thick line). After the first vessel has been exhaustively extracted it is withdrawn from the cycle by valve operation, its pressure is released, it is refilled and re-introduced into the line so that it is traversed by the dense gas as last of the series (dotted line). In this procedure several valves have to be operated at the same time. This is accomplished simply by pneumatically operated and controllable valves.

Aspects of Experimental Planning

The planning of experiments to obtain the best high pressure extraction is aggravated by the large number of factors. As a result of the high expenditure involved it is advisable to perform the experiments on the smallest scale possible. Table 5 contains information about the various parameters and their importance for yield and selectivity of extraction. The table also informs about the most suitable scale for optimising the parameters.

The parameters are listed in the order in which the steps towards the best conditions should be made. The pre-treatment of the sample (shredding, attainment of a definite water content, treatment with acid or alkali etc.) is a factor which influences yield and selectivity of the extraction in decisive fashion. Here, microextraction can yield qualitative information rapidly. An optimisation of the conditioning of the sample is possible only on the preparative scale, however.

The parameters under the heading of the extraction agent can be optimised largely on the laboratory scale. Microextraction provides valuable tips about selection of the gas for extraction, its solvent power and selectivity as functions of pressure and temperature, and about possibilities of fractionation by varying the conditions of state. It is possible to build on this information to ascertain the optimum pressure and temperature conditions for extraction and separation out on the preparative scale.

Table 5. Parameters of dense gas extraction [40]

Extraction Parameter	Influence		Scale of Experiment		
	Yield	Selectivity	Labor.		Technical
			Micro	Prep.	
Sample for Extraction					
Pre-treatment (shredding, water content etc.)	× ×	×	●	●	
Extraction agent					
Type and state of gas	× ×	× ×	●	●	
Pressure/temperature of extraction	× ×	× ×	●	●	
Pressure/temperature of separation	×	×		●	
Gas throughput	× ×	×		●	●
Conduct of the process					
Length of the extraction	× ×	×		●	●
Extraction, multiple	—	× ×	●	●	
Separation, multiple	—	× ×		●	
Form of operation	×	×			●

Under the heading "influence" the meanings are: × × decisive influence; × influence; — no influence ● possible scale for experimental work

The amount of solvent required for extraction and the relative throughput (the ratio between the mass flow of dense gas and the amount of sample taken) are important factors which determine the yield of the extract and the energy and time requirements of the process. To optimise these parameters it is very useful to plot diagrams relating extraction yield to the time of extraction or the gas amount put through. Relative throughput values of 5 to 20 kg gas per hour and per kg sample have proved good in extraction with dense carbon dioxide [39]. A certain saturation of the extraction gas becomes established in this region, i.e., the yield increases approximately proportionally to the amount of gas. The exact value, always to be recommended, of the relative throughput depends on several factors. One may mention: the content of extract in the sample; the solubility of the substances to be extracted, their accessibility for the dense gas (diffusion) and, not to be neglected, the mass transfer kinetics which, for their part, depend on the types of compound and the conditions of state, especially the temperature, of the dense gas in the extraction vessel. It has been found also that a particular value for the equilibrium charging of the gas phase, e.g. of 75%, is more rapidly reached as pressure is raised [41].

One can refer the mass flow of dense gas to the extraction volume also, if the bulk density of the sample is taken into account. Assuming an average bulk density of the sample to be 0.5 kg/l (which is a normal value for dried plant material) it can be deduced that a compressor or pump performance of 2.5 to 10 kg gas/h is suitable per litre extraction volume. This means in turn that, depending on the gas density, the solvent in the extraction vessel is changed about every 10 minutes.

Questions about the conduct of the process must be answered partly in the laboratory, partly in the pilot plant. Variations of procedure for increasing selectivity, such as the multiple stage extraction or separation, can be tested in preparative laboratory installations. However, the duration of exhaustive extraction of a particular amount of crude material can be ascertained only on the technical scale. The gas consumption found in laboratory experiments cannot necessarily be translated to the large scale. The bulk densities of the materials to be extracted may show deviating values as a result of differing charging techniques; and the slenderness of line of the extraction vessel (ratio of height to cross-section) influences the flow conditions of the dense gas. Other factors which can be optimised only in pilot or large scale plants are, notably, the type of operation (e.g. quasi-continuous multiple-vessel extraction in counter-current, or employment of stirring procedures) and economic-technical aspects of the course of the process, where mention must be made of, for example, the extraction dead time for emptying and refilling the high pressure vessel, including pressure and temperature regulation. Even the withdrawal of the extract from the separator can sometimes become a problem on the large scale which requires special adaptation to the nature and amount of the product.

III.3. Extraction Using Entrainers

The dense gases suitable for extraction of natural products are in general weak solvents. This brings on the one hand the advantage of high selectivity but, on the other hand, limitations in that some substance groups of interest are insoluble or can be extracted only under extreme conditions or using an uneconomically large gas flow. For this reason attempts have been made to improve the solvent capacity of dense gases by adding small amounts of an auxiliary. These auxiliaries, known as entrainers or modifiers, are compounds of volatility intermediate between that of the condensed phase to be dissolved and that of the dense gas. For separation of natural products under non-aggressive temperature conditions, entrainers boiling between 20 and 100 °C at ordinary pressure have proved useful. These include classical solvents, such as alcohols, chlorinated hydrocarbons, n-hexane, acetone and water. Sometimes the entrainer can be chosen so that, through special interactions of the extraction phase with the substrate, definite components of the mixture to be separated are preferentially dissolved. The separation factor is, however, generally lowered by addition of the entrainer, particularly as the proportion of the entrainer increases.

As well as increasing the solubility in the gas phase of poorly soluble substances, with the advantage of applying working pressures lower than those with the pure gas, the addition of an entrainer also enables the gas to be regenerated more efficiently by small changes of state or merely through temperature effects. For explanatory purposes Fig. 25 shows the temperature-dependence of a ternary system made up of a poorly volatile substance, A, dense gas, B, and entrainer, C, in a triangular diagram with temperature along the vertical axis and at a constant supercritical pressure of $2p_c$, for example.

A miscibility gap is present in the whole temperature range on the prism surface of the binary system, AB; it is most extensive at intermediate temperatures. Here a second miscibility gap occurs also in system BC, over a limited interval of temperature; the solubility of the poorly volatile substance, A, in the gas phase is accordingly appreciably reduced. Starting from the extraction state with phase diagrams of type 1, in which the binary system entrainer-gas is supercritical and dissolves large amounts of

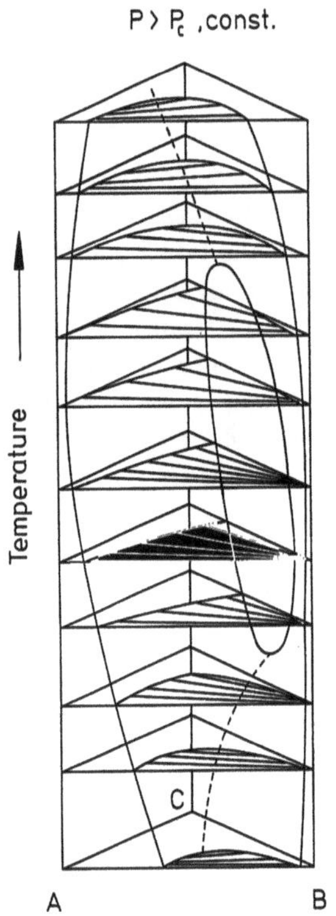

Fig. 25. T-x-diagram of a ternary system consisting of a poorly volatile compound, A, dense gas, B, and entrainer, C, at constant supercritical pressure; – – – – critical solution curve

A, the transition to the state of separation with a type 2 diagram can be achieved simply by changing the temperature. During this transition the critical point moves from the side of the poorly volatile compound to the gas side and the feebly positive tie-line gradient is inverted to a strongly negative slope; the major part of the entrainer thus condenses together with the component A and precipitates.

This complicates the situation markedly if work under reproducible conditions is desired and, as necessary on the preparative scale, one aspires to a solvent circulation as closed as possible. The poorly volatile substance, derived from a plant material for example, must be separated from the entrainer in a second separation step; and the loss in entrainer must be replaced if it is desired to maintain a definite concentration of entrainer in the extraction gas. Further, it must be remembered that every exchange of the extractor contents during extraction of solids in the batch process is also accompanied by losses of entrainer; and the extracted sample must be desolventised and the solvent recuperated, as in methods of classical extraction. For this reason, entrainers are preferably used in the extraction only of liquid samples and especially in combination with the procedure of counter-current extraction in fractionation columns (Chapter III.4.). The employment of organic entrainers creates in any case problems similar to those arising in classical extraction methods if products free of residues are sought.

Zhuze [46] was the first to describe the use of the entrainer principle in dense gas extraction. Elgin and Weinstock [47] pointed out, using triangular diagrams, the marked dependence of the miscibility gaps on pressure and temperature in relevant ternary systems. Entrainer effects were touched on also by Zosel [48] in an application for an extensive patent concerning dense gas extraction, in 1964. Without going into details he claimed, for example in the extraction of chlorophyll from shredded plant parts, that "the amount of a substance taken up by a supercritical gas only with difficulty can be increased by adding a compound taken up more easily by the gas". The entrainer technique in combination with counter-current extraction was described by Panzner, Ellis and Bott [49] and, in an improved version with a second auxiliary column for separating entrainer and product, by Peter, Brunner et al. [50, 51, 52]. Numerous ternary systems with various entrainers were tested and described by Brunner [19].

III.4. Continuous Extraction of Liquids

III.4.a Fundamentals

The extraction of solids with dense gases aims at obtaining substances, that of liquids usually more at refining, for improving the quality of crude products mostly obtained by classical preparative methods. Examples of such separations are the deacidification of fatty oils, removing oil from crude lecithin or separation of undesirable hydrocarbons from essential oils. These procedures can be carried out continuously since no problems are created by the introduction of liquids or liquefiable quantities in the high pressure field.

The separation here depends on the differing partition of the components to be separated between the liquid sample and the dense gas. A theoretical separation stage is attained when the contact between the phases is sufficiently close and maintained long enough to enable a partition equilibrium to be set up between the liquid sample and the solvent. As far as the separation process is concerned the procedures can be stepwise or differential in course. The former group includes the so-called "mixer-settlers" such as containers with stirring devices, in which the two phases are brought to equilibrium by stirring or another type of mechanical mixing, and then separated by settling out. The latter (differential) group is exemplified by the counter-current columns, such as packed or sieve plate columns. The phases are in counterflow to each other in these columns and in permanent contact. Mass transfer takes place in several theoretical plates, giving a better separation performance than in a single mixing.

The amount of extraction agent and the number of theoretical stages required for a separation depend on the difficulty of the problem. These factors of influence can be determined by measuring the partition equilibria. In simple cases the extraction of liquids is described by the partition law of NERNST. An ideal extraction system is characterised by a constant partition coefficient and complete immiscibility of the two phases; the equilibrium curve is then a linear relation between the concentrations of the substance to be extracted in the sample and in the extraction agent. The equilibrium curve is no longer linear in non-ideal cases, where the partition coefficient is not constant and the two phases forming the extraction system are partly miscible. The course of such a curve is obtained from points derived by measuring the concentrations of the two phases yielded by vigorous mixing of sample solutions of various concentrations with the extraction agent. The coordinate system with axes at a right angle suffices no more if it is desired to follow the changes of all three components during extraction.

The equilibrium states of such ternary or quasi-ternary mixtures are best given in triangular diagrams (cf. Chapter II.4.2). The optimum working conditions for pressure and temperature, and the requirement of extraction agent determined by the solvent capacity of the gas, can be, respectively, read or calculated simply from them.

The theoretical number of stages associated with the separation problem can be determined by considering the two phases between which partition of the substance to be extracted takes place. The equilibrium conditions are portrayed in a right-angled coordinate system, with the concentration of the substance to be separated in the sample on the abscissa and its fraction in the solvent on the ordinate (see Fig. 26).

The theoretical number of stages needed for separation can be determined with the help of the graphical procedure of McCabe and Thiele [53]; this was originally developed for rectification but is applicable also to procedures of extraction, in particular counter-current extraction [54]. An advantage of the procedure is that the values sought can be found by drawing in only straight lines, in addition to the equilibrium curve. The enrichment line depends on the ordinate intercept α, that is, on the reflux ratio and the desired purity of the extract (x_t). The stripping line is derived from the

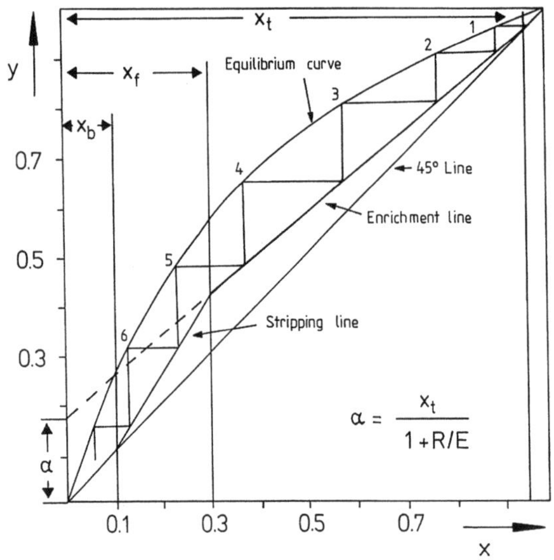

Fig. 26. Determination of the theoretical plate number of a countercurrent column using the equilibrium curve according to the McCabe-Thiele procedure.
x = mole fraction in the sample of the substance to be extracted
y = mole fraction in the solvent of the substance to be extracted
x_t = composition of the extract at the column head
x_f = composition of the feed
x_b = composition of the raffinate
R = reflux; E = extract flow; R/E = reflux ratio

point of intersection of the enrichment line with the feed concentration x_f, and the required composition of the raffinate, X_b on the 45°-line. As α becomes smaller, i.e., as the reflux ratio increases, the operating lines flatten and become identical with the 45°-line at infinite reflux ratio. The complete height between the 45°-line and the equilibrium curve is realised with a single theoretical plate only when the column is used at total reflux, that is, when the complete extract flow separates from the solvent at the column head and flows back into the column. Fig. 26 shows an example of a graphic solution for a separation of a hypothetical charge in which the theoretical number of stages is obtained at a reflux ratio of 6.5. From this it is easy to see that separation becomes more difficult and hence the theoretical number of necessary stages becomes larger, the nearer the equilibrium curve approaches the 45°-line and the greater are the required purities of extract and raffinate. A more comprehensive and detailed discussion of the relations between theoretical plate number, minimum reflux ratio and minimum solvent ratio, which influence the extractive separation of liquid mixtures, is to be found in the literature [54, 55].

The theoretical stages necessary for separation are often attainable only with difficulty in classical liquid-liquid extraction; this is because the two necessary operations – mixing with mass transfer and separation – are in many cases beset with problems and hard to realise in a system of two liquid phases. When the extraction system consists of a liquid and a dense gas the situation is more favourable. The mass transfer takes place markedly faster as a result of the lower viscosity and larger diffusion coefficient of a dense gas (of Chapter II.3). There must be a minimum difference in density between the two phases to ensure a separation without problems. This is more easily possible when dense gases are used since the density and also the dissolving selectivity can be varied within wide limits via the nature of the extraction gas and its pressure and temperature. It must be remembered, however, that the density of the liquid phase becomes smaller with increasing pressure, due to increase in the amount of dissolved gas; and the density of the gas phase increases. In this way the relative positions of condensed phase and dense gas can interchange under particular conditions of pressure and temperature (a barotropic phenomenon [56–58]). This is of vital importance for the course of the process, above all for the direction of flow of the charge and the solvent.

III.4.b Stepwise Separation

A single stage process is the simplest procedure for extracting liquids with dense gases. The phase to be extracted is placed in a high pressure vessel and the dense gas flows through it. Efficient mixing is needed to establish a partition equilibrium between the phases. Mechanical stirring by rotation or piston drive is possible. Because of the necessary high pressure the construction of such apparatus can be very expensive. Two ways are possible for transmitting mechanical energy to the pressure vessel. In one, the driving shaft is actuated directly outside the vessel and a stuffing box used to seal against the high pressure. High rotational moments or piston drives can be communicated to the interior of the vessel but there are limits to the revolution speed or piston stroke frequency imposed by the friction at the stuffing box. Further, the driving shaft usually has to be cooled and lubricated. The solubility power of dense gases must be considered in choosing a lubricant.

As an alternative, a drive without stuffing box is possible. For this, a packing consisting of a permanent magnet is fixed at the end of the mixing shaft. The whole unit is in a pressure-resistant casing and screwed on to the cover of the vessel in a gas-tight way. An external unit, likewise with permanent magnet, is set in motion and actuates the internal unit via magnetic coupling through the pressure-resistant anti-magnetic casing. This mixing arrangement with permanent magnet and no stuffing box normally requires no lubricant and can permit high revolution speeds or piston stroke frequencies; the transmitted power (rotational moment or piston drive) depends on the strength of the magnetic coupling. A disadvantage of vessels with stirrers is that stirring under pressure involves high costs and maintenance. On the industrial scale, where high rotational moments have to be transmitted, only the use of stuffing boxes can be recommended; the magnitudes of the revolution speed and of the extraction pressure are then limited. An example of extraction with single stage mass transfer in an autoclave with stirring is in Chapter IV.2, namely, the removal of oil from crude lecithin, using supercritical carbon dioxide.

Schultz and co-workers [66] described a multistage separation process for extracting organic compounds from aqueous solutions (e.g., from fruit juices) with liquid carbon dioxide, in which the two phases were mixed and brought to separation several times. Based on investigations on the laboratory scale [67], which yielded qualitative information about whether flavouring substances could be extracted with liquid CO_2, the authors described a pilot plant with which quantitative process data could be obtained about flow rates, yields of extract and recoveries of the flavouring substances.

The extraction unit consists of a glass column, an evaporator and a condenser and is built into a pressure-resistant casing fitted with sighting windows for supervising

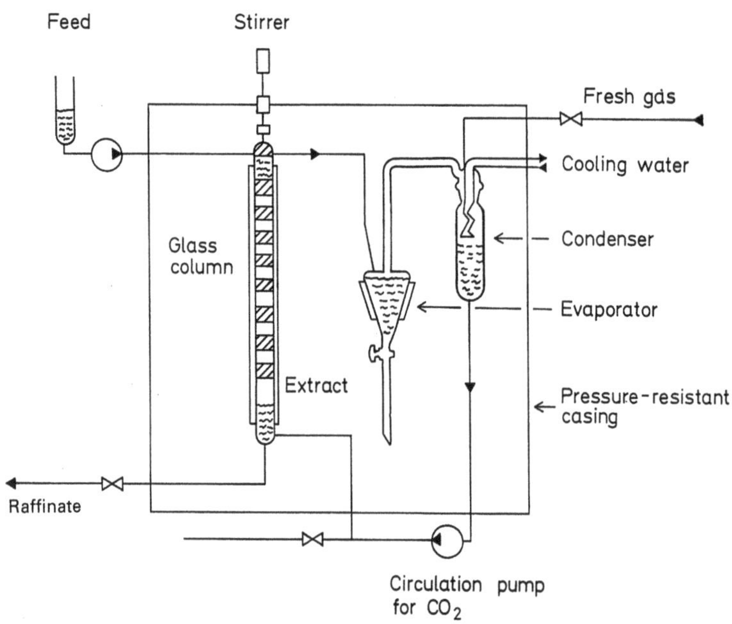

Fig. 27. Flow sheet for multistage liquid-liquid extraction [66]

the course of the process. Liquid carbon dioxide is circulated with the help of a pump. The liquid-liquid extraction proceeds at room temperature and the corresponding saturation pressure of the carbon dioxide (60 bar at 22 °C). Evaporation and condensation of the extraction agent are regulated by heating or cooling.

Fig. 27 gives a simplified flow sheet of the pilot plant. The process consists of a multistage counter-current extraction, in which the charge to be separated is pumped in at the the top of the column and flows down in counter current to the liquid carbon dioxide. The aqueous solution to be extracted is vigorously mixed in ten stages with the extraction agent using a rotating stirrer. Packings of stainless steel wool are between the stirring zones to effect separation of the two phases. The solvent, charged with extract, leaves the column at the head, flowing into the evaporator, where the carbon dioxide is evaporated isobarically by supplying heat, so that the extract separates out. The gaseous carbon dioxide is then cooled and thus liquefied in the condenser and the gas cycle is finally completed by introducing it again into the extraction column from below with the aid of a pump. The extracted aqueous phase is withdrawn at the foot of the column via a valve. This pilot plant for liquid-liquid partition enables about 6 l charge to be extracted per hour using 5 l/h of carbon dioxide.

Coenen and co-workers also investigated the extraction of liquid mixtures with supercritical gases. They described the working up of used oil [68], the refining of crude edible oils, obtained by conventional means [69] and the isolation of colouring matter from paprika oleoresin [70]. The experimental plant used in this work (see Fig. 28) can be employed to extract solids in discontinuous, and liquids in continuous, processes. In the latter, the liquid substrate in the extraction vessel (containing packing material to increase the surface area for mass transfer) is extracted with supercritical

Fig. 28. Experimental plant for separating mixtures with supercritical gases (works photo, Krupp, Essen, Germany)

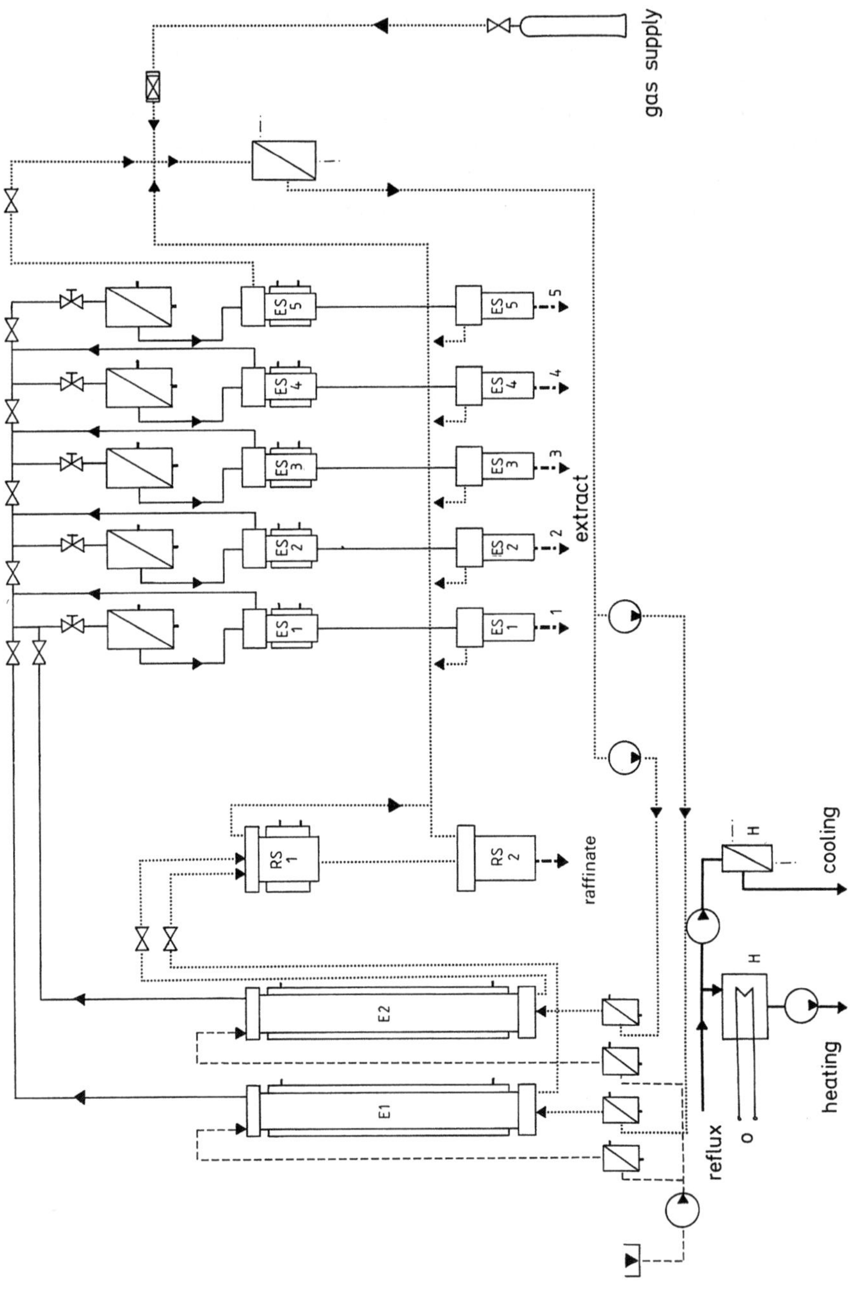

Fig. 29. Flow sheet of the experimental plant of Fig. 28

gas, so that the components of the sample which are insoluble in the extraction agent remain behind. The extracted material is fractionated by stepwise precipitation from the supercritical gas phase.

Fig. 28 is a photograph of the experimental plant in the Krupp research institute. It has two high pressure vessels, each of 7.5 l volume, for extraction; and five extract separators, each of 1 l volume for the separation. Further pressure vessels serve for recovering solvent from the extracts and the extraction residue. The course of the procedure can be explained with the help of the flow sheet (Fig. 29). The liquid mixture is metered into the extraction vessel at the top, using a high pressure pump. It is then brought into counter-current contact with the supercritical gas. The gas phase, charged with extract, is withdrawn at the top of the extractor and its pressure reduced to fractionate the extract in successive separators. The gas, free of extract, from the last separator and the gas from the vessels for solvent recovery are brought again to the conditions for extraction and returned to the extraction vessel.

III.4.c Counter-current Extraction

Several possibilities for separating liquid mixtures with dense gases were given already in the previous chapter. Mainly components of markedly differing solubilities are separated by these methods. Liquid substrates with relatively similar components can, nevertheless be separated by counter-current extraction in mass transfer columns and this is discussed below. Because of the high number of theoretical plates, a satisfactory separation is accomplished even when there are only small differences in solubility in the dense gas. The effectiveness of the separation is increased still further by setting a definite reflux of the product at the column head.

The first investigations of fractionation of liquid mixtures using supercritical gases were described as long ago as 1964 by Zosel in a patent applied for by the Max-Planck-Institute for Coal Research [48]. In the following years further work was published by this institute on the separation of liquid mixtures, notably the fractionation of an α-olefine mixture with 16 to 20 C-atoms and of the triglycerides of fish liver oil [59], the work up of high-temperature tar from brown coal [60] and the concentration of the ester of eicosapentaenoic acid ($C_{20:5}$) in a mixture of ethyl esters of fatty acids from fish oils [61]. An apparatus on pilot plant scale was used in these studies (Fig. 30). The extraction and fractionation part consist of the "destractor" of volume 8.5 l, a 2000 mm long packed column of 2.2 l capacity and a "hot finger" in the column head (content 0.85 l, length 700 mm) which is thermostated independently of the "destractor" and the packed column.

Fig. 30 shows how this rectification apparatus works. The mixture to be separated is placed in a supply vessel (destractor) and supercritical gas led through it from below. The charged gas phase reaches the hot finger via the packed column; some of the dissolved substances separate out on the finger as a result of increase of temperature and run back into the exchange zone. The part still in solution in the supercritical gas is then separated by isothermally lowering pressure. A continuous procedure in not possible with the apparatus, however, because all the components of the mixture are extracted in succession, for which the pressure of the dense gas may have to be increased stepwise. The separation thus depends on a fractionation of the substance mixture with time improved by the column.

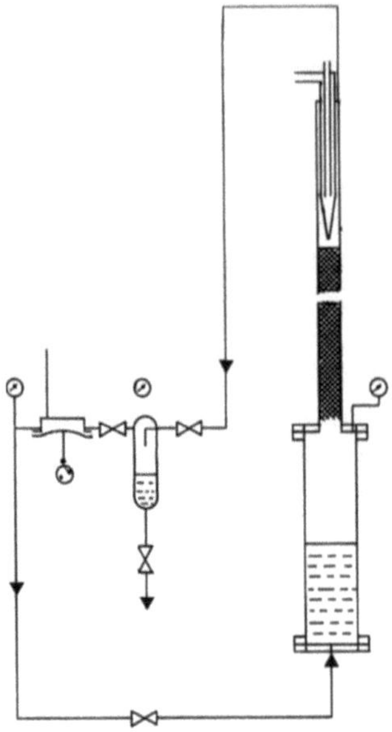

Fig. 30. "Rectification apparatus" of Zosel

In the continuous run the liquid sample must be added steadily to maintain a definite ratio to the flow of extraction agent through the mass transfer column. Care must be taken, however, that the mass flows do not exceed the maximum value permissible by the construction of the apparatus. The counter-current column would then be flooded and the phases forming the extracting system would be carried away without separation having been achieved.

Depending on processing conditions in the column, separation out of the extract can be accomplished by either increase or decrease of temperature at the column head, yielding an internal reflux of product. Establishing a definite reflux ratio through reduction in temperature gives advantages not only in the fractionation of essential oils [62] (cf. Chapter IV.3); this principle can be adopted also for separating mixtures of low volatility, such as fatty oils or mixtures of wax esters [63] (cf. Chapter IV.2).

Fig. 31 shows schematically the construction of a laboratory installation which functions continuously for fractionating liquid mixtures with the help of dense gases, as employed by Stahl and Gerard [62]. The central item of the installation is the mass exchange column. The stainless steel tube-shaped autoclave is 1500 mm long and 25 mm in internal diameter and designed for a maximum working pressure of 700 bar. It is sealed top and bottom with screw fittings and inner-situated O-rings. A water jacket in three parts takes care of the thermostatisation. The lower, central and upper parts of the column can then be held at different temperatures. The material exchange between dense gas and condensed phase in packed columns depends markedly on the moistened surface of the column packing. A sequence has been given for the wetting properties of various column packing materials [64]: glass, ceramic material, stainless steel, polyvinyl chloride (PVC), polypropylene (PP) and polytetrafluoroethylene (PTFE). On account of their ease of wetting, glass spheres of 6 mm diameter

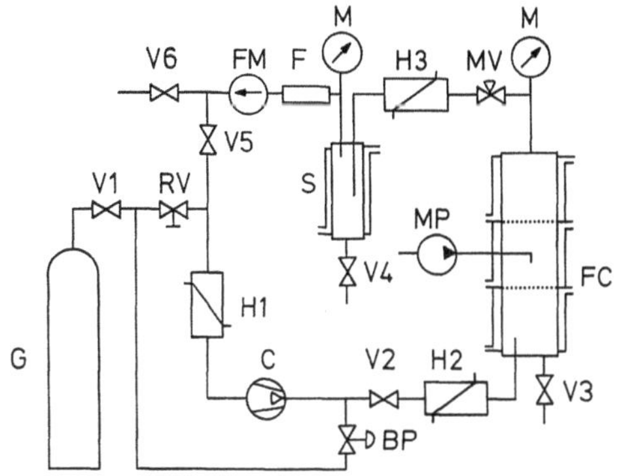

Fig. 31. Diagram of the counter-current extraction installation for fractionating liquid mixtures.
G gas supply; C membrane compressor; FC packed column; MP membrane metering pump; S separator; M strain gauge pressure transducer with digital indicator F active charcoal-blue silica gel filter; FM mass flowmeter; RV reducing valve; BP back pressure regulator; MV fine metering valve; V1–V6 shut-off valves; H1–H3 heat exchangers

were filled into the high pressure column in order to increase the surface area. They yield a geometric surface area of 5,300 cm^2, and the effective volume of the column is reduced by the packing to 340 ml. The maximum possible flow rate of extraction agent in this type of column without impairing the separation performance is 2 to 2.5 kg/h, depending on the separation conditions of pressure and temperature and on the difference in density between the dense gas and the sample.

The generation and regulation of pressure and also the circulation of the extraction agent are carried out analogously to the procedures in the laboratory plant in the preparative extraction of solids (cf. Chapter III.2, Fig. 17). In the fractionation process the dense gas is passed through heat exchanger H2 and led from below into the packed column FC. The liquid charge is fed continuously with the metering pump MP, into the middle of the column. Exchange of material takes place between the extraction agent and the sample mixture to be separated. By rising through the column the dense gas is enriched in the components which are more soluble in it. The internal reflux is regulated by raising or lowering the temperature at the column head, depending on the separation problem and the conditions of the process. The extraction gas, charged with dissolved substance, leaves the column at the head for separator S where the extracted material is precipitated and the gas regenerated. The extract is taken from the separator via the valve V4. The more poorly soluble part of the sample mixture descends through the packed column and is released through valve V3.

A definite reflux of product at the column head is consequently adjustable in the counter-current column via the temperature difference between column base and column head. Another way of doing this is to pump some of the extract separated from the gas stream back into the column as an external reflux. This technique was proposed by Ellis [65] and by Peter, Brunner and co-workers. For fractionating involatile substance mixtures they used entrainers in addition to the dense gas and employed two separating columns in this variation. They described for example the separation of the monoglycerides of oleic acid from a mixture of glycerides of this acid, using acetone as entrainer [50]; the separation of saturated and unsaturated fatty acids from linseed oil, using an extracting mixture of CO_2 and dichloromethane [51]; and the removal of free fatty acids from palm oil, employing ethanol as additional auxiliary [52].

Their installation, half-technical scale, utilised two plate columns, of which the main column served to separate the substances of low volatility and the extra column to remove the extracted components from the circulating gas. These two bubble cap columns (NW 65) are the central parts of the installation. They were designed to withstand a pressure of 150 bar at 180 °C. The main column consists of three, the other, of two stages, each of 17 plates, each plate having a bubble-cap. The plates are 60 mm apart. The column stages (height of 1000 mm) are joined to one another with lenticular gaskets via flange connections. The necessary heat for the process is supplied from heat exchangers at the foot of the main column and at the head of the auxiliary column. The columns are also equipped with filament windings and insulated against losses through heat radiation. Plunger metering pumps are employed to introduce the liquid flow and regulate the reflux. A piston compressor is used to apply pressure to the extraction gas. The gas within the closed column system is circulated isobarically with a membrane gas circulating compressor.

The principle of the process can be explained with the help of Fig. 32: The liquid mixture to be separated is introduced, together with the entrainer, into the middle of the main column. The dense gas traverses the column from below, dissolving the more soluble parts of the sample. The gas enriched with extract leaves the column at the head and enters the auxiliary column, in which the extracted substances separate out. The thermodynamic conditions in the main column must be chosen so that the binary system of dense gas and entrainer is supercritical (lower triangular diagram in Fig. 32) to attain a relatively high solubility of the substances to be separated. The temperature in the auxiliary column is raised isobarically, leading to a pseudo-ternary equilibrium of Type II. In this way the solubility of the poorly volatile compounds is lowered markedly and the entrainer partly condenses. The substances which have separated from the gas run down through the auxiliary column. A part of this stream is returned to the main column as reflux. The circulation gas, now freed from the poorly volatile substances but still containing a definite proportion of entrainer, leaves the auxiliary column at its head and sweeps once more through

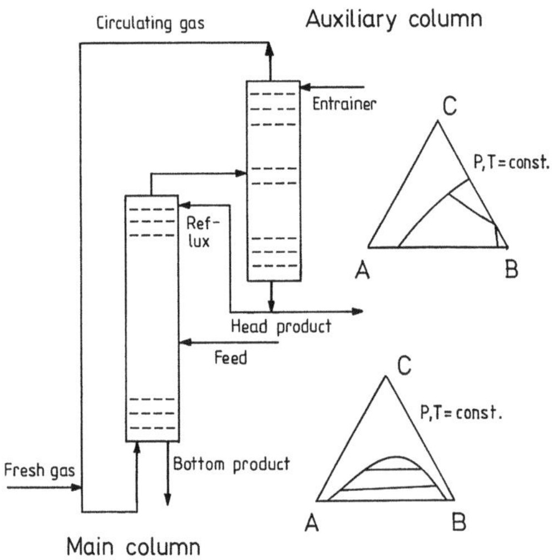

Fig. 32. Scheme of a fractionation process with supercritical gases, using entrainers, and presentation of the thermodynamic conditions in triangular diagrams [52].
A mixture to be separated; B dense gas; C entrainer

the main column from below. The components of the sample mixture which are less easily soluble in the dense gas are taken off at the foot of the main column and the extracted compounds at the foot of the auxiliary column. Entrainer is added at the head of the auxiliary column to give an additional washing effect which improves the separation of the extracted components from the circulating gas stream. Entrainer withdrawn from the cycle with the streams of product must be replaced; this is mostly done via the entrainer portion in the mixture fed in.

Investigations on fractionation of the glycerides of oleic acid using supercritical carbon dioxide in a counter-current column with n-hexane or carbon tetrachloride as entrainer, were published also by Panzner, Ellis and Bott [49]. Separation was effected in a packed column. The extracted compounds were precipitated by reducing pressure to about one-third of that of the extraction procedure.

The counter-current extraction with dense gases in mass transfer columns is an alternative to classical vacuum rectification or molecular distillation, even though a different result can be obtained as a consequence of the different principle of separation. This is explained in the example below: naphthalene is the more volatile component in the vacuum distillation of a mixture with dodecane at 100 °C, and is separated from the mixture. The ratio of the volatility of naphthalene to that of dodecane is about 1.10 to 1.15. If this mixture is separated with supercritical ethylene at 75 °C, dodecane is preferentially taken up by the supercritical phase and thus extracted. A separation factor of 1.58 is attained at a pressure of 80 bar and a gas charging of 1.6% by weight. As pressure is increased, the solubility increases but the separation factor decreases. At 145 bar and 75 °C, 34.3% of the sample dissolve but the separation factor is lowered to 1.28 [65].

III.4.d High Pressure Jet Extraction

The extraction of liquid mixtures of substances demands intensive and uniform contact between solvent and substrate. This causes an increase in the surface area of the liquid phase, for which a definite amount of kinetic energy must be supplied to overcome the surface energy. An example of this is the use of mechanical energy to operate stirring devices. With media of low viscosity, where the surface or cohesive forces are relatively weak, the potential energy of the medium itself suffices, so that the kinetic energy needed to create a large surface is provided by the flow down through a packed column. The solvent suffers a certain loss in pressure on passing through the column, which serves to increase the surface area. Similar considerations apply also to cases in which closer contact is brought about by one phase bubbling through the other and upward- or downward-acting forces, due to differences in density, are effective. Both these last two methods are tied by certain conditions and can be used only with sufficiently low-viscosity media. Vessels with stirring equipment are of more general use but, because of the mechanically operated moving parts, are unsatisfactory in high pressure procedures.

Stahl and Quirin worked out a procedure with which media of low or high viscosity can be brought into energetic contact with dense gases and extracted without employing mechanically moved parts [71]. Substrate and solvent are continuously introduced by jets into a mixing vessel of very small volume. A small fraction of the high pressure of both phases-needed in any case-is converted into the kinetic energy required for creating a large surface and for mixing.

The complete circulatory procedure of the high pressure jet extraction is represented schematically in Fig. 33. The heart of the installation is the mixing and extraction part, which consists of two stainless steel capillaries, concentrically placed one above the other. The fluid mixture to be extracted is led through the inner jet. The second capillary, slightly larger in diameter, is pushed over the exit of the substrate capillary so that it overlaps for a short distance. The dense gas flows parallel to the sample mixture through the confined, ring-shaped space between the two jets and attains thus a high speed. At the end of the substrate jet, turbulence phenomena cause vigorous mixing of the substances introduced each time with fresh solvent.

If conditions are favourable a short distance within the jet is sufficient for complete separation of the soluble components from the small particles of substrate created. The insoluble components land in a collecting vessel, the extracted ones, dissolved in the dense gas, are transferred to a separator and collected after decompression. The regenerated gas is brought again to the conditions for extraction and this completes the cycle.

This dual mixing jet described works more efficiently than an assembly where only the sample is introduced into the gas phase or only the extraction gas into the sample mixture. It profits from two effective mechanisms for increasing surface area. First, the sample mixture is forced through a fine capillary, where it assumes a thread-like structure. One of the substrate jets proposed by the authors with, for example, an internal diameter of 0.2 mm, generates in this way from 1 l of starting material a thread structure of about 20 m^2 area. This structure is then deformed through turbulence in the fast flowing gas at about 5 m/sec and splits up into small segments

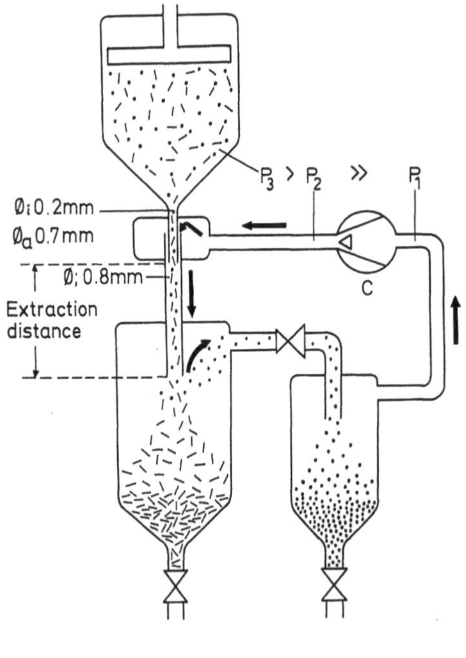

Fig. 33. Scheme for continuous high-pressure jet extraction of flowable masses, with an example of jet design.
p pressure; C compressor [71]

with still larger surface area, which permit good mass transfer. Since the extraction proceeds in a very small volume and extremely small amounts of substrate come in contact with fresh solvent in constant sucession, there is no extraction gradient in time as far as amount and composition of dissolved components are concerned.

The separation performance of the arrangement with jets is lower than that from mass exchange columns but a definite number of separation stages can be attained within the extraction zone through repeated contact of the substrate with gas already charged with extracted material. Decisive factors here are length and cross-section of the extraction path and the magnitude of the mass flows of substrate and of solvent and their ratio to one another. The jet procedure is always to be favoured when the components to be separated differ largely in solubility, as for example, in the separation of crude lecithin into neutral lipids and phospholipids, or the separation of pepper oleoresin into pungent materials (piperine) and flavouring substances (essential oil).

In contrast to the case with most other procedures of refining, there are no limitations in the high-pressure jet extraction concerning the consistency and composition of the sample mixture, provided that it flows at the extraction temperature. Emulsions and dispersions can be extracted just as can solutions of various materials which, after the separation, can become partly solid. The purification of crude lecithin is an example of an application of a viscous substrate, solid raffinate and liquid extract.

The dimensions of the jets must be accomodated to the separation problem to be solved. Table 6 shows the different jet parameters which have to be considered. The substrate jet, D1, should be as short as possible to keep to a minimum the drop in pressure of the substance mixture introduced. The overlapping part of the two capillaries, otherwise not of significance, should thus be short. The remaining jet parameters are not fixed irrevocably but must be optimised and adjusted to each other. The internal diameter of the substrate jet, D1, determines the primarily produced surface of the sample mixture and its exit speed. The wall thickness of D1 determines the volume expansion available to solvent and substrate after their encounter at the end of the jet, D1. The intermediate space formed by the overlapping of the two capillaries is decisive for the speed of the introduced gas and hence also the formation of turbulences at the edge of the substrate jet. The internal diameter of the capillary D2, its length and segmentation determine the extraction volume and, in this way, the time of contact and extraction. The jet D2 can, for example, be designed as a smooth tube or widen in steps, as a result of which the velocity of flow falls in stages and the time of contact increases.

Table 6. Jet parameters

D1

D2

1) length (D1)
2) overlap (D1/D2)
3) ϕ_i (D1)
4) ϕ_a (D1) $-$ ϕ_i (D1) = wall thickness
5) ϕ_i (D2) $-$ ϕ_a (D1) = gas passage
6) ϕ_i (D2) } = extraction volume
7) length (D2) }

The condition of state of the solvent and the mass flows of substrate and of solvent also determine the result of the extraction. The amounts and solubilities of the substances to be separated or the desired degree of extraction of the substrate have thus to be taken into consideration. A rapid throughput of material is important for economic success, even when combined with an inferior utilisation of the solvent capacity. If the optimum operation has been ascertained in accordance with the requirements, the jet procedure permits an advantageous and usually continuous high pressure extraction of flowable masses without having to resort to mechanically operated parts.

References

1. Hála, E., Pick, J., Fried, V., Vilim, O.: Gleichgewicht Flüssigkeit – Dampf, Akademie Verlag, Berlin (1960)
2. Schneider, G. M.: Experimental Thermodynamics, Vol. II, Experimental Thermodynamics of Non-Reacting-Fluids, (LeNeindre, B. und Vodar, B., Ed's), Butterworths, London, 1975, p. 787
3. Tsiklis, D. S.: Techniques for Physicochemical Studies under High and Superhigh Pressure, 3rd ed., Khimiya, Moscow, 1965
4. Tsiklis, D. S.: Phase Separations in Gas Mixtures, Khimiya, Moscow, 1969
5. Young, C. L.: Specialist Periodical Reports, Vol. 2 (Chemical Thermodynamics), p. 71, The Chemical Society, London (1978)
6. Hicks, C. P., Young, C. L.: Chem. Rev. *75*, 119 (1975)
7. Schneider, G. M.: Ber. Bunsenges. Phys. Chem. *70*, 10 (1966)
8. Alwani, Z., Schneider, G. M.: Ber. Bunsenges. Phys. Chem. *73*, 294 (1969)
9. Lentz, H.: Rev. Sci. Instrum. *40*, 371 (1969)
10. Steiner, R., Schadow, E.: Z. Phys. Chem. (Frankfurt) *63*, 297 (1969)
11. Buback, M., Franck, E. U.: Ber. Bunsenges. Phys. Chem. *76*, 350 (1972)
12. Schröder, E., Arndt, K.-F.: Z. Polymerforsch. *27*, 135 (1976)
13. Tsiklis, D. S., Maslennikova, V. Ya.: Dokl. Akad. Nauk. SSSR *157*, 426 (1964)
14. Takenouchi, S., Kennedy, G. C.: Amer. J. Sci. *262*, 1055 (1964)
15. Tödheide, K., Franck, E. U.: Z. Phys. Chem. (Frankfurt) *37*, 387 (1963)
16. Streett, W. B., Hill, J. L. E.: J. Chem. *54*, 5088 (1971)
17. Gerritsen, H. G., Hartmann, H.: Chem.-Ing.-Tech. *51*, 303 (1979)
18. Maier, M., Stephan, K.: Chem.-Ing.-Tech. *56*, 222 (1984)
19. Brunner, G.: DSc Thesis, Erlangen, 1978
20. Tsang, C. Y., Streett, W. B.: J. Chem. Eng. Data *26*, 155 (1981)
21. Kuk, M. S., Montaga, J. C., in: Chemical Engineering at Supercritical Fluid Conditions, eds. Paulaitis, M. E., Penninger, J. M. L., Gray, R. D., Davidson, P., Ann Arbor Science (1983), p. 101
22. Radosz, M.: Ber. Bunsenges. Phys. Chem. *88*, 859 (1984)
23. Reid, R. C.: Supercritical Fluid Extraction, A Perspective, Hougen Lecture Series, University of Wisconsin, 1981
24. Kurnik, R. T., Holla, S. J., Reid, R. C.: J. Chem. Eng. Data *26*, 47 (1981)
25. Stahl, E., Gerard, D.: Perf. Flavor. *10* (2), 29 (1985)
26. Michels, A., Dumoulin, E., van Dijk, J. J. T.: Physica, *27*, 886 (1961)
27. Czubryt, J. J., Myers, M. N., Giddings, J. C.: J. Phys. Chem. *74*, 4260 (1970)
28. Van Leer, R. A., Paulaitis, M. E.: J. Chem. Eng. Data *25*, 257 (1980)
29. McHugh, M., Paulaitis, M. E.: J. Chem. Eng. Data *25*, 326 (1980)
30. Johnston, K. P., Ziger, D. H., Eckert, C. A.: Ind. Eng. Chem. Fundam. *21*, 191 (1982)
31. Kurnik, R. T., Reid, R. C.: Fluid Phase Equil, 8, 93 (1982)
32. Stahl, E., Willing, E.: Mikrochim. Acta *1980 II*, 465
33. Quirin, K.-W.: Fette, Seifen, Anstrichm. *84*, 460 (1982)

34. Ohgaki, K., Tsukahara, I., Katayama, T.: Kagaku Kogaku Ronbunshu *10* (3), 372 (1984)
35. Kwiatkowski, J., Lisicki, Z., Majewski, W.: Ber. Bunsenges. Phys. Chem. *88*, 865 (1984)
36. Stahl, E., Schilz, W.: Z. Anal. Chem. *280*, 99 (1976)
37. Stahl, E., Quirin, K.-W., Mangold, H. K.: Chem. Phys. Lipids *31*, 313 (1982)
38. Gährs, H. J.: Ber. Bunsenges. Phys. Chem. *88*, 894 (1984)
39. Schütz, E.: Thesis, Saarbrücken, 1979
40. Stahl, E., Schütz, E.: Chem. Ing. Tech. *52*, 918 (1980)
41. Stützer, D.: Thesis, Erlangen 1983
42. Stahl, E., Schütz, E.: Arch. Pharm. *311*, 992 (1978)
43. Stahl, E., Gerard, D.: Parfuem. Kosmet. *63*, 117 (1982)
44. „Discussion Paper" of NOVA SWISS, Effretikon/Schweiz (1984)
44a. Lack, E. A.: Kriterien zur Auslegung von Anlagen für die Hochdruckextraktion von Naturstoffen (Thesis), dbv-Verlag für die Techn. Universität Graz, 1985
45. Körner, J. P., Saamer, P.: Inf. Chim. *243*, 269 (1983)
46. Zhuze, T. P. et al.: Maslob. Zhir. Prom., *24*, 34 (1958)
47. Elgin, J. C., Weinstock, J. J.: J. Chem. Eng. Data *4*, 3 (1959)
48. Zosel, K.: Germ. Pat. 1 493 190 (Studiengesellschaft Kohle mbH), filed 1964
49. Panzner, F., Ellis, S. R. M., Bott, T. R.: Inst. Chem. Eng. Symp. Ser. *54*, 165 (1978)
50. Peter, S., Brunner, G., Riha, R.: Fette, Seifen, Anstrichm. *78*, 45 (1976)
51. Brunner, G., Peter, S., Retzlaff, A., Riha, R.: High Pressure Science and Technology, Sixth AIRAPT Conference, Timmerhaus, K. D., Barber, M. S., ed's., Plenum Press, New York, 1979, Vol. 1, p. 565
52. Brunner, G., Peter, S.: Chem. Ing. Tech. *53*, 529 (1981)
53. McCabe, W. L., Thiele, E. W.: Ind. Engng. Chem. *17*, 605 (1925)
54. Ullmanns Enzyklopädie der technischen Chemie, 4. edition, Bd. 2, Verlag Chemie, Weinheim 1972
55. Kirschbaum, E.: Destillier- und Rektifiziertechnik, 2. edition, Springer, Berlin–Göttingen–Heidelberg 1950
56. Cailletet, L.: Compt. Rend. *90*, 210 (1880)
57. Kuenen, J. P.: Z. Phys. Chem. *11*, 38 (1893)
58. Kritschewski, I. R., Bolschakow, P.: Acta Physicochim. USSR *14*, 353 (1941)
59. Zosel, K.: Angew. Chem. *90*, 748 (1978)
60. Eisenbach, W., Niemann, K.: Erdöl und Kohle *34*, 296 (1981)
61. Eisenbach, W.: Ber. Bunsenges. Phys. Chem. *88*, 882 (1984)
62. Gerard, D.: Chem. Ing. Tech. *56*, 794 (1984)
63. Leiner, S.: Thesis, Saarbrücken 1986
64. Shi, M. G., Mersmann, A.: Chem. Ing. Tech. *56*, 404 (1984)
65. Ellis, S. R. M.: Brit. Chem. Eng. *16*, 358 (1971)
66. Schultz, W. G., Schultz, T. H., Carlson, R. A., Hudson, J. S.: Food Technol. *88*, 32 (1974)
67. Schultz, W. G., Randall, J. M.: Food Technol. *24*, 1282 (1970)
68. Coenen, H., Rinza, P.: Chem. Ing. Tech. *54*, 386 (1982)
69. Coenen, H., Kriegel, E.: Chem. Ing. Tech. *55*, 890 (1983)
70. Coenen, H., Hagen, R.: Gordian *83*, 164 (1983)
71. Stahl, E., Quirin, K.-W.: Fette, Seifen, Anstrichm. *87*, 219 (1985)

IV. Applications of Dense Gases to Extraction and Refining

IV.1. General Survey

Schultz and Randall published in 1970 the results of qualitative investigations into the extractability of lipophilic natural substances with liquid carbon dioxide [1]. They described the influence of molecular weight and functional groups of aroma compounds from fruit juices on the solubility in the liquid carbon dioxide. In 1976 Stahl and Schilz studied the extractability with supercritical carbon dioxide of numerous individual compounds from various classes (e.g. polyaromatics, phenols, aromatic carboxylic acids, anthraquinones, pyrones, hydrocarbons and other lipids [2]; this furnished the first clear picture of the solubilites of natural substances in dense carbon dioxide. This information, together with experience from recent years, has enabled the rules of thumb given below to be suggested for extraction with dense carbon dioxide:

1) easily extractable (up to 300 bar) are lipophilic compounds with a molecular mass of up to 300–400, such as hydrocarbons, ethers, esters, ketones and similar compounds.
2) the presence of polar functional groups lowers, and may even completely prevent, extraction.
3) not extractable are polar substances, such as sugars, glycosides, amino acids, lecithins, etc.; and polymers, including proteins, cellulose, polyterpenes and plastics. Non-polar oligomers are only sparingly soluble.
4) water is poorly soluble in liquid carbon dioxide (ca. 0.1% by weight, at 20 °C) but shows increasing solubility in supercritical CO_2 with temperature increase (ca. 0.3% by weight at 50 °C).
5) fractionation is possible when the substances display differences in molecular mass, vapour pressure or polarity.

It is difficult to forecast extractabilities of lipophilic natural substances with other gases since only few relevant literature data are available. Thorough investigation of solubility in various gases has been carried out on the steroids [3]. In addition to carbon dioxide the gases dinitrogen monoxide, trifluoromethane, ethane and ethene were tested for their extractive power. The results of this work permit the order of extractability of steroids to be given below:

$$C_2H_6 > N_2O > CO_2 = C_2H_4 > CHF_3$$

In every case the extractive power of the gases increased with increase in pressure and – at pressures above the intersection point of the isotherms – also in temperature. Further, the general solubility sequences of variously substituted steroids were the same as in the investigations with carbon dioxide [4]. The solubility in all the gases studied diminishes with increasing polarity of the solute (substance).

Comparison of the different supercritical gases shows that each has a selective action, due to specific interactions between dense gas and condensed phase. The result of this is that the solubility sequence, mentioned above for steroids, is not generally valid. Thus, for example, opium alkaloids, especially thebaine, are appreciably more soluble in trifluoromethane than in supercritical carbon dioxide [5, 6]. Steroids and some fatty oils are about equally soluble in carbon dioxide and in ethene [7] whereas pure hydrocarbons or aromatics, such as naphthalene or phenanthrene, dissolve better in ethene [8, 9].

The different solubility powers of the various gases cannot be explained directly through physical properties, such as the gas density, critical point or dipole moment. Simple and generally valid rules about the solubilities of lipophilic natural substances can be no better established for the various dense gases than for the classical solvents.

The field of application of dense gases as solvents and extraction agents is extensive. Methods of substance separation have become highly important in all spheres and consequently this new separation procedure has led to many-sided research activities. Most of the applications are still in the development stage and are being carried out only on the laboratory or pilot plant scale to explore procedural techniques. However, intensive efforts are being made in connection with industrial applications. This marked interest is demonstrated by the considerable number of patent applications during the last few years.

The technical proposals published so far can be summarised in various groups. Table 7 contains a compilation of some interesting applications of dense gases to substance separation. The table does not claim to be complete. It contains some fundamental work and some newer results involving the individual compound classes. The practical applications are discussed in detail in the following chapters.

The extraction and refinement of edible oils is a prominent domain of application of high pressure technology. The use of dense carbon dioxide offers here the chance of combining several complicated stages of classical chemical engineering, leading to an essentially simplified overall procedure.

The sphere of flavouring materials (aromas, spices and essential oils) is of great practical interest in the world of nutrition and also for the pharmaceutical domain and the perfume and cosmetic industries. The situation is favourable here for extraction with dense carbon dioxide. Concentrates can be obtained under mild thermal conditions with a physiologically unobjectionable solvent, a procedure which is usually superior to classical extraction or steam distillation. Further, most of the substances concerned are highly soluble even at medium gas pressures of ca. 100 bar. They are therefore attainable with a comparative minimum of effort and possess, in addition, a high market value. The number of patents or patent applications in this sphere is correspondingly large. It must be mentioned here, however, that the patent literature, in accordance with its own aims and fields of interest, often contains no detailed investigations of special problems and data which do not always refer to the best conditions for separation.

Table 7. Literature Survey of Separation of Substances with Dense Gases

Gas	Conditions of Extraction	Separation- or Extraction Problem	Reference	Year
Edible Oils, Fats and Waxes				
C_3H_6, C_3H_8	100–115 °C, 60–110 bar	Extraction of lanolin from wool fat	10	1958
CO_2	45–50 °C, 280–350 bar	Extraction of the fatty oil from ground coconuts, peanuts, soybeans and sunflower seeds	11	1971*
CO_2	90–200 °C, 200–245 bar	Deodorisation and deacidification of soybean oil and palm oil by countercurrent extraction in a column, adsorption of the extracted material on active charcoal	12	1973*
CO_2/H_2	190–210 °C, 200—235 bar	Simultaneous hydrogenation and deodorisation of nut oil, sunflower oil, spermaceti in a packed column with finely divided nickel catalyst	13	1974*
CO_2 + acetone	69.6 °C, 132.5 bar	Separation of oleic acid from a glyceride mixture through countercurrent extraction in a bubble cap column	14	1978
	80 °C, 135 bar	Separation of monoglycerides from a glyceride mixture		
C_2H_4	27–50 °C, 100–160 bar	Fractionation of codfish oil	15	1978
CO_2 + CCl_4 CO_2 + hexane	74—135 °C, 80—100 bar	Extraction of triglycerides, followed by fractionation in a packed column	16	1978
CO_2	20–50 °C, 200–700 bar	Extraction of the seed oil from soybeans, sunflower seeds and rape	17	1980
CO_2	31–150 °C, 72–1000 bar	Extraction of carnauba wax, beeswax, wool wax and montan wax and of jojoba oil	18	1982*
CO_2	25–80 °C, 90–2600 bar	Solubility behaviour of soybean oil under high pressures	19	1982

Table 7. (continued)

Gas	Conditions of Extraction	Separation- or Extraction Problem	Reference	Year
CO_2	80 °C, 200 bar	Fractionation of butter fat, concentration of short-chain triglycerides in the extract, long-chain triglycerides in the residue	20	1982
CO_2	50–80 °C, 500–1000 bar	Extraction of cotton seed	21	1984
CO_2, C_2H_4 CHF_3 $CClF_3$ $CBrF_3$	20–100 °C, 150–500 bar	Extraction of crude montan wax	22	1984
CO_2	90 °C, 900 bar	Removal of oil from crude lecithin through high pressure jet extraction	23	1985
Flavours, Spices and Essential Oils				
CO_2, C_3H_8	Liquid	Extraction of essential oils from various plants; carbon dioxide showed higher selectivity than propane	24	1968
CO_2	22 °C, 63 bar	Flavouring essences obtained from freshly homogenised fruits (apples, oranges, pears) and fruit juices	1	1970
CO_2 $CO_2 + H_2O$	45–50 °C. 315–400 bar	Extraction of resin, essential oil, α- and β-acids from airdried hops, then extraction of the tannins with moist CO_2	25	1971*
CO_2 N_2O	50–60 °C, 280–400 bar 50 °C, 200 bar	Extraction of odour substances with dry CO_2, flavouring agents with moist CO_2, from ground spices, piperine from black pepper, eugenol and acetoeugenol from cloves, vanillin from vanilla pods, cinnamaldehyde and eugenol from cinnamon sticks	26	1971*

Table 7. (continued)

Gas	Conditions of Extraction	Separation- or Extraction Problem	Reference	Year
CO_2	Liquid	Extraction of mainly easily volatile spices and flavouring substances from 20 different natural products, e.g. calamus, anise, cloves, ginger, coriander, bay leaves, parsley, chamomile, eucalyptus leaves	27	1974
CO_2	40 °C, 70–400 bar	Microextraction with direct coupling with thin-layer chromatography, numerous examples	2	1976
CO_2, N_2O	40 °C, 70–300 bar	Extraction of chamomile flowers with fractionation of extract by stepwise separation	28	1978
CO_2	20–40 °C, 90–250 bar	Extraction of the thermolabile pyrethrins from pyrethrum flowers	29	1980
CO_2	0–40 °C, 80–200 bar	Preparation of concentrated extracts from fresh plant material, especially from kitchen herbs, e.g. dill, estragon, onions, blackcurrant buds, mimosa blossoms	30	1981*
CO_2	34 °C, 90 bar	Extraction of lilac flowers	31	1982
	40 °C, 300 bar	extraction of lemon peel		
CO_2	40 °C, 100 bar 20 °C, 80 bar	Extraction of the poisonous thujone from common wormwood, without reducing the content of valuable bitter principles	32	1983
CO_2	0–10 °C, 80 bar	Extraction of juniper berries, hop flowers and ginger roots	33	1984
CO_2	40–80 °C, 70–100 bar	Continuous removal of terpenes from citrus oils by countercurrent extraction in a packed column	34	1984

Table 7. (continued)

Gas	Conditions of Extraction	Separation- or Extraction Problem	Reference	Year
CO_2	10–100 °C, 30–150 bar	Solubility and separation behaviour of essential oils and possibilities of fractionation	35	1985
Further Isoprenoids				
CO_2 + ethanol CO_2 + acetone	40–50 °C, 80–200 bar	Preparation of very finely divided carotinoids	36	1979*
CO_2	40 °C, 70–300 bar	Solubility behaviour of carotinoids, microextraction of carrots, tomato pulp, algae and paprikas, directly coupled with thin-layer chromatography	37	1980
CO_2	40 °C, 81 bar	Extraction of pine- and fir-wood, extract separated into 3 fractions, resin acids, fatty acids and turpentine oil	38	1980*
CO_2	40 °C, 120/320 bar	Extraction of paprika in two stages: 1) 120 bar: extraction of pungent substances (capsaicine) and the free fatty acids 2) 320 bar: extraction of the colouring materials (capsanthine)	39	1983
CO_2	40 °C, 80–200 bar	Quantitative determination of the solubility of steroids, search for connections between chemical structure and solubility	4	1984
Alkaloids				
CO_2 N_2O	50–70 °C, 250–1000 bar 28 °C, 65 bar	Extraction of nicotine from tobacco of water content brought to 15–25%	40	1970
CO_2 + H_2O	40–80 °C, 120–180 bar	Decaffeination of coffee beans with CO_2, saturated with water, adsorption of the caffeine on active charcoal in liquid CO_2	41	1970*

Table 7. (continued)

Gas	Conditions of Extraction	Separation- or Extraction Problem	Reference	Year
CO_2 $CO_2 + H_2O$	45 °C, 400 bar	Extraction of flavouring materials from tea with dry CO_2, then extraction of the caffeine with moist CO_2, flavour transferred to tea leaves already decaffeinated	42	1972*
CO_2, N_2O, CHF_3	20 °C, 40 °C, 70–150 bar	Quantitative determination of the solubility of opium alkaloids by the dynamic method, using a microextraction apparatus	6	1980*
CO_2, CHF_3	20 °C, 70–200 bar	Extraction of the baine from *Papaver bracteatum Lindl.*; CHF_3 showed better solvent power and higher selectivity than CO_2	5	1980
Polymers				
CO_2	40–80 °C, up to 1000 bar	Solubility behaviour of polyvinyl chloride, polymethyl methacrylate and polystyrene	43	1976
CO_2, C_2H_4, C_3H_8	40–160 °C, up to 2000 bar	Extraction of polyethylene, polystyrene and polyethylene glycol, consideration of solubility parameters	44	1984
CO_2, C_2H_4 CHF_3, $CClF_3$	40–130 °C, 150–400 bar	Extraction of polyacrylamide, polystyrene, polyoxymethylene, polypropylene and polyethylene, separation of monomers, oligomers, water and solvents	45	1984
Wood				
Acetone	250–340 °C, 250 bar	Thermal degradation of cellulose and chitin	46	1978
Toluene	340 °C, 80 bar	Pyrolysis/extraction of spruce wood	47	1978
n-Pentane, diethyl ether, acetone + various alcohols	250–280 °C, 100 bar	Thermal degradation of birch wood with supercritical organic solvents	48	1979

Table 7. (continued)

Gas	Conditions of Extraction	Separation- or Extraction Problem	Reference	Year
Acetone, methanol	260–360 °C, 100–280 bar	Obtaining liquid fuels and chemicals (phenolic components) from cedar wood	49	1983
CO_2 N_2O, C_2H_4, C_3H_8	35–60 °C, 100–620 bar T_C + 9 °C, 210–620 bar	Extraction of resins and fatty acids from pine wood, and of wax from fir bark		
Coal and Petroleum Products				
Toluene	350 °C, 100 bar	Extraction/liquefaction of coal, obtaining aliphatics, aromatics and oxygen-containing compounds	50	1975
C_3H_8	140 °C, 130 bar	Deasphalting of the top residues from petroleum distillation	15	1978
C_2H_4	45 °C, 60–110 bar	Fractionation of α-olefines (C_{14}–C_{20})		
Pentane Benzene	140–290 °C, 20–83 bar 160–360 °C, 20–58 bar	Stepwise extraction of tar sands and peat with succeeding adsorption column with active charcoal for fractionating the extract	51	1980
C_2H_6 CO_2	55 °C, 200 bar 55 °C, 135 bar	Extraction of brown coal tar with ethane yielded creosote oil (substituted phenols + carboxylic acids) and paraffins; in a high pressured column carbon dioxide separated this extract selectively into creosote oil + lower paraffins and higher paraffins + 1-olefines	52	1981
C_3H_8	110 °C, 200 bar	Extraction of a dark red oil (paraffinic compounds) from Athabasca-tar sand, asphaltenes remained in the residue		
CO_2, C_2H_4 C_2H_6, C_3H_8	$T \gg T_C$, $P \gg P_C$	Work up of used oil to reusable lubricants by counter-current extraction in packed column and multiple stage separation	53	1982

Table 7. (continued)

Gas	Conditions of Extraction	Separation- or Extraction Problem	Reference	Year
C_3H_8 + toluene	120 °C, 80 bar	Separation of fine solid particles from a dispersion in viscous liquids, e.g. shale oil	54	1983
Aromatics + hydrogenated aromatics	420–480 °C, 50–200 bar	Extraction/liquefaction of coal	55	1983
Miscellaneous				
C_2H_6 + C_3H_8	12 °C, 27 bar	Desalting of sea water (3% NaCl)	56	1960*
CO_2, CCl_2F_2	liquid	Impregnating foodstuffs (e.g. carrots, potatoes, crisps) with antioxidants (e.g. butylhydroxyanisole tocopherols, ascorbyl acid palmitate) which were dissolved in the liquid gas and the foodstuffs immersed in the solutions	57	1973*
CO_2	55.5 °C, 170 bar	Regeneration of adsorbents, e.g. removal of phenol from an active charcoal column	58	1974*
CO_2 C_2H_4	35–65 °C, 103–172 bar 35 °C, 103–172 bar	Separation of mixtures of ethanol and water	59	1981
CO_2	100 °C, 120–280 bar	Extraction of PCB from transformer oil (silicone oil)	60	1982
	50–120 °C, 90–280 bar	extraction of alachlor adsorbed on active charcoal, extraction of lipophilic organic compounds from aqueous solutions		
CO_2	60 °C, 100–350 bar	Removal of pesticide residues from drugs, e.g. of DDT (and BHC) from senna leaves	61	1984
CO_2	95 °C, 155 bar	Separation of ethanol from aqueous solutions	62	1984
	60 °C, 150 bar	Separation of butanol, glycerol, ethylene glycol and acetic acid from aqueous solutions		

* patent or patent applied for

The increasing demands of the food and cosmetic industries for natural plant colouring materials renders interesting the extraction of red and yellow carotenoids from plant material. Through their high extractive selectivity, dense gases can yield carotenoid fractions of great purity, largely free of undesirable material. There is an increasing demand also for steroid-containing medicaments (corticoids, contraceptives); these can be half-synthesised from naturally occurring sterols, so that sterol extraction from, for example, fatty oils or tall oils plays an important part. These plant starting materials generally have only low sterol amounts; this creates difficulties of separation but these can be surmounted by using dense gases.

A number of publications, mainly in the patent literature, has been devoted to the "detoxification" of tea, coffee and tobacco; individual variations are distinguished through details in the technical procedure. The selective extraction of the alkaloids can be favoured by a particular water-content of the plant material; examples are the extraction of caffeine from green coffee beans or tea leaves and of nicotine from tobacco leaves. Many other alkaloids can also be extracted quantitatively, e.g. thebaine from *Papaver bracteatum*, after having freed the alkaloid bases from the plant material by pre-treatment with alkali.

The centre of interest with polymers concerns separation of monomers, oligomers and solvents. Polymers in which food articles are packed or which are used in medicine are allowed to contain low molecular weight material only in small, legally established amounts.

Coal, tar sand and petroleum products are usually extracted by classical solvents at high temperature under conditions above the critical points. The aim is to obtain as large as possible a part of the material in liquid form. Another task in this connection is the work up or fractionation of already liquid starting materials or residues. The problems of wood extraction are similar; important here are procedures for obtaining liquid fuels by pyrolysis/extraction, or for separating particular chemicals under mild conditions.

In addition to these domains there are many special uses, e.g. removal of alcohol from aqueous solutions, regeneration of adsorbents, separation of pesticide residues from drugs etc.

IV.2. Fatty Oils and Waxes

The most important procedures for obtaining fatty oils are pressing out mechanically and extracting with solvents. In both cases the optimum pre-treatment (conditioning) of the starting raw material improves both the yield attainable and the colour and sensory quality of the crude oil. Starting material of fat content exceeding 30–40% is, for economic reasons, subjected to as combined treatment of pressing, followed by extraction. The pressing generally lowers the oil content to about 20%. Using more complicated pressing procedures, more oil can be removed but even in the most favourable case more than 5% oil remain in the pressed cake. Total removal of oil can be achieved only with lipophilic solvents which enable fat contents of less than 1% to be obtained in the extracted material (grist).

The commercial extraction of oil is usually performed in a continuous countercurrent process with technical hexane as solvent; which has high stability, little corrosive activity and good solvent properties. The solvent enriched with extract (miscella) attains an oil content of 25 to 35% during the procedure. The extract is separated from the solvent by distillation and evaporation; solvent is recovered from the grist in special desolventing-toaster-installations. These last two stages can be circumvented by the use of dense gases as solvent.

The crude oils obtained by classical methods are usually cloudy and give an unsatisfactory impression to the senses. They are therefore submitted to a procedure of refining before use. This consists of removal of slime, acid, colour and off-odour. These stages also can become largely superfluous or greatly simplified with oils obtained by high pressure extraction, especially with dense carbon dioxide. The greater ease of these procedures, together with some difficulties associated with the hexane extraction hitherto employed (high solvent costs, danger of fire, low selectivity, problems of residues and emission) raise the question of whether the high pressure extraction is not an alternative for obtaining fatty oils and waxes. The considerations below can serve as a basis for discussing this.

IV.2.a Solubility Behaviour

Knowledge of the phase behaviour of a substance is a basic item of information for judging and designing a suitable procedure for extraction. Investigations have been carried out on the solubilities in dense carbon dioxide of triglycerides and wax esters over a wide range of pressure and temperature [19, 63]. In this, a mixture of natural composition was studied each time, rather than a single model substance; the natural oil solubility was considered to yield information of more value for practice since with a mixture of substances the interactions between similar components can lead to an increase of concentration in the dense gas phase [64, 65]. However, quotation of solubility data for a mixture of substances is justified only when the individual components behave so homogeneously that no alteration of composition occurs during extraction and hence no change in the amounts of substances dissolved. This requirement is fulfilled by the triglyceride (TG) or wax ester (WE) mixtures chosen, so that the system natural oil/CO_2 can be regarded as pseudo-binary in each case.

Soybean oil was studied as the naturally occurring triglyceride mixture. As seen in Table 8 it contains about 80% of esters of oleic and linoleic acids. Soybean oil is

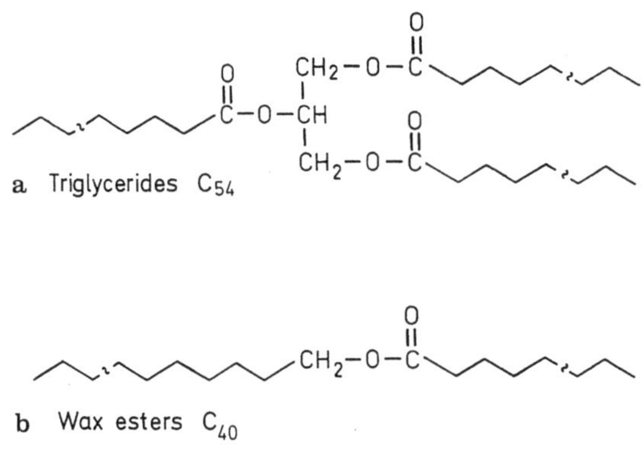

Fig. 34a, b. Structure of the compound classes investigated

Table 8. Composition of soybean and jojoba oils as percentages

	$C_{16:0}$	$C_{18:0}$	$C_{18:1}$	$C_{18:2}$	$C_{18:3}$	$C_{20:1}$	$C_{22:1}$	$C_{24:1}$
Soybean oil [67] (acids)	10	3.5	21	56	8			
Jojoba oil [68] (acids)	1		11			71	14	1
Jojoba oil (alcohols)			1			44	45	9

outstanding in its worldwide production and nutritional significance. According to information of the U.S. Department of Agriculture (USDA), the world production of soybean oil in the season 1984/85 amounted to 90.3 million tonnes. As a representative of the wax ester class, a cold pressed jojoba oil was investigated; it differed from the triglycerides in the fatty oils in that it consisted of monoesters of long-chain unsaturated alcohols (notably eicos-11-enol and docos-13-enol) with the very same fatty acids. More than 80% of the liquid wax esters consist of C_{40}- and C_{42}-compounds [66]. Jojoba oil is obtained from the nut-like fruits of *Simmondsia chinensis*, a desert bush. It is a valuable wax with properties similar to the sperm oil obtained from the whale and has a correspondingly wide range of uses. The average molecular mass of jojoba oil is 600, in comparison with 900 for soybean oil.

The solubilities were measured using the analytical dynamic method. A fast, simple method to quantify the substance mixtures investigated is to weight the amounts of substance dissolving in the carbon dioxide. The desired solubility result is then obtained directly by relating this to the amount of gas used. The data are given in the units of mg dissolved substance per g solvent (mg/g = g/kg = % by weight $\times 10^{-1}$), suitable for practice.

Fig. 35 is a sketch of the experimental set-up for measuring solubilities. The gas supply unit for generating and regulating pressure, functioning up to a pressure of 3000 bar, was described in Chapter III.2. For the experiment, the oil is introduced into the autoclave E in an available form on a carrier. The dense gas under definite conditions of pressure and temperature is then passed slowly through the sample. It becomes charged, according to its state, with part of the oil and carries this into the separation autoclave, A. Separation of the dissolved material is brought about by

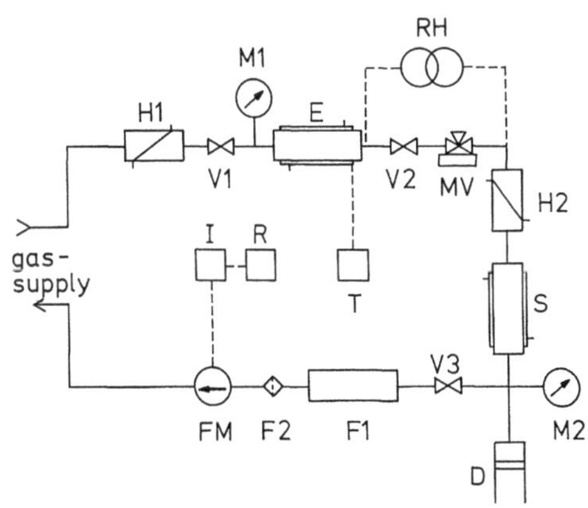

Fig. 35. Experimental assembly for measuring solubilities using the analytical dynamic method.
H heat exchanger; V cut-off valve; M manometer; E equilibrium cell; HV heated precision metering valve; RH resistance heating; S separating vessel; D safety disc; F1 active charcoal-silica blue gel filter; F2 sintered filter; FM mass flowmeter; I integrator; R recorder; T device for measuring temperature [19]

depressurising the miscella at the heated precision metering valve, RV. The gas itself continues through the mass flowmeter, D, which is joined to the integrator, I, and the recorder, R, to yield the exact amount of gas; this then returns to the supply unit, thereby completing the cycle. The amount of oil dissolved is removed from the separator at the end of the experiment and weighed.

The pressure-temperature-volume behaviour of the solvent carbon dioxide is shown in Fig. 36. If the gas is at a temperature below the critical value (31 °C) its density rises abruptly when the liquefaction pressure is reached (e.g. 35 bar at 0 °C). Since liquids have low compressibilities the density of the condensed gas then increases only slightly with increase in pressure. The gas/liquid phase boundary is no longer traversed above the critical temperature. As pressure rises the gas density increases continuously; a constant transition takes place from gas sensibly without solubility power to supercritical gas with good solvent properties. However, even here the real increase of density to values similar to those of liquids occurs in a restricted range of pressure from the critical value (73 bar) up to about 200 bar, provided that the temperature is not too much above the critical value. A slow and steadily weakening increase of density is then observed at high pressures with both the supercritical and the liquefied gases.

Fig. 37 shows the phase behaviour of soybean oil at different temperatures in the pressure range 100 to 2600 bar. The isotherms indicate first an increase of solubility with increasing pressure, the gradient of the isotherms becoming more and more steep with rise in temperature. It is interesting that the solubility does not level off at pressures up to 500 bar and above. This behaviour of fatty oils, also of wax esters and sterols (as will be seen later) in dense carbon dioxide differs from that of numerous other substances, e.g. naphthalene [69], phenols [70], opium alkaloids [6] etc., which display a flattening of solubility with increasing pressure already from 200 bar and hence display a relative course of solubility parallel to the density isotherms of carbon dioxide.

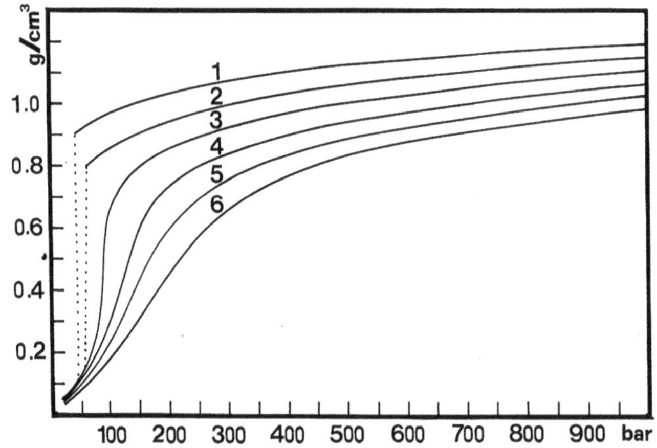

Fig. 36. pVT-behaviour of carbon dioxide, 1–6 = 0,20, 40, 60, 80, 100 °C

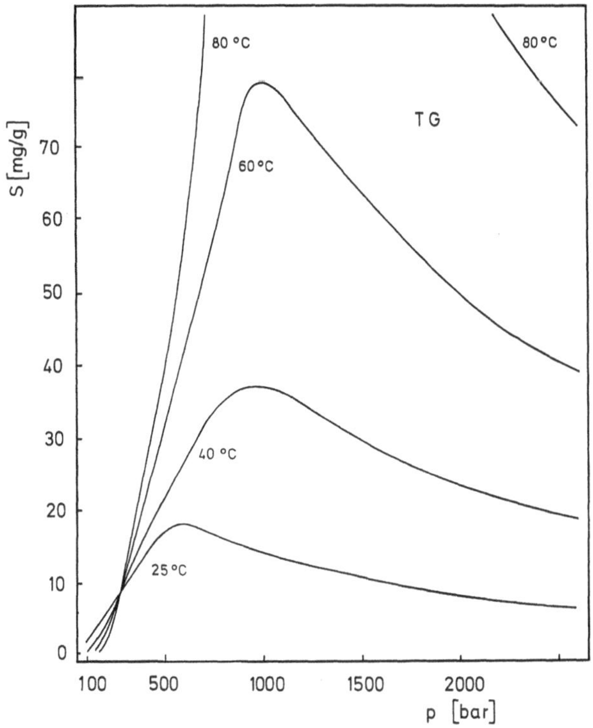

Fig. 37. Solubility isotherms of soybean oil in dense carbon dioxide, as a function of pressure (smoothed curves) [19]

As the pressure is increased still further, the solubilities at 25, 40 and 60 °C finally pass through a maximum which is more distinct the higher the temperature. Whereas the solubility maximum in liquid carbon dioxide is reached at 600 bar, the maxima in the supercritical phase region occur at a higher pressure, of the order of 1000 bar. The solubility between 800 and 1000 bar at 80 °C is so large that it can no longer be measured using the analytical dynamic method (limit of ca. 8–10% by weight). The tendency to higher solubilities continues as the temperature rises beyond 80 °C, provided that the system is at a sufficiently high pressure. This is shown in the solubility curves calculated by Eggers, Sievers and Stein [71] and seen in Fig. 38. The experimental data could be satisfactorily correlated with the help of the Redlich-Kwong equation, as modified by Hederer [72], the parameters of which were adapted to the values found-even though the solubility maxima could not be taken in.

The solubility isotherms of soybean oil intersect in the vicinity of 300 bar pressure and at a relatively low concentration in the gas of 0.9% by weight. This is a consequence of the differing temperature effects on the solubility, depending on pressure. In the pressure region below this inversion point, solubility decreases with rise of temperature; in the region above, solubility increases. The intersection point of the isotherms is encountered only within a limited temperature interval; a region of overlapping is observed above this. The differing influence of temperature on the solubility can be understood when it is remembered that an increase in temperature lowers the density of the compressed gas and increases exponentially the vapour pressure of the weakly volatile component: The fall in density with increasing temperature is especially

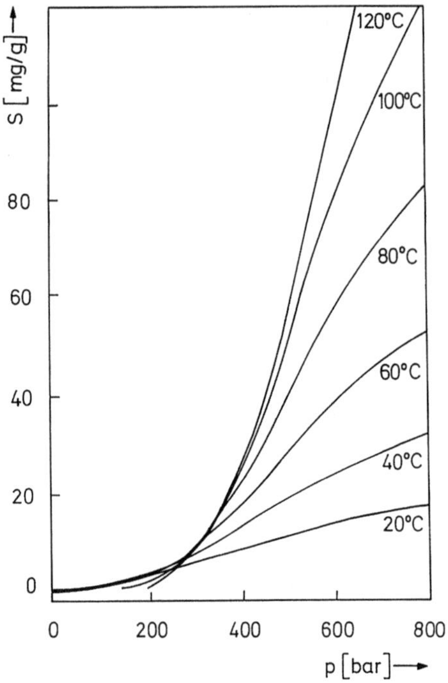

Fig. 38. Solubility curves calculated for soybean oil in CO_2 as a function of pressure at various temperatures [71]

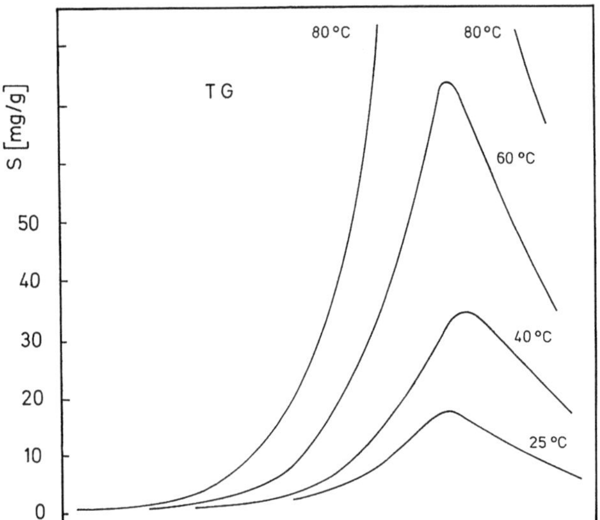

Fig. 39. Solubility isotherms of soybean oil in dense carbon dioxide, as a function of the gas density (smoothed curves) [19]

marked in the low pressure region of 1–$2p_C$ (Fig. 36) and leads here to a fall in solubility. At higher pressures the density of the solvent changes only little with temperature and is of less importance; the vapour pressure of the substrate now plays the important part and is responsible for the positive solution enthalpy.

The intersection point of the isotherms does not appear if the charging of the gas phase is plotted against the gas density, as in Fig. 39. At constant density an increase in temperature increases solubility in every case, occasionally appreciably. In the region of the solubility maximum, the curves show an approximately symmetrical parabola-like course. The maxima are at a gas density of about 1.06 g/cm^3. Temperature increase shifts their position slightly towards lower density values.

A study of the solubility of jojoba oil for comparison shows that the isotherms of the wax esters take a form similar to those of the triglycerides. They likewise pass through almost symmetrical maxima, when regarded as a function of the gas density, and suffer a slight displacement, from 1.07 to 1.03 g/cm^3 on increasing the temperature from 20 to 60°. If the solubility isotherms are plotted as a function of the pressure, a representation more appropriate for extraction practice, an isotherm intersection point reappears, for jojoba oil at about 250 bar and a concentration in the gas of 0.45% by weight (Fig. 40).

From Fig. 40 it can be seen also that the solubility of the wax esters at low temperatures, in liquid CO_2 at 20 °C for instance, is likewise little influenced by pressure whereas at higher temperatures, especially above 60 °C, increase in pressure can bring

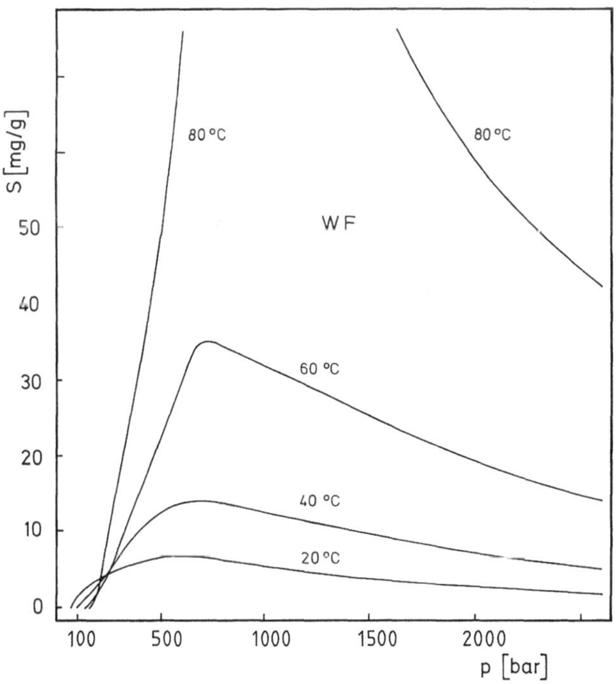

Fig. 40. Solubility isotherms of jojoba oil in dense carbon dioxide, as a function of the pressure (smoothed curves) [63]

about a marked increase in solubility. The steep isotherm gradient in this temperature region permits an effective separation of the dissolved oil even through small changes in pressure. For example at 80 °C it is not normally necessary to lower the pressure to below 150 bar, to precipitate the wax esters; this corresponds to a residual solubility of 0.02% by weight and can effect considerable saving of energy costs because a large part of the volume work for recompression of the circulating gas has to be performed at lower pressures where the gas is highly compressible.

Comparison of the absolute values of solubility reveals that, at 20, 40 and 60 °C, the triglycerides are more soluble in dense carbon dioxide than are the wax esters. At 40 °C for instance, soybean oil attains a maximum solubility of 3.6% by weight compared to only 1.4% of jojoba oil. The solubility maximum of the triglycerides in the supercritical gas is at a rather higher pressure than that of the wax esters, the solubilities of which begin to fall again above 700 bar. In contrast to the data at the lower temperatures, the solubilities of the wax esters in the isotherm climb reach those of the triglycerides at 80 °C and are, in fact, slightly higher. The solubility of jojoba oil is evidently more strongly influenced by temperature than that of soybean oil. The temperature behaviour is best seen when the experimental data are plotted in the form of solubility isobars. This is done in Fig. 41 and the course of the curves shows clearly the differences between soybean and jojoba oil.

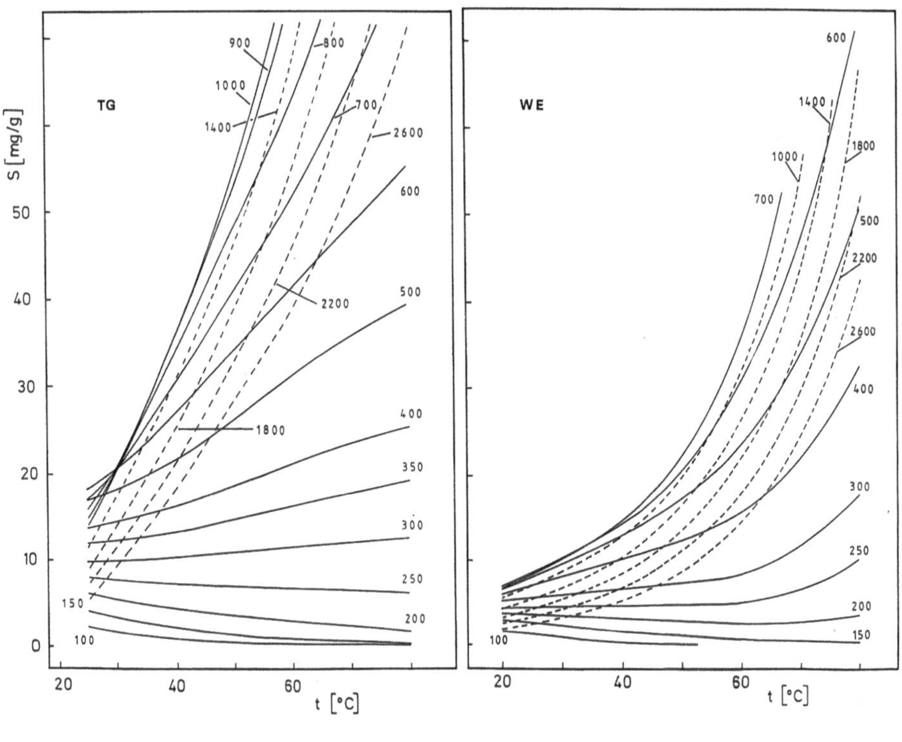

Fig. 41. Solubility isobars of soybean (TG) and jojoba (WE) oils in dense carbon dioxide, as a function of the temperature; pressures in bar written on the curves

A family of solubility isobars is drawn in for soybean and jojoba oils. It can be seen that a weakly negative solution enthalpy is found only at low pressures. At the pressure of the inversion points (300 bar for soybean, 250 for jojoba oil) the solubility is independent of temperature within certain limits. The solubility increases with rise in temperature for the majority of isobars, at higher pressures. This increase of the isobars becomes steeper as pressure is increased and the solubilities attained in the supercritical carbon dioxide continue to rise up to the pressure corresponding to the solubility maximum, i.e. up to about 1000 bar for the triglycerides and 700 bar for the wax esters. Increase of pressure still further leads to the isobars shown by dotted lines which are almost parallel and displaced towards lower solubilities again. In comparison to the triglycerides the wax esters show a more marked rise of the isobars, especially above 60 °C. The rise can be observed already at lower pressures and has the consequence that the isobars pass through a flat minimum at 200 bar. Here it is evident that, with rising temperature, the increase in vapour pressure of the condensed phase dominates the effect of lowering of the density of the solvent. The cohesion of the condensed phase of the smaller and less polar molecules of the wax esters thus appears to decrease more appreciably under the influence of pressure and temperature than that of the triglycerides. This means that carbon dioxide can dissolve better in the condensed phase and, in contrast, a certain amount of the wax esters is more readily taken up in the gas phase and solvated [73].

The course taken by the solubility isobars enables a clear comment to be made about the problem of extraction: liquid or supercritical? Just in the case of the wax esters it is evident that the solubility is very small and can scarcely be influenced by pressure on the liquefied gas below 31 °C. The wax esters and also the triglycerides show, already at 400 bar in the supercritical region, solubilities so high that they could never be reached at any pressure used on the liquefied gas. For instance the solubility of the wax esters increases by a factor of 8 when the temperature is increased from 20 to 80 °C. Accordingly the extraction at high pressures and temperatures offers the interesting chance of an effective and economically favourable separation through isobaric lowering of temperature alone.

When a definite amount of a condensed phase, for example soybean oil, dissolves in a dense gas such as CO_2, the process can be regarded as "evaporation" because the substrate becomes part of the gas phase. This actual solubility, based almost entirely on intermolecular interactions, deviates considerably from the ideal solubility which is attained only through the vapour pressure of the condensed phase, without the assistance of solvation effects. The ratio of the real, much higher solubility to that calculated for ideal conditions is termed the "enhancement factor" (E) [74]. This factor is thus a measure for the apparent increase in vapour pressure of a poorly soluble substance as a result of interactions with the dense gas. Fig. 42 portrays the enhancement factors of soybean oil, expressed as a function of the gas density. The values for the ideal solubilities were calculated with the help of the law of partial pressures, taking an average molecular mass of 900 for soybean oil and the temperature-dependence of its vapour pressure to be correctly expressed by the equation $\lg p = -8700/T + 14$ [75].

It is seen that soybean oil has especially high enhancement factors. The content of oil in dense CO_2 can be up to 18 powers of ten greater than would be expected from the vapour pressure. The largest values are reached at high pressures and low tem-

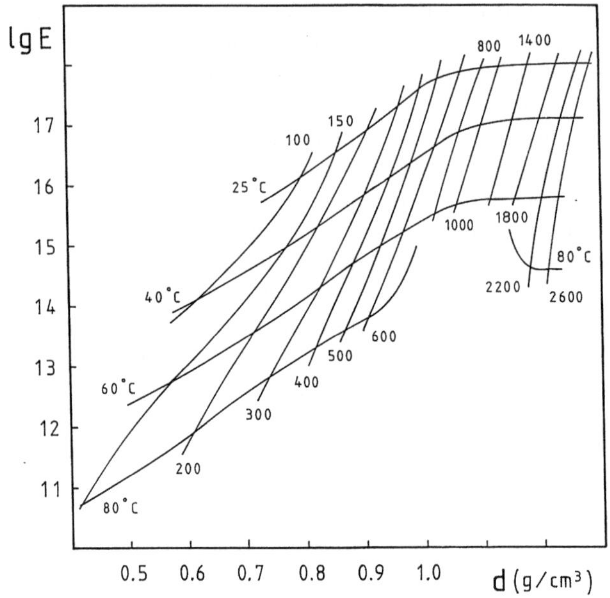

Fig. 42. Enhancement factors of soybean oil in dense carbon dioxide, expressed as a function of the gas density (pressure in bar) [19]

peratures. The slope of the isobars at falling temperatures is most markedly defined at high pressure. The increase of the different isotherms is characterised by almost the same slope. When the density is reached which corresponds to the solubility maximum, the isotherms take a horizontal course; this means that the enhancement factor keeps its attained value when pressure or density are still further increased.

The course of the isotherms can be explained in that the calculated relative amount of substance in the gas phase decreases continuously as pressure rises whereas in reality, as a result of interactions, a rise in solubility is first observed. After the solubility maximum has been reached, the observed reduction of solubility agrees roughly with the calculated diminution of substance amount in the gas phase. The enhancement factor then remains unchanged and the solvation power of the dense gas has reached a saturation value which can no longer be exceeded.

Enhancement factors have been determined also in the system oleic acidethylene, with values reaching 10^{10} in the region of 100 bar at 20 °C [76]. An investigation of the relative concentration enhancement of oleic acid in various gases at the same pressure yielded a series in ascending order of: carbon dioxide < ethylene, chlorotrifluoromethane < bromotrifluoromethane < propane [77]. Data on enhancement factors with other systems, e.g. caffeine/CO_2 and anthracene/CO_2 have been quoted also [78].

When the solubility data just discussed are applied to the extraction of oil seeds it must be borne in mind that the analytical measurements were carried out on a small scale under well controlled conditions and using a carrier material on the surface of which the fatty oil was present as a thin film; the extraction conditions in practice deviate considerably from these.

Diffusion effects play a part during extraction from a natural product, i.e., oil transport within the seed particle. Full profit from the dissolving capacity is hindered also by the restricted mass transfer of lipid molecules into the dense gas phase in connection with the high relative flow speeds encountered, sometimes of over 20 kg CO_2 per h and per kg of seed. Systematic studies of the solvation kinetics of fatty oils have not yet been performed but some experience has been had of the influence of particle size and shape on the yield of oil and rate of extraction; this is comparable to classical oil extraction [17, 71, 79]. The best results are obtained with oil seeds that have been flattened into thin flakes of 0.2 mm wall thickness. Powdered material of 0.2 mm grain size is less satisfactory in comparison, and extraction is clearly lower from larger-sized fragments. No oil can be extracted from whole seeds, even when these are very small.

The natural moisture content of the oil seeds of ca. 10% scarcely affects the solubility of a fatty oil in dense carbon dioxide. It does play some part, however, by increasing the permeability of the cell membrane through swelling effects, and hence improving the accessibility of the fatty oil.

If the degree of extraction of the oil seed exceeds a definite value which depends on the gas flow, on the particle size and on the geometry of the autoclave, the charging of the solvent falls steadily. A uniformly good extraction efficiency is accomplished only when the solvent is conveyed in counter-current; this can be achieved in the high pressure procedure by using several extraction vessels (Chapter III.2).

These effects can lead to maximum attainable concentrations in the dense gas being up to 40% lower than the solubility values; the variation of the gas charging with pressure and temperature remains unchanged, however. The available solubility data are thus useful criteria for judging and designing processes for extraction of triglycerides and wax esters from natural raw materials and also for refining crude oils.

The results obtained for soybean oil should be applied only to comparable oils within the extensive oleic acid-linoleic acid-group (olive, ground nuts, sesame, corn germ, sunflower, lupin and cottonseed oils). Any larger differences in the molecular mass and/or polarity of the triglyceride molecules can lead to considerable differences in solubility under one and the same set of conditions.

Table 9. Maximum attained concentration in the gas in the extraction of various oil seeds with CO_2

Oil extracted	Fatty acid group	Concentration in the gas at 600 bar/40 °C	Ref.
Coconut fat (copra oil)	Lauric, myristic (C_{12}, C_{14})	3.2% by weight	[80]
Palm oil	Palmitic (C_{16})	— —	
Soybean oil	Oleic, linoleic ($C_{18:1}$, $C_{18:2}$)	2.0% by weight	[17]
Cacoa butter	Stearic (C_{18})	—	
Rapeseed, colza oil	Erucic ($C_{22:1}$)	— —	
Castor oil	Ricinoleic ($C_{18:1\,(OH)}$)	0.3% by weight	[81]

Table 9 gives examples of the maximum attained solubilities of various triglycerides in the extraction of oil seeds with carbon dioxide at 600 bar and 40 °C. Comparison of coconut fat and soybean oil shows clearly the influence of the molecular mass. The latter contains C_{18}-acids, whereas the former contains C_{12}, a little C_{14} but also C_8 and C_{10}-acids. Consequently soybean oil has mainly C_{54}, copra oil C_{36}-triglycerides to be dissolved. Under conditions where both oils are liquid, the smaller molecules give the expected noticeably higher solubilities. The differences within a fatty acid group are smaller, as shown by the calculated solubility values of Eggers et al. [71] who assumed smaller differences in the molecular masses. A hydroxyl group in the hydrocarbon chain of the fatty acids exerts a far more powerful effect than does the addition of a CH_2-group. The drastic reduction of solubility in dense carbon dioxide caused by this group is not surprising when one considers the divergent behaviour of castor oil in classical solvents, such as lower solubility in hexane but greater solubility in ethanol.

An interesting aspect of the discussion of solubilities is the comparison of dense carbon dioxide with the alcoholic solvents proposed as a substitute for hexane in extracting fatty oils. Fig. 43 shows the solubilities of an oil of the oleic-linoleic acid-group in absolute 2-propanol, in ethanol, in the azeotropes of each alcohol with water, and in dense carbon dioxide at 600, 800 and 1000 bar. The values are as percents by weight, referred to each solvent and plotted against temperature.

It can be seen that the solubility increases in all cases with rise in temperature. The alcohols named, in the absolute state, are easily the best solvents for fatty oils. However, they are uninteresting in practice because the seed sample has to be subjected to a complicated preliminary drying and the solvent must be kept completely anhydrous. Uptake of water up to the azeotropic point, 4.4% for ethanol and 12% for 2-propanol, lowers considerably the solubility of the oil. The minimum concentration of oil in the extracting agent normally required for economic extraction, 10%, is, in fact, reached with the 2-propanol-water azeotrope clearly below the boiling point (80.4 °C) but the quality of the crude extract obtained is lowered; the high content

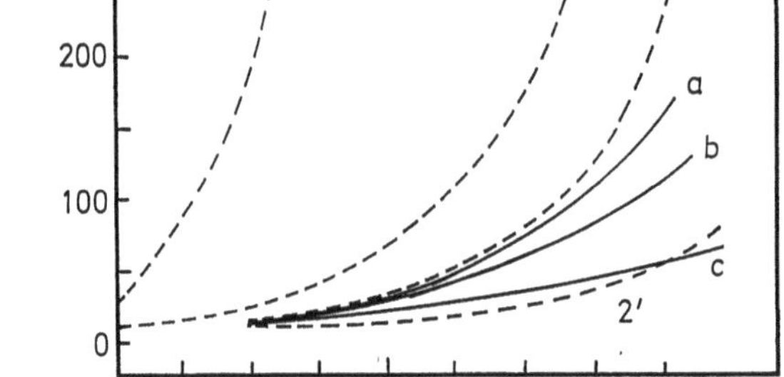

Fig. 43. Solubility of soybean oil in various solvents, as a function of temperature.
1 2-propanol; 2 ethanol; 1′ azeotrope of 2-propanol-water; 2′ azeotrope of ethanol-water [82, 83, 84]
a CO_2-1000 bar; b CO_2-800 bar; c CO_2-600 bar [19]

of water increases the polarity of the solvent, leading to an increased co-extraction of non-lipids. In the case of the ethanol-water azeotrope, only half the required solubility is reached even at the boiling point (78.5 °C) and work must be carried out under pressure at higher temperature to attain a suitable solubility value. Further, the complete separation and recovery of the alcohols from oil and residue is demanding and difficult.

As far as the solvent capacity for fatty oil is concerned, dense carbon dioxide behaves at 600 bar like the ethanol-water azeotrope and at 1000 bar like the 2-propanol-water azeotrope.

Solubilities of 25% by weight can be obtained in carbon dioxide already at 800 bar when the extraction temperature is raised only to 100 °C. This has no detrimental effect on the quality of the oil under the conditions of dense gas extraction (exclusion of oxygen). The marked dependence of dissolving power on temperature enables the phases largely to be separated into oil and solvent merely by cooling the miscella.

Friedrich also carried out studies of the solubility behaviour of the triglycerides in carbon dioxide [85]; the vegetable oil was not brought on to an inert carrier but directly extracted from the plant raw material (soybeans, cottonseed). The values found between 40 and 80 °C in the pressure region up to 1000 bar showed no solubility maxima but, in agreement with the results discussed, showed very clearly that the simultaneous use of high pressures (above 550 bar) and high temperatures (above 60 °C) effects a marked increase in solubility of the oil in CO_2, well over 3% by weight, which can be of importance in practice.

Extensive studies of phase equilibria, especially of the binary systems from oleic acid or glyceride mixtures, and dense gases were published by Brunner [86]. Ternary systems, with an added classical solvent, and also four-component systems made up of two substances of low volatility, an entrainer of medium volatility and a dense gas, were also investigated. The experiments were performed mainly with the help of an analytical static method in the pressure range below 400 bar. Data on the gas solubility in the condensed phase and also solubilities of the condensed phase in the dense gas were obtained and could be mathematically correlated. Brunner and Peter [73] published results on the solubility of palm oil in various gases and of these gases in palm oil, as well as solubility data of palm oil in carbon dioxide using ethanol as entrainer. The authors also measured the solubility of oleic acid in dense ethylene at various temperatures and pressures up to 350 bar.

The system oleic acid/CO_2 and binary mixtures of CO_2 and monoolein were investigated by King et al. [87] although only up to a pressure of 200 bar; they studied in particular the gas side which is important for the extraction. Chrastil described investigations of the solubilities of some pure substances of the fatty acid, triglyceride and wax ester classes, likewise up to pressures of only 250 bar [88]. De Filippi quoted values for the solubility of a vegetable oil at 20 °C in liquefied, at 55 °C in supercritical, carbon dioxide, in the pressure region up to 500 bar [60].

Czubryt et al. carried out at an early date investigations of the solubilities of, among other compounds, stearic acid and 1-octadecanol at 40 °C and under high pressures from 300 to 1900 bar; they employed analytical dynamic method [89]. Solubility maxima were found for both substances at about 400 bar. The dissolved compounds were determined with a flame ionisation detector only indirectly after passage through a pyrolysis unit, so that inference of absolute solubility values was problematic.

The data obtained were correlated according to the regular solution theory with the help of solubility parameters.

A mass of additional information of a more qualitative character about the extractability of fatty oils and waxes can be obtained from the numerous publications on the fractionation and preparative isolation of these products from natural materials.

IV.2.b Extraction of Oil Seeds

A substantial advantage of the dense gas procedure for obtaining oils and waxes from naturally occurring substrates is the high quality of the extracts. This applies in particular to extraction with dense carbon dioxide. It is of interest that carbon dioxide is an isostere of carbon disulphide, the earliest used solvent for extracting oil seeds. Oils obtained according to the classical procedure are suitable for human consumption only when undesirable accompanying substances have been removed; comprise phospholipids, polysaccharides, free fatty acids, waxes, substances imparting odour and taste, colouring material products of autoxidation and pesticide contaminants. The work demand of this refining process to yield a product fit for marketing is appreciably reduced with oils extracted by dense carbon dioxide (Table 10). Since its selectivity can be regulated through the parameters of pressure and temperature, dense carbon dioxide is suitable even for refining crude oils obtained by classical procedures.

Such refining and fractionating procedures under high pressure can be carried out continuously-but this does not yet apply to the direct extraction of oils from crude plant materials. However, fatty oils such as from soybeans, rapeseed, sunflower seeds etc., are mass produced and are prepared by conventional methods using continuous procedures in amounts exceeding 1000 tonnes daily. The advantages of dense gas extraction can thus be fully exploited only when solid materials can be continuously supplied to, and removed from, the extraction vessel.

Apart from this general engineering problem, on which work towards a solution is already being carried out, the following pages are devoted to considering experience and possibilities of the high pressure method for the extraction of various starting materials and the solution of special problems of fractionation in the lipid branch.

The extraction of fatty oils with carbon dioxide and other dense gases was proposed in the patent literature around 1950 [90, 91, 92]. This idea was taken up later in further patents by Zosel [93] and Vitzthum and Hubert [11] who described the extraction of copra, sunflower seeds, soybeans and ground nuts with supercritical carbon dioxide at 280–350 bar. Only in recent years, however, have detailed and comprehensive investigations been undertaken by numerous teams (Stahl, Friedrich, Peter, Eggers, Marr etc.). These studies are unanimous in revealing the following general features of a fatty oil extracted by dense carbon dioxide: The yield of oil is very slightly less than that from extraction with hexane. One important reason for this is that the phospholipids and glycolipids are extracted only in traces or not at all. Oils extracted with carbon dioxide are thus already devoid of lecithin and contain less than a tenth of the phosphorus amount of usual crude oils. The loss through refining is then proportionately less. The content of fatty acids is roughly the same as in oils extracted with hexane. During batch extraction the free fatty acids and some odoring components are extracted more at the beginning whereas pigments, other nonsaponifiable material and traces of phosphorus-containing çomponents prefentially towards the end. To some small extent, therefore, a fractionation with time occurs; as expected, substances of lower volatility, higher molecular mass and/or polarity appear more towards the end of the extraction. On the whole, the contents of free fatty acids and of non-saponifiable material, and also the peroxide number, are of the same order as in the hexane extract; the tocopherol content can be higher. Dense gas extracted oils contain appreciably less heavy metal, especially iron, depending on the type of equipment. In general the colour is lighter than that of oil extracted by hexane. This applies especially to the extracts obtained with liquid carbon dioxide.

The odour is mostly milder, the taste less harsh than displayed by the classically prepared crude oils. Comparable products result in both cases after refining, of good quality taste and which show no differences even after longer storage. There may be departures from this general characterisation in the one or the other feature if the starting materials are unusual.

Stahl, Schütz and Mangold were among the first to make a more accurate study of the extraction of oil seeds with dense carbon dioxide; they described quantitatively the influence of pressure, temperature, gas flow and particle size of the crushed seeds on the extraction of oil from soybeans, sunflower seeds and rapeseed [17]. Oil samples, differing clearly in colour, odour and taste were obtained by fractionated separation.

Friedrich, List and Heakin also investigated exhaustively the extraction of soybeans with supercritical carbon dioxide under various conditions. Their comparison of the high pressure and hexane extracts agreed with the description just given above [94].

Table 10. Refining procedures of edible oils

Conventional Chemical Refining		Refining with Dense Carbon Dioxide	
	Crude oil		
Removal of lecithin	+ 2% water, 80 °C, lecithin swells up; separation in high-speed separators	Removal of lecithin, slime and acid, deodorisation	Extraction of conventionally obtained crude oil, or, better, direct extraction of the oil seed with fractionated separation
Removal of slime	+ 0.1% phosphoric acid: proteins and carbohydrates precipitate Filtration or centrifugation		
Deacidification	Neutralisation of the fatty acids with NaOH (losses through hydrolysis)		Drying and stabilising
	Separation of the "soap gel", washing with water, centrifuging (losses through emulsion formation)	Bleaching if necessary	If necessary, after-treatment with Fuller's earth Filtering
	Stabilisation and drying in vacuo, 100 °C		Refined oil
Bleaching	Treatment with activated aluminium silicates, in vacuo Filtering of the adsorption agent (Fuller's earth)		
Evaporation, deodorisation	Steam distillation in vacuo, 5 mbar, 250 °C Refined oil		

Table 11. Comparison of crude oil extracted from soybeans by hexane and carbon dioxide [94]

Solvent	Yield %	Residual Oil %	Free Fatty Acids %	Peroxide Number m equiv./kg	Non-saponifiable matter %	Contents (ppm)	
						Fe	Phosphorus
Hexane	19.0	0.7	0.6	<0.1	0.6	1.45	505
CO_2	18.3	2.1	0.3	<0.1	0.7	0.3	45

They found, too, time-dependent extractions of accompanying substances such as free fatty acids, non-saponifiable material and phosphorus-containing compounds. In Table 11 data are given for crude oils extracted from soybeans with carbon dioxide and with hexane.

Corn (maize) germ oil (USA production of 400,000 t in 1982) belongs to the same group as soybean oil. Maize contains only 4% of oil so that extraction is not economical. But the separated germ-buds contain 40–50% fatty oil. The corn flour residue from oil extraction is also valuable for human nutrition. The carbon dioxide extraction yields a qualitatively superior flour residue compared to that from conventional hexane extraction [96]. The reason for this is the more extensive separation of the oil contained in the germs (residue of only 1% instead of 2%) and better removal of the bound lipids and bitter principles (residue of 2% instead of 8%) by the dense gas extraction. At the same time it was found that the CO_2-extracted germ flour was less sensitive to oxidation and the peroxidases were seven times more deactivated. It seems improbable, however, that these have been denatured as they are stated to be very stable to heat; furthermore, studies of seed residues from high pressure extraction have hitherto yielded no signs of appreciable protein changes. In any case the taste of the CO_2-extracted flour is outstandingly neutral (only half so bitter) and it is stable for a long period; the flour has 20% protein, of balanced amino acid composition, and is thus a food additive of high nutritional value.

In contrast the germ oil extracted by carbon dioxide (the solubilities agree well with those of soybean oil – 1.2% (wt.) at 345 bar and 2.7% at 550 bar, 50 °C in each case, our own investigations giving 1% at 350 bar/40 °C and 1.8% at 450 bar/60 °C [97]) shows the advantages mentioned, such as lighter colour, lower refining loss and even a smaller fraction of free fatty acids, than a germ oil obtained by pressing, but it is clearly more sensitive to oxidation. The proportion of tocopherol and of non-saponifiable material was found to be the same in both oils. It seems reasonable to assume that other antioxidants, native to the germs (among others, phospholipids) are insoluble in carbon dioxide and are therefore not found in the oil extracted under high pressure but remain in the residual flour, thereby increasing its stability. The same authors describe elsewhere [98] the characterisation of the CO_2-extracted germ oil in comparison with pressed oil and hexane-extracted oil from pre-pressed material and also the working up of these crude oils using classical alkali refining, bleaching and deodorisation.

The fatty oil from cottonseed is among the most important representatives of its class. A world production of 4.2 million tonnes was expected for 1985 (principally in China, India, Pakistan, the USSR and USA) [99]. The direct use of the oil and protein residue is made difficult, however, by the presence of 0.4–1.7 % of yellow gossypol, a phenolic, dimeric sesquiterpene derivative which is toxic in larger amounts. Procedures of adsorption and sedimentation and also treatment with alkalies are the usual ways of removing this poison from the oil. Another method is to heat the moist seed to 100 °C under pressure before extraction; part of the gossypol is then bound on the proteins [100].

Cottonseed oil extracted with hexane is cloudy and extremely dark coloured on account of the content of gossypol and other pigments. Particularly demanding refining is therefore needed. Oil extracted with carbon dioxide is appreciably lighter in colour and is characterised by a lower content of gossypol. List, Friedrich and Pominski quoted a content of 0.02 %, independent of the extraction conditions at 550 or 1000 bar at a temperature of 50 or 80 °C; classical crude oils, even from seed treated by heat, always contained 0.2 % [101]. Leiner also could establish no significant influence of pressure on the gossypol content of cottonseed oil in extraction with carbon dioxide [81]. In work with untreated seed at extraction temperatures above 60 °C he found a decrease of gossypol but an increased content of pigments. A moisture content of less than 15 % of the seed did not affect the result of extraction but an enhanced co-extraction of gossypol was observed with seed of higher moisture content.

According to List et al. the crude oils extracted with CO_2 require less alkali for working up and suffer less loss on refining. Further, the amount of pigment precursors, which could not be removed during refining and bleaching and which caused the oils to darken after long standing, was substantially lower.

Leiner tested various ways of acquiring a still better quality oil directly through the high pressure extraction. In multistage separation of the oil, gossypol and other pigments always separated in the first separator as the difficultly soluble components; the gossypol content of the main fraction of the oil could thus be reduced to 100 ppm under favourable conditions. Increased concentration of the free fatty acids and steroids could be observed in the last separator but no fractionation of triglycerides could be observed. Accompanying components were more efficiently separated by lowering of pressure than by reducing solubility through change of temperature. Counter-current extraction of the crude oil in a high pressure column also enabled pigments and gossypol to be removed. It was also possible to lower the gossypol content of the oil from 2600 to 10 ppm for example, by counter-current washing of the oil-charged gas phase with a 1 % sodium hydroxide solution. Finally a successful attempt was made to refine the cottonseed oil by passing the extract-laden gas stream through a pressure vessel filled with an adsorbent. Basic adsorbents, such as polyamide and aluminium oxide, had the strongest affinity for gossypol and other pigments; these enabled a light coloured oil, practically free of gossypol to be obtained. Part of the fatty acids and steroids was retained at the same time, whereas the triglycerides were able to pass this filter unimpaired. A relatively large amount of adsorbent was required but could be regenerated. The best procedure was given as a combined pre-washing with alkali and then adsorption on aluminium oxide, in which all the steps of the procedure could proceed on-line.

Lupins are gaining increasing importance as oil seed in Europe, Africa, South America and Australia. Their seeds yield a valuable vegetable oil and contain high-grade protein. Both these constituents are comparable to soybeans as far as the com-

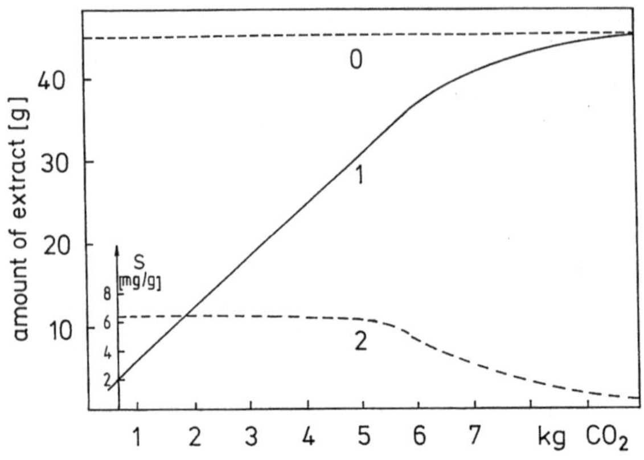

Fig. 44. Extraction of 250 g lupin seeds with supercritical carbon dioxide at 300 bar and 40 °C, expressed as a function of the solvent throughput [103]: 0: yield of oil in the Soxhlet extraction with hexane (15.8%); 1: yield in the CO_2-extraction; 2: concentration of oil in CO_2

position is concerned [102]. The utilisation of oil and protein for human nutrition is, however, restricted by bitter tasting and poisonous quinolizidine alkaloids, mainly lupanine, hydroxylupanine, sparteine and tetrahydrorhombifoline, present in the seed in amounts up to 4%.

Stahl, Quirin and Mangold described the extraction of a faintly yellow-green oil, almost free of phospholipids, from lupin seeds, using supercritical carbon dioxide at various pressures up to 300 bar [103]. The course of the extraction, shown in Fig. 44, is characteristic of oil seed. The curve of the yield shows at first a linear ascent, corresponding to the maximum gas charge of 0.6% by weight at 300 bar and 40 °C, reached already at the beginning. When about 80% of the oil has separated, extraction becomes less and the yield approaches asymptotically a final value which very nearly corresponds to the hexane extract of 15.8%. A 12 g amount of CO_2 per g seed is needed to extract half of the oil and 36 g for exhaustive extraction. The loss of efficiency at the end of extraction makes clear the advantage of a quasi-counter-current extraction with several extractors.

About 10% of the alkaloids present in the seed pass into the oil fraction in the process with carbon dioxide, just as in the hexane extraction, by preference the more lipophilic tetrahydrorhombifoline, whereas the strongly basic lupanine tends to remain in the grist residue. The bitter quality of the oil is normally removed by washing with dilute mineral acid. However, the alkaloids can also be washed out by a countercurrent of water in a column situated after the extractor in the carbon dioxide extraction [81]. This lowers the alkaloid content to below the limit of detection of 25 ppm so that the bitter taste completely disappears, whereas the non-washed oil has a bitter value of 2000 (according to the DAB = German Pharmacopoeia).

The removal of bitter taste from the lupin seeds using acids, alkalies or alcoholic solutions [104, 105] is always accompanied by considerable losses of sugars and protein. Accordingly the possibilities of separating all the alkaloids from the seed with the help of dense carbon dioxide was investigated [97]. In this way two things were shown: first, that all the free alkaloid bases in the lupin seeds are easily soluble in the dense carbon dioxide; secondly, that only part of the alkaloids can be extracted from the powdered seed samples. The influence of different factors was studied and it was discovered that a higher moisture content did not improve alkaloid extraction as in other cases (caffeine from green coffee beans [41] or nicotine from tobacco leaves [40]). A slightly better extraction amount was found only by raising the temperature, although in no case could more than 15–20 % of the alkaloids be separated. About 30 % of the alkaloids could be removed after pre-treating the powdered seeds with excess 7 m aqueous ammonia solution, and about 50 % after treatment with concentrated (1.3 m) alkali hydroxide solution if the lupin powder was redried before the extraction. This suggests that the alkaloids are firmly fixed to acid groups in the sample matrix which is understandable when the basic dissociation constants are regarded [106, 107]; these show the lupin alkaloids, in contrast to many other alkaloids, to be strong bases which can be liberated only with alkali solutions. The high buffer capacity of the seed proteins, which constitute 40 % of the seed, is an obstacle, however [108]. Since strong alkalies reduce the availability and quality of the proteins a better alternative to be recommended is the use, as hitherto, of polar aqueous solvents to remove the alkaloids as salts.

Bunzenberger, Lack and Marr described the extraction of edible oil from rapeseed, where the pivotal part of the work was the use of carbon dioxide in the liquefied state [109]. The maximum attainable charging of the gas, derived by drawing tangents to the curve of yield in the first phase of extraction, showed higher values in liquid than in supercritical carbon dioxide in the pressure range studied, of 100–250 bar, agreeing with the solubility data for fatty oils. For example, at 250 bar, the concentration in the gas fell in the sequence 0.44–0.39–0.35 % by weight as the temperature rose from 20° to 30° to 40 °C, respectively. The authors observed a drop in the charging of the gas relatively early, after only 40 % of the oil present had been extracted. This may have been due to an unsuitable autoclave shape or, more likely, to inadequate pulverisation of the seeds rich in oil (particle sizes of 1 mm and less were quoted), so that the extraction of oil was determined at an early stage by diffusion effects.

Eggers, Sievers and Stein also studied comprehensively the extraction of oil from rapeseed [71, 110]. They first described a modified Redlich-Kwong equation for calculating the quasi-binary phase equilibrium of oil-carbon dioxide but then reached the conclusion that, even when solubility parameters were adapted, only limited information could be derived for the design in practice of the extraction procedure. The study of the extraction course of seeds, mechanically preconditioned in various ways, yielded information about the kinetics of mass transfer, important for the extraction process. Seeds which were only peeled and those which in addition were subjected to cell cracking by sudden lowering of pressure, were extracted with low efficiency. With seeds made into thin platelets of 0.1–0.2 mm layer thickness, the maximum attained gas charging remained constant until 80 % of the oil had been extracted. In all experiments with flaky material less than 2 % hexane-soluble constituents remained in the residual grist. High extraction was obtained also from seeds, the oil content of which had been first reduced from 42 to 22 % by cold pressing with a filter press and the pressed cake had been loosened up by mechanical pulverisation before extraction with carbon dioxide.

A certain proportion of water is co-extracted in each oil extraction. It is largely independent of pressure and constitutes an especially large fraction in proportion to the amount of oil obtained at the end of the extraction. The peroxide number of the rapeseed oil extracted with carbon dioxide was lower than that of the hexane extract; the fraction of free fatty acids fluctuated and always fell towards the end of the extraction. The quality of the grist was very high and differed from the starting material only in having a slightly lower protein solubility.

The concentrations attained in the gas in a mass flow of 10 kg CO_2/h and kg seed were distinctly lower than the calculated solubilities. The concentration was 2.4 instead of 3% at 750 bar/40 °C and 4.4 instead of 8% at 750 bar/80 °C. Our own investigations gave a 2.8% solution in CO_2 with colza oil at 600 bar/60 °C. Mixing propane with the carbon dioxide clearly increased the loading of the gas compared to pure carbon dioxide and enabled the amount of solvent needed for exhaustive extraction to be reduced. A propane content of 9.5%, below the flash point, had a dissolving power at 350 bar corresponding to that at 750 bar with pure carbon dioxide at the same temperature.

A comparison was made of the energy requirements of the conventional hexane extraction with those of various CO_2 process-cycles (compressor-, pump-operation, various separation conditions) to assess the economic situation. The high-pressure extraction does not cut a good figure if the CO_2-circulation is considered without recovery of heat and if the falling gas charge at the end of extraction is considered. If, however, the saturation solubility is continuously maintained in the counter-current procedure, the energy requirement of the hexane extraction can be undercut under conditions of high solubility (750 bar/80 °C). Finally, developments in the sphere of continuous high-pressure extraction are reported in which the solid is introduced into the pressure vessel with the aid of a cage screw extruder and removed again after counter-current extraction has been performed.

Bully et al. described likewise the extraction of the fatty oil from Canadian rape (*Canola* seeds [111]), free of glucosinolates and erucic acid with supercritical carbon dioxide at 40 °C [112]. They used a modified high pressure liquid chromatograph, fitted with a 10 ml extractor. The extract phase was analysed with the help of a flame ionisation detector. A gas charge of 0.54% by weight was quoted at 300 bar and 40 °C; this rose to 0.87% by weight at 350 bar but only to 0.72% when an increased flow of carbon dioxide was employed (27 instead of 6–10 g CO_2 per g seed per h). The authors also studied the gas charge as a function of the oil content of the seed and quoted extraction equations; with the help of the equilibrium charge concentration and the mass transfer coefficient these equations enabled the course of the extraction to be described. This is of importance for designing a counter-current extractor for continuous operation.

Brannolte, Mangold and Stahl investigated on a preparative scale the extraction of the fatty oil from copra, the dried fruit pulp of the coconut palm which contains 63.5% of triglycerides [80]. They showed that only one tenth of the amount of carbon dioxide and hence only one-tenth of the time, was required to separate completely the oil present at 900 bar and 60 °C compared to 300 bar and 40 °C. They concluded from their results that a minimum pressure of 600 bar and a minimum temperature of 60 °C are to be recommended, especially in the extraction of materials rich in oil, using dense carbon dioxide.

Leiner described the extraction of the fatty oil from evening primrose (*Oenothera biennis*) seeds [81]. This high grade oil contains essential polyunsaturated fatty acids, the physiological activity of which is determined by ω-6,9-double bonds (linoleic acid family), among which γ-linolenic acid, $C_{18:3}$ (6c, 9c, 12c) is of particular interest. A light yellow, high grade oil, practically free of phospholipids, was obtained in the extraction. Neither displacement of double bonds, nor isomerisation of cis-substituted fatty acids to trans-, nor autoxidation of the oxygen-sensitive, unsaturated molecules could be observed. Our own studies showed that a concentration of 2.7% by weight in the gas is attained at 800 bar and 40 °C. For the purpose of enrichment of the active γ-linolenic acid the fatty acids of the evening primrose oil were converted into their ethyl esters and extracted in counter-current in a mass transfer column, using carbon dioxide. Although the esters of the mixture (consisting of ca. 70% linoleic acid and 10% of each of linolenic, oleic and palmitic acids) scarcely differ in their physical properties, a fraction was obtained with nearly twice as much palmitic acid, the complementary fraction then containing a correspondingly higher fraction of unsaturated fatty acids.

Stahl, Quirin and Mangold exploited the fast and simple method of micro-extraction coupled with thin-layer chromatography (FE-TLC, cf Chapter III,2) to separate unusual lipids from special animal and plant tissues [113]. The various constituents of different seeds were compared (Fig. 45) and the effect of pressure and temperature on the extraction of individual components was studied taking castor seeds as example. Further, the intensities of the substance zones obtained enabled the effects of different solvents added to the CO_2 on the extraction of lipids from castor seeds to be estimated.

The extraction and the possibilities of fractionating castor oil with dense carbon dioxide was studied by Leiner on the preparative scale [81]. The hydroxy-triglycerides of castor oil were much less soluble in the CO_2 than those of the fatty acids without hydroxyl groups. In the extraction from the seed, concentrations in the gas of 0.25%, 0.30%, 0.32% and 0.77% (all by weight) were found at 400 bar/40 °C, 400 bar/80 °C, 600 bar/40 °C and 600 bar/80 °C, respectively. The alkaloid ricinine, present in up to 0.1% in castor seeds, was partly co-extracted. The CO_2-extract contained 300 ppm, whereas less than 10 ppm were found in the hexane extract. This may be due to the higher temperature chosen for the high pressure procedure (80 °C), compared to that in the soxhlet extraction with hexane, which facilitated liberation of the alkaloid from the plant matrix.

An attempt was made to separate the undesired triglycerides containing one or two hydroxyl groups from the desired ones with three or four, by counter-current extraction in a high pressure column. The ratio of desired to undesired fatty acids was 2.9 in the starting oil product; this was reduced to 1.4 in the head product and increased to 5.9 in the product withdrawn at the foot of the column, the ratio of the amounts being 35:65 (top:bottom).

The defatting of snacks is an interesting suggestion for using dense gas extraction of lipids [114]. Potato crisps contain, for instance, 40–45% frying fat. Extraction with dense carbon dioxide enables this fat to be removed down to the desired residual content without problems as a result of the large surface and without physiological risk. The snacks thus treated are more digestible, lower in calories and keep better.

The dried yolk of hens' eggs has also been extracted with supercritical CO_2 [81]. The powder was from fresh yolks which had been warmed for a few days at 55 °C for pasteurisation and prepared by vacuum spraying; this reduced the water content from 50% to less than 5%. Extraction of this egg powder with carbon dioxide at 300 bar/40 °C gave a 40% yield of a lipid extract, coloured bright orange, cloudy through solid lipid material and with egg odour. The extract contained triglycerides, free fatty acids, waxes and cholesterol. The phospholipids, mainly phosphatidyl-

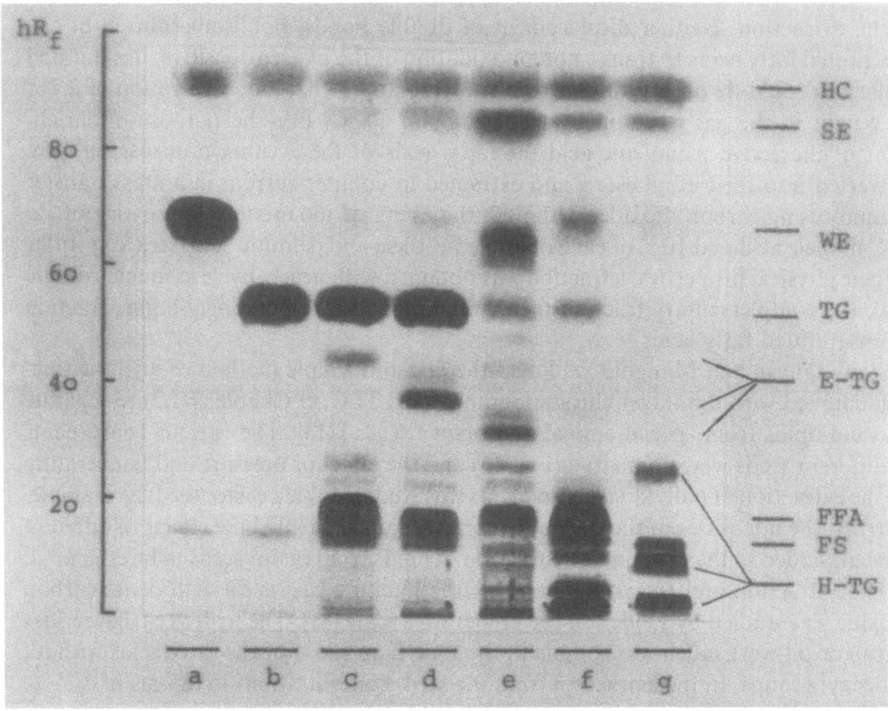

Fig 45. FE-TLC of various oil seeds at 200 bar/40 °C [113].
a *Simmondsia chinensis*; **b** *Glycine max*; **c** *Malope trifida*; **d** *Artemisia absinthium*; **e** *Vernonia anthelmintica*; **f** *Dimorphoteca aurantiaca*; **g** *Ricinus communis.*
E-TG triglycerides with epoxy-fatty acids; FFA tree fatty acids; H-TG triglycerides with hydroxy-fatty acids; FS free sterols; WE wax esters; SE sterol esters; HC hydrocarbons
Mobile phase: hexane-diethyl ether (40 + 80) (15 cm), then carbon tetrachloride (17 cm)

choline, present in 20% amount in the egg powder, remained in the residue which is nearly colourless and odourless after the treatment. The typical egg odour returns partly after a longer period of storage. The proteins were neither denatured nor coagulated by the extraction and the powder could be regenerated by adding water.

A high pressure procedure for obtaining anhydrous animal fats from butcher shop residues has also been described [115]. In this so-called cold melt procedure the fat is pressed out of, and extracted from, the material by static contact for 10–60 min with a supercritical gas, e.g. carbon dioxide (320 bar/45 °C), without high temperature being necessary. The fat collects in the lower part of the pressure vessel, and is clearly separated from the residue. It can be taken off in liquid form by warming the pressure vessel, or separated mechanically from the residue after removing the total contents of the vessel.

IV.2.c Extraction of Waxes

The oldest work of preparative high pressure extraction is the removal of asphalt from petroleum [116, 117] and the extraction of high molecular paraffin waxes from bitumen, peat and appropriate deposits (ozokerite) [118, 119]. These processes work at high temperature, using liquid hydrocarbons under supercritical conditions as solvents and are not considered further here.

Zhuse, Jushkevic and Gekker described a similar procedure for obtaining lanolin from wool or crude wool fat [10]. Propane or a propane/propene mixture was used at 100–110 °C, just above the critical temperature. The solubility of the lanolin could be increased by adding a subcritical C_4-hydrocarbon fraction. Depending on the extraction pressure (which was 110 bar at the most), fractions of different colour, liquefying point and molecular mass were extracted and separated at 40–50 bar. Albumins and impurities remained in the residue. Valteris gave a detailed account of the procedure [120].

An additional example of wax extraction is furnished by bee-glue (propolis), a dark green to red-brown, sticky material of hyacinth-like, aromatic odour, which becomes brittle after some time. Propolis contains 10–20 % real wax together with resins and small amounts of volatile oils, and also a large proportion of flavonoids, intensely coloured yellow-black. The constituents determining the value of propolis are the flavonoids (e.g. galangin, pinocembrin, pinobanksin, pinobanksin-3-acetate) and esters of caffeic acid, which possess a marked antimicrobial activity [121, 122]. These antimicrobial, antimycotic and antiphlogistic properties are utilised, for example, in the use of propolis in toothpastes and mouth washes for prevention of tooth decay, parodontosis and bad breath. Propolis is difficultly soluble in virtually all solvents. This is due to high-molecular solid waxes and resins, which prevent the liberation and emulsification of enough active principles; this interferes especially in the use in aqueous systems.

The facts that the undesirable non-polar components are soluble in dense carbon dioxide and the polar, antibiotic flavonoids are, in contrast, insoluble, has been exploited for the fractionation of propolis. It was possible using CO_2 at 600 bar/40 °C to separate a light yellow, solid wax fraction in 20 to 27 % amount by weight from non-purified, powdered samples of propolis from various sources [123]. A distinct gradient of colour, consistency and amount of extract was observed in the course of the extraction (beginning: yellow, half-solid, spicy, balsam-like odour; end: light, solid, odourless). There is thus no purpose in quoting solubility data. Under the conditions given, 20 kg CO_2 were employed per kg propolis. The valuable flavonoids could be obtained in a highly pure state and in an amount of about 60 % by weight of the crude propolis by carrying out cold percolation of the extraction residue with an alcoholic solvent. About 15–20 % material remained of an insoluble residue of impurities to be discarded.

The extraction of liquid wax esters from jojoba seeds with dense CO_2 [97] corresponds essentially to deriving fatty oils from oil seed. Concentrations in the gas of 1.0, 1.2 and 3.1 % by weight were obtained at 400 bar/40 °C, 400 bar/50 °C and 660 bar/80 °C, respectively, in the extraction of powdered seeds, which contained 52 % oil. These values are only slightly lower than the data derived from quantitative measurements. It may be concluded from this that the mass transfer of the smaller wax ester molecules into the dense gas phase is kinetically less inhibited that that of the triglycerides.

A high grade oil could still be extracted with CO_2 from jojoba press cake residues which had been stored for some time and contained 16.6 % oil. (analytical data in

Table 12. Analytical data of an oil, extracted with CO_2 (400 bar/40 °C) from jojoba press cakes [97]

Density d_{20}^{20}	Refractive Index n_{20}^{20}	Iodine Value	Saponification Value	Acid Value	Gardner Number (colour)
0.872	1.4659	84	90	4.0	6

Table 12); for use in cosmetic products this extracted oil needed only a reduction of the acid number to below 0.4. A hexane extract made for comparison was cloudy, relatively dark coloured and of unsatisfactory odour. The best results however were obtained in direct extraction of freshly powdered seed which, as a result of its low water content, yielded immediately a clear oil of light yellow colour, odourless and needing sensibly no refining.

The seed residue contains 80% protein, 8% sugar and about 4.5% simmondsin, a cyanomethylene-cyclohexyl-glucoside which is not acutely toxic but nevertheless restricts the use of jojoba residue as feeding stuff for animals. Methods are being tested out for an economic separation of this compound which remains untouched during the CO_2-extraction [124].

A total extract of 3.5% could be obtained in the extraction of wheat bran with CO_2 at 600 bar/60 °C; it was of partly liquid, partly solid consistency [125]. On reducing the pressure to 350 bar an extract, waxlike and solid at room temperature, was obtained in 3% yield; it contained less fatty oil, was almost entirely soluble in alcohol and had a pleasant fruity smell. The concentration in the gas was 1.6% by weight at 350 bar/60 °C. After the extraction the bran showed for its part, an imbibition value 15% higher, was practically sterile and freed from its bitter after-taste.

IV.2.d Influence on the Seed Proteins

Fatty oils can be extracted with dense gases from plant material under non-aggressive conditions of temperature without having to remove solvent from the grist by heating. The further use of the grist residue to yield soluble protein is especially advantageous when the oil removal has caused no denaturation of the seed proteins and nothing happens to be present in the extracted meal which impairs digestion and has to be inactivated by heat treatment.

Stahl, Quirin and Blagrove carried out a study on whether the proteins of powdered soybeans, lupin-, cotton- and jojoba seeds, which naturally contain about 10% of water, underwent changes during a two-hour extraction with supercritical carbon dioxide at 350 bar and 40 °C [126]. Intact reference samples were obtained by cold percolation with hexane. The proteins were obtained from the extracted meal in various ways and studied by sodium dodecylsulphate-polyacrylamide-gel electrophoresis. No difference in the relative proportions of individual protein subunits could be found between the CO_2-treated seed powder and the reference samples, with or without reducing the disulphide bonds. Further, no oligomerisation through intermolecular bridging via disulphide groups was observed. The same results were obtained in degradation experiments with trypsin of globulins isolated from lupin seed treated with carbon dioxide and with hexane. The composition of the amino acids also showed no changes related to the pre-treatment. This suggests the conclusion that the native conformation of the proteins and their nutritive value is not impaired by extraction with carbon dioxide at 350 bar and 40 °C.

Weder investigated the influence of moist supercritical CO_2 on ribonuclease and lysozyme as model substances [127, 128]. Treatment for two or six hours at 300 bar and room temperature or 80 °C had no effect on the amino acid composition. Slight oligomerisation through formation of intermolecular disulphide bridges could, however, be demonstrated with both substances, using gel electrophoresis. Fission of some peptide bonds within the disulphide loops was recognisable at the same time. Treated samples were found to be slightly more easily cleavable than untreated ones in degradation experiments with trypsin; this is probably explained through the unfolding of the protein molecules, caused by water. These changes are brought about also by treating with moist nitrogen under the same conditions, or under normal pressure by heating to 80 °C with water; neither the pressure nor the carbon dioxide can thus have played a part.

L-glutamic acid, L-glutamine, L-methionine, L-leucine, L-alanine, β-alanine and L-lysine were also treated with moist CO_2 for 6 hours at 300 bar and 80 °C. Analysis of the amino acids showed a loss only with glutamine, of 15–23%, due to conversion into 2-pyrrolidinone-5-carboxylic acid; however, this reaction occurs merely by warming. According to Weder, neither this result nor the trifling protein alterations described, which are primarily due to the influence of heat and water, have a detrimental effect on the nutritive value of protein-containing natural products extracted under high pressure.

Friedrich described the preparation of defatted soybean flakes of high grade protein quality [129]. The fresh bitter taste of the grist from conventional extraction with hexane presents a problem in the subsequent use in food. In addition autoxidation or lipoxygenase-catalysed degradation can impair the flavour of the grist and reduce its food value. Inactivation of the enzyme activity by toasting lowers the protein solubility expressed as NSI (nitrogen solubility index) from 85 to the unacceptable level of less than 60.

The stability of the grist towards oxidation could be improved and the undesired flavouring components removed by extraction with carbon dioxide without reducing the protein solubility too much when certain conditions of extraction are respected. The standard of quality of the grist demands a residual oil content of less than 1%, an NSI-value over 60, a taste quality over 6 on a ten-point scale, and a taste intensity below 1.5 on a three-point scale. A somewhat larger amount of solvent may be necessary for removing the interfering flavourings than for oil removal. A gas stream of 5–12 kg CO_2 per h and per kg seed is recommended in order to avoid too long an extraction time and temperature strain of the proteins. The seed should contain 9 to 12% moisture, the pressure be at least 690 bar to guarantee high solubility, and the temperature be between 81 and 100 °C; below 81 °C an unattractive taste resulted; above 100 °C, the protein was too sparingly soluble. The defatted soybean flakes could be easily milled to a fine, dispersable flour which contained 50% protein and was suitable, without further treatment, for adding to beverages, dough and pastry. Alternatively high grade protein concentrates could be prepared from the CO_2-extracted soybean flakes according to known procedures.

IV.2.e Fractionation of Fatty Oils and Waxes

Dense gas extraction can be used as an alternative to the classical methods of vacuum distillation, thin film evaporation or solvent extraction for separation of, in particular, mixtures of high-boiling substances without excessive temperatures. Care must be taken in the extraction of liquid charges that the system does not enter the domain of barotropic phase inversion, in which fractionation effects can no longer be exploited as a result of inadequate phase separation. This danger arises especially with fatty oils, which on account of their high molecular weights require high gas densities to attain the desired solvent capacity. Thus, it may sometimes be necessary to replace carbon dioxide by hydrocarbons, such as ethylene, ethane or propane, which have lower densities. It must be remembered in batch procedures that the volume of the liquid substrate phase can increase appreciably on dissolving the dense gas; this can lead to an overflow of the vessel containing the mixture to be separated.

Two suggestions for working up crude vegetable oils are described in the patent literature. The method of Coenen and Kriegel permits the complete refining of a crude oil, since the whole amount of triglycerides and the more easily soluble accompanying substances, such as free fatty acids, aldehydes, ketones, terpenes and other flavouring substances, are taken up in the gas phase and carried further [130]. The dense gas used as solvent should thus have a high capacity for triglycerides, for example, propane at 150 bar and 130 °C. Mucous material and other polymeric compounds are not soluble under these conditions and remain behind as residue; the fatty oil and the other components can be separated in a fractionation process by stepwise pressure reduction to 7.4 bar, for instance, and/or change of temperature. In this way the contents of free fatty acids in the triglyceride fraction could be reduced below the required value of 0.04% by weight, and the amounts of aldehydes and ketones below the limit of detection [131]. Another method, which omits the stages of lecithin and slime removal, and hence requires a partly refined oil, was described by Zosel [12]. His procedure accomplishes deacidification and deodorisation through selective extraction. Since the fatty oil is not taken up in the gas phase, a solvent was used that dissolved the accompanying matter, present only in small amounts, for example, CO_2 at 200 bar and 150 °C. The extraction was performed in counter-current in a packed column which rendered possible good contact and mass transfer between crude oil and gas. The gas stream with the dissolved components leaving the head of the column, was regenerated without change of state, using a solid adsorbent (active charcoal). The treated oil was free of odour and taste and its content of free fatty acids had been reduced from 0.4 to 0.02%.

The high pressure procedure described combines various quite different classical procedural steps in refining, such as deacidification through treatment with alkali or vacuum distillation, and the especially laborious deodorisation by steam distillation at high temperatures and low pressures, in one single step carried out under mild conditions and accompanied by only small refining losses.

Peter, Brunner and Riha investigated the separation of the monoglyceride of oleic acid from a mixture containing 50% mono-, 35% di- and 12% triolein and also 3% glycerin [14]. It was carried out in a counter-current column using dense carbon dioxide and acetone as entrainer. The entrainer had several functions: it improved the solubilities of the poorly volatile components; bettered the separation factor through special interactions; and, in this particular case, helped to "wash out" the dissolved components from the circulating gas in an auxiliary column inserted after, thereby saving de- and re-compression.

In order to be able to make a better choice of the working conditions, the systems CO_2/acetone and CO_2/glyceride mixture were studied separately and the results plotted in p-x-diagrams. The quasi-ternary system CO_2/acetone/glycerides at 130 bar and the temperatures 70, 100 and 110 °C was plotted in a triangular diagram; the partial system CO_2/acetone was supercritical at 70 and 100 °C but subcritical at 110 °C, then showing a miscibility gap. A supercritical extraction phase CO_2/acetone is required to concentrate effectively the weakly volatile substances; accordingly, the operating conditions of the separating column were chosen as 130 bar at 80 °C at the column bottom and 70 °C at the top. A 10% amount by weight of entrainer was used in the liquid phase, which was introduced into the middle of the column. Under these conditions, the gas phase took up 5–10% of components of low volatility, the monoglycerides with separation factors of 1.4 to 3, being concentrated with respect to the diglycerides. An increased fraction of entrainer led to a higher charge of the gas but a lower separation factor. The poorly volatile components obtained from the head of the main column were separated in an auxiliary column at the same pressure but 110 °C; under these conditions, the system gas/entrainer was subcritical, so that some acetone condensed. Using a reflux ratio of 1.5 it was possible to enrich the monoglycerides in the head product to 95%. After the separation the entrainer was removed from the products by distillation.

Brunner and Peter described a similar procedure for deacidifying edible oils and fats [7]. As a result of the differing vapour pressures and solubilities triglycerides and free fatty acids can be separated with supercritical carbon dioxide alone. Addition of ethanol as entrainer, however, enabled the separation factor to be doubled, the pressure necessary for the same solubility to be halved, and the poorly volatile components to be separated from the circulating gas by change of temperature only. Carbon dioxide with 10% by weight of ethanol was used at 70 °C and 135 bar in the separating column to deacidify palm oil; 1% of substances of low volatility, principally fatty acids, was taken up in the gas phase. Increase of temperature to 110 °C effected phase separation leaving a residual solubility of only 0.1%. Estimates of cost on the basis of prices in Sept. 1977 gave 0.30 DM/kg for 3000 t/a and 0.10 DM/kg for 15000 t/a for the deacidification of crude oil without recovery of the entrainer from the products. The operating pressure could be reduced to 55 bar and the dissolving capacity increased to 6.8% by using propane as dense gas and hexane as entrainer. However, the separation factor free fatty acids-triglycerides was then appreciably lower, imposing higher demands on the separation performance of the column.

Panzner, Ellis and Bott described the separation of the triglyceride of oleic acid from a mixture consisting of 40% mono-, 50% di- and 10% tri-olein [16]. As pure carbon dioxide displayed no measurable dissolving capacity for the glycerides in the pressure range available from 80 to 100 bar, carbon tetrachloride or n-hexane was added as entrainer to improve the glyceride solubility. The use of carbon tetrachloride led, however, to extensive corrosion, which was explained through the presence of moisture under the conditions of the procedure and subsequent formation of a strong inorganic acid. The fractionation was carried out as a batch process. The glycerides were first mixed with 50% by weight of entrainer and placed in the extraction vessel. Carbon dioxide was then passed through from below and the extract phase led through a fractionating column mounted above it.

A product containing up to 98% trioleate (which has the highest molecular mass but the lowest polarity), could be taken from the column head, whether hexane or carbon tetrachloride had been added to the glyceride mixture. The extraction temperature was between 80° and 120 °C and fell about 20 °C from column foot to column head. The extracted triglycerides were recovered by lowering the pressure to 30 bar. The concentration of entrainer in the bottom product fell from 50 to 35% in the course of the extraction. The system was so selective for trioleate that only a short fractionating column was needed. Here, too, however, the necessity to separate the entrainer from the separated glyceride fractions is irksome.

Dense gases have been used also to fractionate fish oils, which, in comparison to vegetable oils, usually possess a broader spectrum of fatty acids. Zosel described the fractionation of fish liver oil in extraction time gradients [15]. The liquid substrate mixture was percolated by a stream of liquefied ethane in the extraction vessel at 27 °C with increasing pressure from 100 to 160 bar. The gas + extract was led without change of pressure but at the supercritical temperature of 55 °C through a column mounted on the extraction vessel and carrying a "hot finger" (90 °C) at its head to cause a partial reflux. The saponification index of the extract, which is inversely proportional to the molecular mass of the triglycerides, fell during the extraction, whereas the iodine value, a measure for the number of double bonds, rose (Fig. 46).

Eisenbach employed the same experimental procedure to separate a fraction rich in eicosapentaenoic acid ($C_{20:5}$) from cod oil [132]. For this, the fatty acids, which had chain lengths from 14 to 22 C atoms and up to 6 double bonds, were converted to their ethyl esters and these were extracted with CO_2 at 50 °C and 150 bar. In the course of the extraction the fatty acid spectrum of the substances dissolved shifts continuously from the smallest to the largest chain length. The initial fractions containing the C_{14}–C_{18}-esters and the final fractions containing the C_{22}-esters were

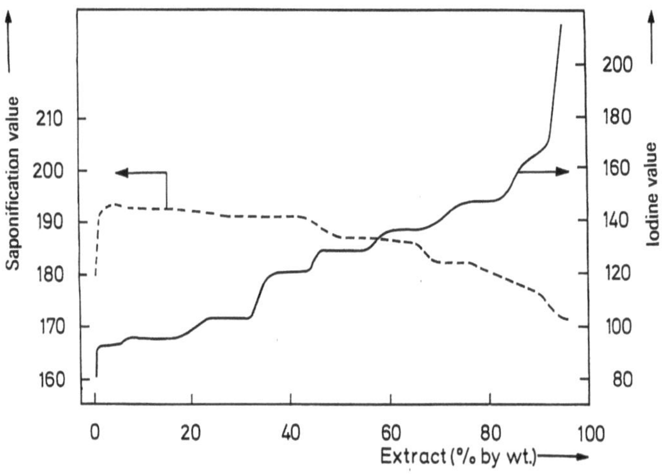

Fig. 46. Variation of the saponification and iodine values of the extract in the course of fractionotion of fish liver oil with ethane at 100–160 bar and 55 °C column temperature [15]

discarded. The C_{20}-ester fraction was collected as long as its content exceeded 90%. The previous ester fraction, C_{18}–C_{20} and the following one, C_{20}–C_{22} were mixed and extracted again according to the same scheme in a second stage under the same conditions. In this way it was possible to obtain 68 or 80% of the C_{20}-esters in the starting material in a purity of 96 or 92%; however, the desired $C_{20:5}$-esters were only slightly enriched with respect to the $C_{20:1}$-esters, also contained in the oil.

Leiner worked on the separation of "Orange roughy oil", a special oil from the deep sea fish *Hoplostetus atlanticus* [81]. The yearly production of this fish oil amounts to 2500 tonnes and it consists of wax esters with 3% undesired triglycerides in addition. He succeeded in reducing the proportion of triglycerides in the head product (Fig. 47b), which comprises 95% of the sample charge, to less than a quarter of the starting amount through counter-current extraction with dense CO_2 in a separating column at 310 bar and 70 °C foot-, 95 °C top-temperature. In the bottom product (Fig. 47c) the triglycerides were enriched by a factor of ten to over 35%. The sterols of the fish oil were found in the head product, along with the wax esters. It must be mentioned that the wax esters are not always to be found in the head product in the wax ester/triglyceride separation. The reverse situation can prevail, depending on the temperature conditions and, especially, on the chain lengths of the molecules of both these lipid classes.

The fractionation of butter fat by simple extraction with supercritical carbon dioxide was described by Kaufmann et al. [20]. The short chain fatty acids in the milk fat are physiologically important in nutrition. In the methods hitherto used for fractionating milk fat, advantage has been taken of the differing melting points of individual components and separation was not according to molecular masses. In the present case, however, 20% of a short chain triglyceride fraction could be separated from the starting material in a single stage extraction at 200 bar and 80 °C and then

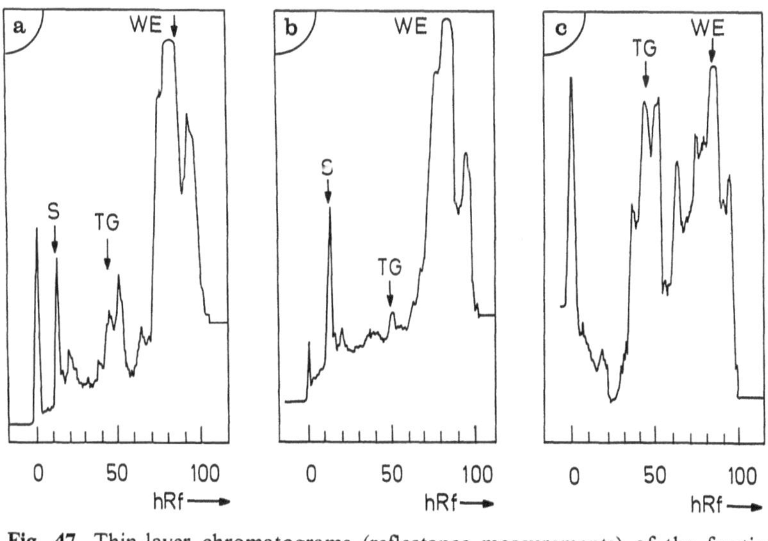

Fig. 47. Thin-layer chromatograms (reflectance measurements) of the fractions of Orange roughy oil after separation in a column with supercritical CO_2 [81].
a sample feed; **b** head product (95%); **c** bottom product (5%)

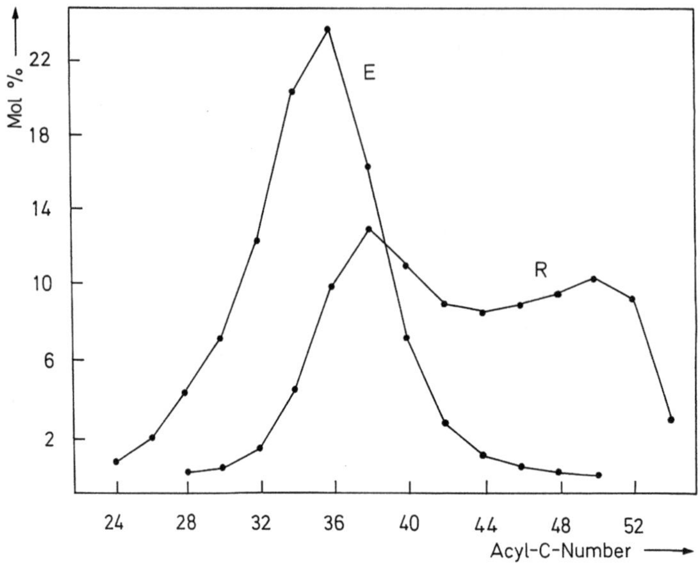

Fig. 48. Molar distribution of the triglycerides of different acyl-C-numbers in the extract (20% by weight) and residue (80% by weight) during fractionation of butter fat with supercritical carbon dioxide at 200 bar and 80 °C [20]

recovered at 30 bar/30 °C. The average concentration in the gas under these conditions was 0.15% by weight. In comparison to the starting fat, the extracted fraction had a lower iodine value (19.6) and a higher content of free fatty acids (1.51%) and of cholesterol (0.55%).

Investigation of the fatty acids in the extract showed an increase of 81% in the short chain molecules from C_4 to C_{10} but a reduction of 44% in oleic acid. Saturated fatty acids had a greater tendency to pass into the extract than unsaturated acids of the same chain length. This is a fundamental difference from fractional crystallisation where the short chain and unsaturated long chain fatty acids are concentrated together as soft fats. Analysis of the triglycerides showed also that the amount of short chain molecules with acyl-C-numbers from 24 to 38 in the dense gas extract was clearly larger than in the starting fat material, whereas the long chain triglycerides with C_{46} and more were almost completely missing (Fig. 48).

A study of the melting behaviour using differential calorimetry showed that the extract had already fully melted at 17–20 °C whereas the starting fat material melted at 37 °C. Still better results could be expected in counter-current extraction of butter fat in a fractionating column.

Stahl and Quirin described a procedure for fractionating extraction of natural waxes [18]. The fractionation resulted here also from extraction gradients in time through successive dissolving of the compounds of a relatively inhomogeneous starting material. There is also the possibility of increasing the gas density stepwise during extraction and of a fractionated separation. The dense gas extraction replaces here the classical purification procedures for crude waxes as primary extracts obtained in the usual way, such as washing with acids, alkalies or solvent mixtures. This is an advantage when the wax products are employed in the cosmetic or food industries. Certain properties of the waxes, e.g. the dropping point, can be influenced within limits during fractionation.

If crude, powdered beeswax, which has a brownish colour, is extracted at 400 bar/60 °C for example, a first extract is obtained of ivory colour, mild, honey-like odour and of sticky, waxy consistency. As extraction proceeds the colour pales to pure white, the odour disappears and the extract acquires a powdery consistency. The dropping point is first below, then above that of the starting product. Wool, carnauba and montan waxes can all be broken down into different fractions in the same way. The simultaneous use of high pressures and temperatures is recommended for obtaining easy solubility in carbon dioxide of these last-mentioned hard waxes in particular.

Braun and Schmidt studied the extraction and refining of crude montan wax, a toluene extract of brown coal which contains resins and asphalt-like products in addition to wax components [22]. The high pressure extraction with CO_2 at 40 °C and lower temperatures gave only low yields, below 10 %, which were hardly affected by pressure. At higher temperatures an increase of yield with rise in pressure was observed. A yield of 38 % by weight was attained at 100 °C and 500 bar, for instance. The series below of increasing yields of extract obtained compares carbon dioxide with other gases:

$$CHF_3(3.0) < C_2H_4(20.0) \sim CO_2(21.5) < CClF_3(56.2) < CBrF_3(95.0)$$

(The figures in brackets denote % extract by weight at 300 bar and 80 °C)

Hydrocarbons and wax alcohols entered preferentially the carbon dioxide extract and increasing amounts of free fatty acids dissolved as temperature was increased. Under the conditions of low solubility, low molecular compounds and hardly any esters were dissolved; the saponification value of the extract and its melting range were appreciably below the values for the starting material. As the dissolving capacity of the gas increased, resin components and simple wax esters went into the extract whereas the asphalt-like pigments and higher molecular, complex wax esters were concentrated in the residue. Differentiation chiefly according to molecular mass

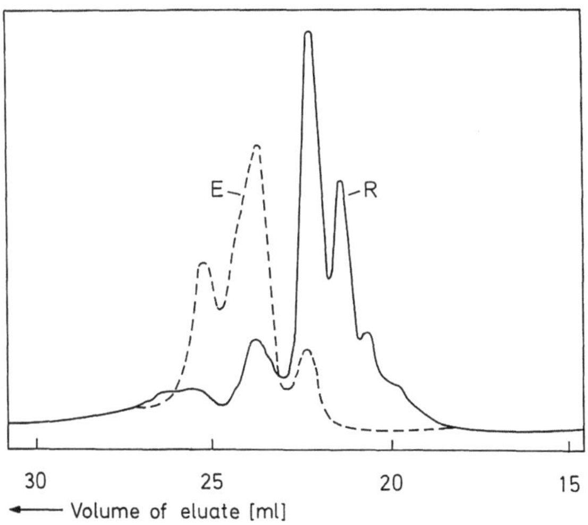

Fig. 49. Gel permeation chromatogram of extract E (8 % by weight) and residue R (92 %) from the extraction of crude montan wax with carbon dioxide at 300 bar/40 °C [22]

was confirmed by gel permeation chromatographic studies of the extract and residue (Fig. 49). It was not possible to effect a complete separation of montan wax according to compound class into wax-components, resin components and an asphalt-like pigment fraction because each of these groups has such a broad spectrum of substances, also as far as molecular mass is concerned, that solubilities unavoidably overlap.

IV.2.f Removal of Oil from Crude Lecithin

If crude oils, obtained by classical hexane extraction, are hydrated, the phospholipids precipitate as so-called wet sludge which is separated from the vegetable oil in special process steps and dried. The product thus obtained is crude lecithin, a brown, sticky mass. Crude lecithin from soybeans contains 65% phosphatides, 35% neutral lipids and traces of water and impurities. If the amount of neutral lipids of the product is reduced to below 5%, so-called pure lecithin is yielded which has a solid consistency and light colour, provided that it has not been subjected to heat strain. About 4000 tonnes of pure lecithin are produced annually in the world; the increase in value through oil removal is ca. 12 DM/kg (figures for 1985). The classical method of oil removal from crude lecithin is through extraction with acetone, although several laborious operations are needed to recover the solvent from the separated fractions. Problems are caused by refining losses, deterioration of quality through subjection to heat and the presence of solvent residues in the pure lecithin.

Dense gas extraction, especially with carbon dioxide, offers an alternative, since the neutral lipids, such as tri- and diglycerides, sterols, sterol esters, fatty alcohols, fatty acids, wax esters and hydrocarbons are soluble in carbon dioxide; in contrast, the various phospholipids are insoluble. The difficulty in practice is to obtain thorough and uniform contact, under high pressure, between the viscous substrate and the dense gas. The situation is aggravated by the increasingly rigid consistency of the crude lecithin especially on the surface, during oil removal; this prevents complete extraction in the interior of the mass. Since the usual advantages of the high pressure extraction – no solvent residues, no endurance of high temperature, no oxygen access – are especially relevant in the oil removal from lecithin, several patents have been applied for.

Heigel and Hüschens described a procedure for extracting crude lecithin, preferably with supercritical CO_2 at 200–500 bar and 35–80 °C [133]. The viscous crude lecithin was placed in a pressure vessel and the dense gas passed through it. A sieve device was situated at the exit of the pressure vessel to prevent solid material from being carried over. A special procedure for improving the contact between the crude lecithin and the solvent was not described. The oil removal would probably be soon retarded by diffusion effects and virtually cease before all the phosphorus-free lipids were separated. The regeneration of the cycled gas does not appear to be an optimum at sub-critical temperatures of preferably 15–25 °C and pressures which have to be below the corresponding limit of liquefaction.

The procedure of Coenen and Hagen functioned at sub-critical temperature with liquefied gases, best with ethane and carbon dioxide [134]. The extraction pressure was between $2\,P_C$ and 350 bar for ethane and 500 bar for CO_2. The extraction temperature was 0 °C to T_C, with ethane preferably at 15–31 °C, with CO_2, 5–30 °C. The low extraction temperature could be a disadvantage in view of the viscosity of the substrate and the kinetics of mass transfer of the lipid molecules into the liquid solvent. Further, studies of the solubilities of triglycerides (soybean oil), wax esters (jojoba oil) and sterols have shown that the dissolving capacity of liquid carbon dioxide is relatively poor and little influenced by pressure. Superior solubilities would be obtained at supercritical temperatures and the highest pressures quoted. In a special version of the procedure, ethanol was added to the liquefied gas as entrainer which leaves no detrimental residues and hence did not have to be quantitatively removed from the lecithin. Here, too, no precise details were provided about the way of mixing the viscous crude lecithin and the liquefied gas. The crude lecithin was fed into the middle of a pressure vessel and the gas led in below and taken off at the side of the head. The extracted oil was precipitated by lowering the pressure to 30–40 bar with ethane and 10–30 bar with CO_2, and, if necessary, increase in temperature, to, at the highest, 40 °C with ethane and 30 °C with carbon dioxide. The lecithin product accumulated as a block or powder and contained at least 90% of acetone-insoluble matter.

Various methods of high pressure extraction of crude lecithin were discussed by Stahl and Quirin [23, 135], particularly in connection with producing an adequate surface of contact, indispensable for effective and complete removal of oil. One possibility is to apply the crude lecithin to a CO_2-insoluble carrier in powder form, for example dextrose, yielding a virtually free-flowing material which can be per-

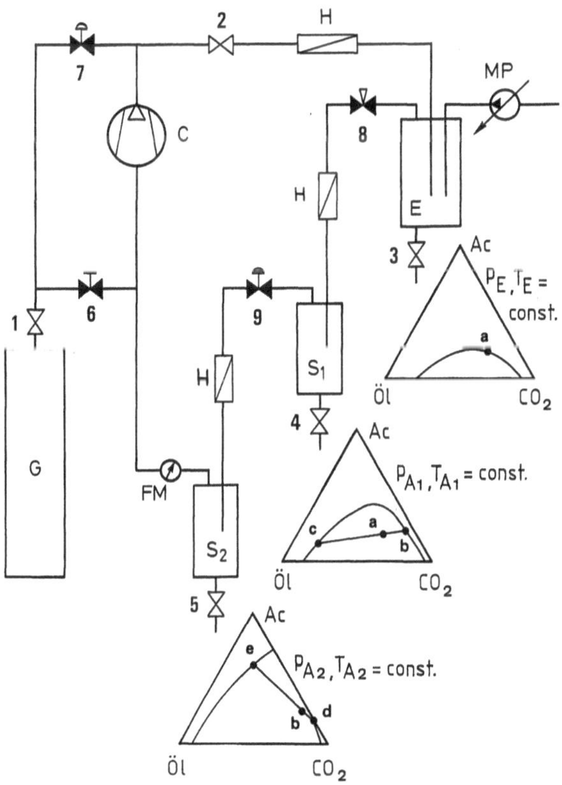

Fig. 50. Schematic representation of the extraction of crude lecithin/acetone (Ac) mixtures, quoting the phases of the quasiternary system Oil-Ac-CO_2 in the extraction vessel E and the separators S1, S2 [23].
G gas supply; C compressor; H heat exchanger; MP metering pump; FM flow meter; 1–5 cut-off valves; 6 reducing valve; 7, 9 back pressure regulators; 8 fine metering valve

colated. The method permits good oil removal but is expedient only when pure lecithin and carrier material can be used together as a product. Besides it must be considered that a large proportion of carrier can lower markedly the bulk density of the crude lecithin in the extraction vessel.

In a quite different procedure the crude lecithin is diluted with a classical solvent before extraction, so as to reduce its viscosity and improve the oil extraction. The best solvent for this purpose was acetone in which, like in CO_2, phospholipids are practically insoluble. These can thus be regarded as insoluble during the whole process which simplifies the situation considerably. Crude lecithin was mixed with six times its amount of acetone, giving a fine precipitate of lecithin. The oil dissolved in acetone can be quantitatively removed from this suspension by high pressure extraction. A suitable scheme for this procedure is sketched in Fig. 50.

The mixture of crude lecithin and acetone is placed in the extraction vessel, E, or introduced with a metering pump MP. The dense gas is sprayed in through a dip pipe, pressure and temperature being adjusted so that the system CO_2/acetone is supercritical, i.e., is completely miscible and has a high solvent capacity for neutral lipids. Suitable conditions are 160 bar/40 °C. The greater part of the extracted oil is precipitated in the first separator under conditions where the system CO_2/acetone is just still supercritical but the two-phase region appreciably extended. The extract phase then breaks down along the drawn-in tie-line to give a liquid phase, rich in oil, which can be taken off via valve 4, and a gas phase of composition b, which flows into the second separator. Here the conditions are chosen so that a miscibility gap occurs in the system entrainer/CO_2; a large part of the acetone then condenses with the remainder of the oil. The residual gas phase, which contains only little acetone, is compressed again and returned to the extraction vessel.

Pure lecithin of light colour and powdery form is obtained in E at the end of extraction. An advantage over the classical de-oiling procedure is that considerable acetone is saved and that there is a chance of a rough separation of oil from acetone by profiting from a miscibility gap in the ternary system oil/CO_2/acetone. Nevertheless, both the oil and the pure lecithin unavoidably contain a certain residual amount of acetone which has to be removed in additional steps of the procedure. Another disadvantage is the unfavourable operation feature of the second separator which imposes a demanding recompression of the circulating gas.

Mixing with acetone can be circumvented by dispersing the crude lecithin directly in dense CO_2 and extracting. The viscosity of the crude lecithin makes mechanical stirring necessary, which has to function under pressure. Fig. 51 shows the extraction diagram for oil removal from crude lecithin in a stirring autoclave using CO_2 at 350 bar and 40 °C [23].

The division into three stages of the course of the extraction can be clearly recognised. A constant extraction of oil is seen in the first stage; the maximum concentration attained in the gas of 0.45% is, however, considerably lower than the equilibrium value measured for soybean oil. Once 50% of the oil has been removed, extraction of the oil in the second and third stages is accompanied with increasing work of diffusion and falling efficiency, despite the stirring. The use of liquefied CO_2 at 350 bar/20 °C gives almost exactly the same results. The de-oiled lecithin thus obtained is brittle rather then powdery and somewhat darker than a product de-oiled in a classical procedure; this is because the pigments are more poorly soluble in CO_2

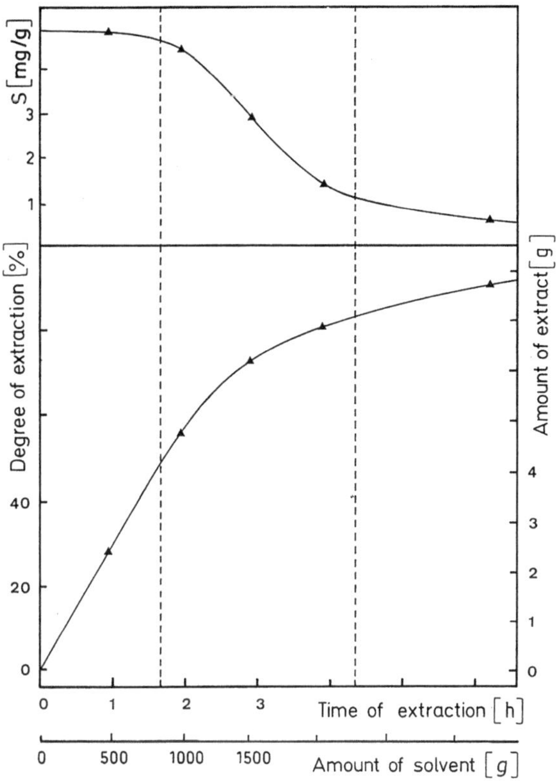

Fig. 51. Course of the de-oiling of crude lecithin (24.3 g containing, 8.5 g oil) in a stirred autoclave, using CO_2 at 350 bar and 40 °C [23]

than in acetone under the given conditions. It must be borne in mind also, that, just on a larger scale, stirring under pressure is expensive, requires comprehensive maintenance and has limitations as far as the extraction pressure is concerned. This method also, like those previously described, can in principle be operated, if desired, quasicontinuously.

A continuous and complete oil removal from crude lecithin was achieved with the help of the high pressure jet extraction already discussed (Chapter III.4) after it had been shown that the viscous mass could be forced with slight excess of pressure through fine capillaries of 0.2 mm internal diameter for example, and that the powdery pure lecithin formed could be easily removed from the collecting vessel via a valve at its foot [135]. Fig. 52 shows a section of the experimental installation for removing oil from crude lecithin and the construction of the jet arrangement.

The jet procedure yielding a residue of less than 3.5% oil in pure lecithin was possible without problems by using as solvent CO_2 at 900 bar and 90 °C. The sample jet was 0.7 × 0.25 mm (external diameter × wall thickness) and the gas jet, 1.6 × 0.4 mm and of 150 mm length. The mass flow of the lecithin feed was ca. 60 g/h, of the carbon dioxide 2.5–3 kg/h. The concentration of lipids in the gas was between 0.7 and 1% by weight. Based on the equilibrium concentration of soybean oil, this means that CO_2 took up less than one tenth of the theoretical amount of neutral lipids, under the conditions of the jet extraction. This is related to the inertness towards solution of the large lipid molecules and to their being fixed within lecithin structures.

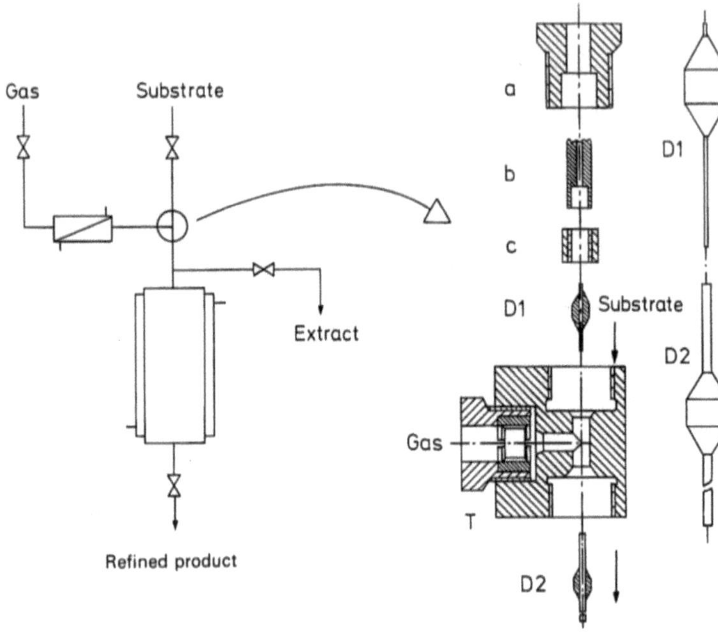

Fig. 52. Experimental assembly for high pressure jet extraction with particulars of the jet arrangement used [23].
a pressure screw; b high pressure tube, 1/4″; c pressure ring T; T-piece; D_1, D_2; stainless steel capillaries with double-conical form

The extraction pressure of 900 bar was attained without technical problems because all volumes in the part of the apparatus under highest pressure could be kept small for the continuous operation. The energy requirement for compressing the CO_2 was in the range of that for distillation processes. When suitable temperatures of 60 or 80 °C were used in the oil separator, pressure needed to be lowered only to 150 bar to regenerate the circulating gas, so that energy for re-compression could be saved. Any loss in quality of the lecithin at the extraction temperature of 90 °C could also be excluded since CO_2 was present as a protective gas and the temperature was high for only a short time as the crude lecithin was brought to the temperature just before extraction and it was in the extraction region for only ca. 0.05 sec.

The extraction conditions described enabled also components normally poorly soluble in CO_2, such as carotinoids and bitter principles, to be separated. The refined product was accordingly a light-coloured, powdery pure lecithin of good odour and taste properties. The grain size of the powder was below 0.1 mm on the average; the tiny particles showed surfaces of irregular structure, confirming the intensive contact with the extraction agent (Fig. 53).

From 1.5 to 3.5% of aceton soluble matter was found in the pure lecithin, corresponding to a qualitativel high-grade commercial product. The fact must be remembered that the jet method permits the separation of crude lecithin into neutral lipids and pure lecithin in one single continuously executed step without secondary procedures, such as evaporation of solvent residues, drying, pulverisation, purification of outgoing air, etc.

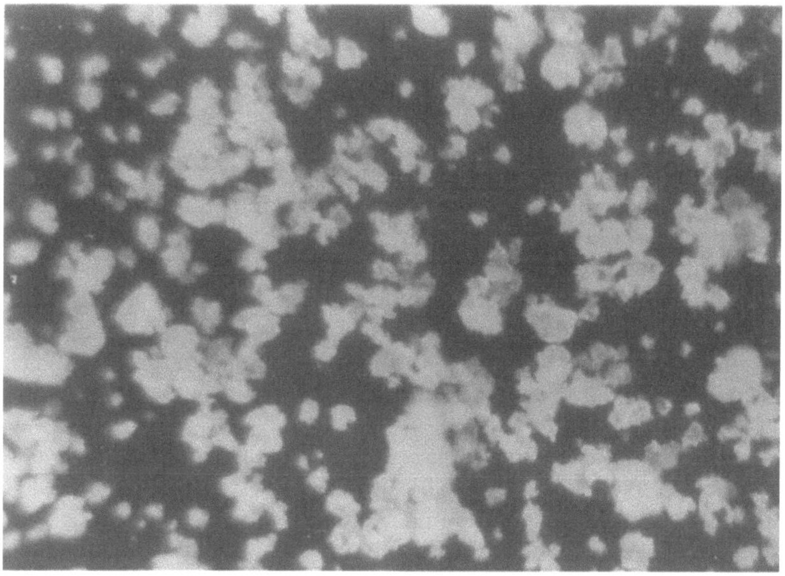

Fig. 53. Microscope view of pure lecithin powder, obtained by de-oiling crude soybean lecithin using high pressure jet extraction with CO_2; the length of the horizontal side of the picture is equivalent to 2 mm [23].

The method has so far not been optimised and increased in scale. It may, however, be expected that, for example, the addition of a small amount of propane, keeping below the explosion limit, to the CO_2 will permit marked lowering of the extraction pressure and further improvement of the gas charging, yet retaining the selectivity as far as the phosphorus-free and phosphorus-containing lipids are concerned.

Another interesting and also continuous way to oil removal from crude lecithin would be available if the solid refined pure lecithin could be converted into a sufficiently fluid state under the conditions of the extraction. The de-oiling could then be performed by counter-current extraction in a fractionating column, a method which was studied by Peter et al. [136]. Lecithin forms a liquid phase under the action of propane, depending on pressure and temperature, both the lecithin and the oil being equally soluble. The desired selectivity can be obtained by adding CO_2, although there is a risk that the phospholipids will solidify again and precipitate. The added CO_2 also depresses the solubility of the neutral lipids. This is significant because the pressure, and hence the density and solvent capacity of the gas phase, cannot be increased indefinitely. Separation is attained only when two phases are yielded and with a sufficiently large density difference between solvent and substrate. The factors of selectivity, viscosity and capacity are thus to be rated as functions of pressure, temperature and composition of the gas phase for demarcation of the possible sphere of work. It is of prime importance to ensure that the limiting solubility curve of the lecithin is traversed downwards at no point in the separating apparatus; this would lead instantly to a blockage. Finally, the separating performance of the column must be coordinated with the chosen operating point.

It was possible to describe the quaternary system, soybean oil-lecithin-propane-carbon dioxide, with the help of a computer programme and experimental solubility data of the binary and ternary sub-systems. These data were used for initial fractionation experiments in a discontinuous laboratory installation. With the help of a model calculation from plate to plate and a balance programme, these results enabled the operation point for experiments in a pilot plant to be established.

This was given as 80 bar, 50 °C and 80% propane in the mixture with carbon dioxide. The concentration of low-volatility material in the head gas at the column exit was ca. 4% by weight; the solvent ratio was 10. Practical characteristic values, such as the flooding point, number of separation stages, etc. were measured experimentally. The residual oil content in the regeneration of the gas was best kept below 0.2% in order to obtain a high quality lecithin. For this the pressure of the gas phase had to be lowered to 60 bar at 80 °C. Regeneration was carried out in two stages, with the help of an auxiliary column.

Under the best operating conditions a pure lecithin containing 91.5% acetone-insoluble material could be obtained, which does not exactly correspond to the quality of conventionally obtained products. The pure lecithin was free from triglycerides but contained still a small amount of sitosterol and, surprisingly, a noticeable amount of free fatty acids. The part of the lacithin extracted with the soybean oil, to be regarded as lost product, could be reduced from an initial value of 32 to 0.3% by weight by operating the separating column with a temperature gradient from 42 °C at the bottom

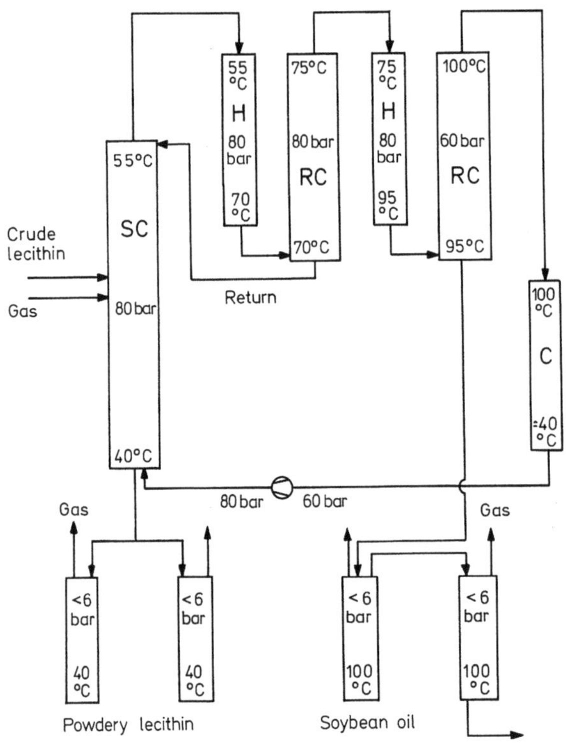

Fig. 54. Schematic process flowsheet of a pilot plant for high pressure extraction of crude lecithin using propane/CO_2, operating pressure and temperatures are drawn in; daily throughput of 1 tonne [137].
SC separating column; RC regeneration column; H heat exchanger; C cooler

to 54 °C at the top instead of isothermally at 50 °C. Moreover, the lecithin-rich fraction appearing as bottom product in the first regeneration column was returned to the separating column.

The information acquired in this smaller pilot plant served as a basis for the design of a pilot plant for an annual throughput of 250 tonnes crude lecithin. The schematic process flowsheet with indication of pressures and temperatures is in Fig. 54. The plant consists of one separating and two regenerating columns with preceding heat exchangers. Supercritical solvent of 80.5% propane and 19.5% CO_2 (both by weight) was passed through the separating column from below. The liquid mixture lecithin/CO_2/propane arriving at the bottom of the separating column was continuously depressurised in vessels for collecting lecithin. From these the powdery lecithin product was periodically removed.

Problems which had to be solved were the removal of lecithin without choking, and withdrawing the separated soybean oil which formed a stable foam with the gas phase. Liberated amounts of gas had to be collected and burnt so that no explosive mixtures with air could be formed.

Protection against explosion was necessary. On the other hand fresh gas had to be introduced to compensate for gas losses, and its composition had to be accurately made up so as to avoid any interference with the process. Finally it must be noted that crude lecithin may contain small amounts of constituents which are not soluble even in pure propane and could pollute the packing of the separating column in prolonged operation.

The estimated costs of preparing pure lecithin according to the column method described are, for a yearly production of 1200 tonnes of pure product, about 40–50% of those of the conventional acetone procedure (price levels of 1985).

IV.3. Essential Oils and Flavours

Essential oils, scented and flavoured materials are widely used. They play an important bart in foodstuffs as flavouring agents and as stimulants of appetite and aids to digestion. They are vital basic materials for scented products in the perfume and cosmetic branches and are processed directly in their pure state to yield high grade natural perfumes. A further application is in the pharmaceutical domain. Many essential oils possess pharmacological activity and thus find interesting application in therapy, e.g. as carminatives, antiphlogistics, sedatives etc. Vapour pressure and polarity are properties which have a special influence on the solubility in dense gases. The lower the polarity, the smaller the molecular mass and the higher the vapour pressure is, the better they can be extracted. Fig. 55 gives the vapour pressure curves of some typical constituents of essential oils as a function of the temperature.

Monoterpene hydrocarbons with 10 carbon atoms, such as α-pinene and limonene, are easily volatile compounds. They have vapour pressures of a few millibar even at room temperature and these reach about 100 millibar at 100 °C. Sesquiterpene hydrocarbons, of higher molecular weight, such as caryophyllene and cadinene, are C_{15}-compounds and have only one-tenth to one-twentieth of this volatility. Oxygen-containing monoterpenes, of which carvone and geraniol have been chosen as examples, and the phenylpropane derivatives such as anethole and eugenol, have

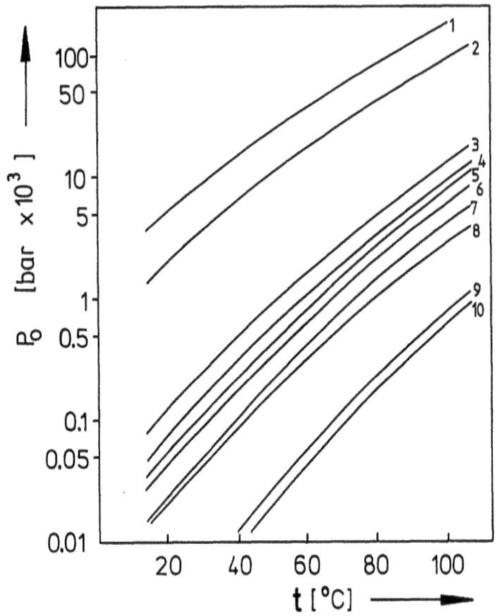

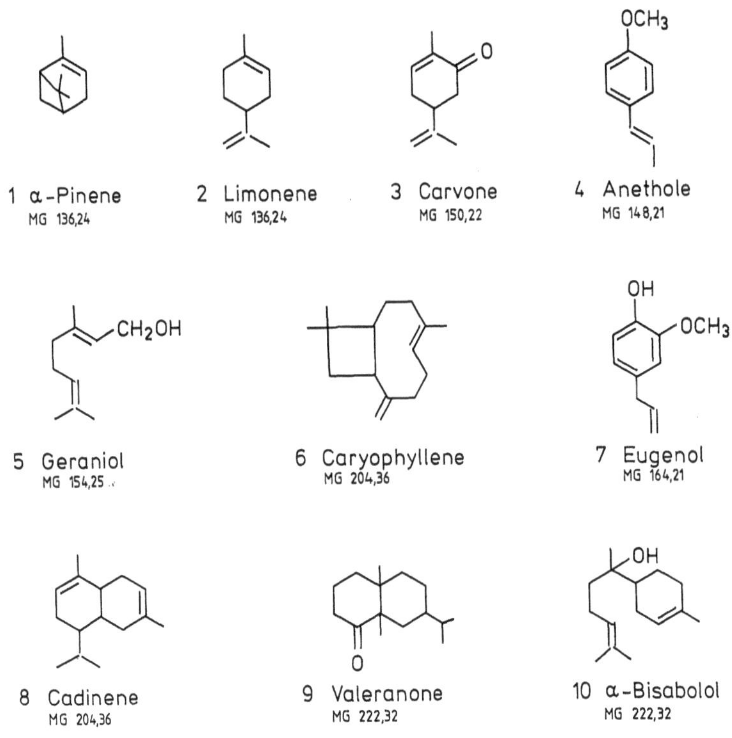

Fig. 55. Vapour pressures of constituents of essential oils, plotted against temperature [138, 139]

vapour pressures of the same order as those of the C_{15}-hydrocarbons. Much less volatile are sesquiterpenes containing oxygen (e.g. valeranone, α-bisabolol); even at 100 °C their vapour pressures reach only about 1 millibar.

Fig. 55 shows the temperature-dependence of the vapour pressure at atmospheric pressure. If high pressure is applied by means of an inert gas to a condensed phase at a given constant temperature the vapour pressure of the condensed phase rises. In the ideal case the pressure-dependence of the vapour pressure is given by the Poynting equation [140]. The dependence of the vapour pressure of, for example, limonene and of anethole on the inert gas pressure at a constant temperature of 40 °C is shown in Table 13. The vapour pressures of these components increase by a factor of about 3 under an inert gas pressure of 200 bar, as a result of the pressure effect alone.

Table 13. Vapour pressures of limonene and anethole at 40 °C and various pressures of inert gas

Inert gas pressure (bar)	1	60	200
P_0 (limonene) (mbar)	6.1	8.8	21.5
P_0 (anethole) (mbar)	0.3	0.47	0.96

It must, however, be stressed that the Poynting equation is based on ideal conditions and does not take into consideration interactions between the molecules of the extraneous gas and those of the condensed phase. It is not thus possible to calculate solubilities with the help of this equation.

A further factor determining the solubility behaviour is the polarity of the compounds. This is governed approximately by the dielectric constant (DC). The lowest DC-values among constituents of essential oils are possessed by the hydrocarbons (limonene, 2.3; values always for 20 °C). Introduction of functional groups raises the DC-value (1-decanol, 8.1; carvone, 13.13; cinnamaldehyde, 15.7). The dielectric constants of essential oils lie between values of 2.9 for oil of angelica and 16.56 for mustard oil [141]. In every case they are therefore lipophilic, easily volatile substance mixtures which can be extracted by unpolar dense gases, such as carbon dioxide. Polar functional groups render extraction of substances of higher molecular weight impossible (4), but their influence on lower molecular essential oils is weaker. High DC-values of the compounds to be extracted can effect, at the most, a small reduction of solubility.

Since essential oils, scented and flavoured materials find application in high grade products in the perfume and cosmetic spheres, in the pharmaceutical branch or in the food industry, only the physiologically unobjectionable carbon dioxide has been used in the work hitherto carried out on high pressure extraction.

As low molecular, lipophilic substances of high vapour pressure, essential oils are easily soluble in dense carbon dioxide. Problems arise, however, in connection with the quantitative separation of the highly volatile components from the circulating gas stream, and with the fact that undesired accompanying substances belonging to other compound classes can be co-extracted. A condition for the best adjustment of the parameters of pressure and temperature is thus a knowledge of the solubility behaviour of essential oils in dense carbon dioxide.

IV.3.a Solubility Behaviour

Liquid carbon dioxide dissolves readily essential oils and flavour substances. Francis described the phase behaviour of a large number of volatile compounds in liquid carbon dioxide as long ago as 1954 [142]. Further investigations of the solubilities of flavour substances from fruit juices were published by Schultz [143]. Some values from these works are listed in Table 14.

Constituents of essential oils, such as limonene or camphor, are completely miscible with liquid carbon dioxide. The same applies to low molecular alcohols. aldehydes, ketones and esters. The solubility diminishes as molecular weight or polarity increases. Although hexanol is miscible with liquid CO_2 in all proportions, only 1.0% of n-decanol as a C_{10}-compound dissolves in it.

Using micro-extraction coupled with thin-layer chromatography, Stahl and Schilz showed in 1976 that essential oils are extracted in detectable amounts by supercritical carbon dioxide at 40 °C under pressures from as low as 70 bar upwards [2]. Stahl and Gerard carried out a comprehensive study of the solubility behaviour of volatile oils and characterised the range of state of dense carbon dioxide in which these substances could be obtained selectively [35]. All essential oil components are freely soluble in compressed carbon dioxide, even of relatively low gas density (Fig. 56). Their solubilities rise exponentially between 70 and 100 bar at 40 °C as a result of the rapid rise in density of the extraction agent. At a pressure of 90 bar, corresponding to a gas density of only 0.5 g/cm^3, solubilities of 50–60 mg/g CO_2 are attained. At higher gas densities, i.e., in more strongly compressed, supercritical CO_2 or in liquid CO_2, miscibility is total.

As seen in Fig. 56, the monoterpene hydrocarbons (e.g. limonene) are the most soluble. This is due to their higher vapour pressure, low polarity and smaller molecular weight compared to the other components of essential oils. Carvone, an oxygen-

Table 14. Solubilities in liquid carbon dioxide

Completely miscible [142]	Fraction dissolving (% by weight) [142]		Insoluble [143]
Stannous chloride	Water	0.1	Urea
Benzine	Iodine	0.2	Glycine
Benzene	Naphthalene	2.0	Phenylacetic acid
Pyridine	Aniline	3.0	Oxalic acid
Acetic acid	o-Nitroanisole	2.0	Succinic acid
Caprylic acid	Oleic acid	2.0	Malic acid
Ethyl lactate	Lactic acid	0.5	Tartaric acid
Amyl acetate [143]	Butyl stearate	3.0	Citric acid
Glyceryl triacetate	Ethyl anthranilate	6.0	Ascorbic acid
Ethanol	Glyceryl monoacetate	1.0	Dextrose
Hexanol	Glycerol	0.05	Sucrose
Benzaldehyde	n-Decanol	1.0	
Camphor			
Limonene			
Thiophene			

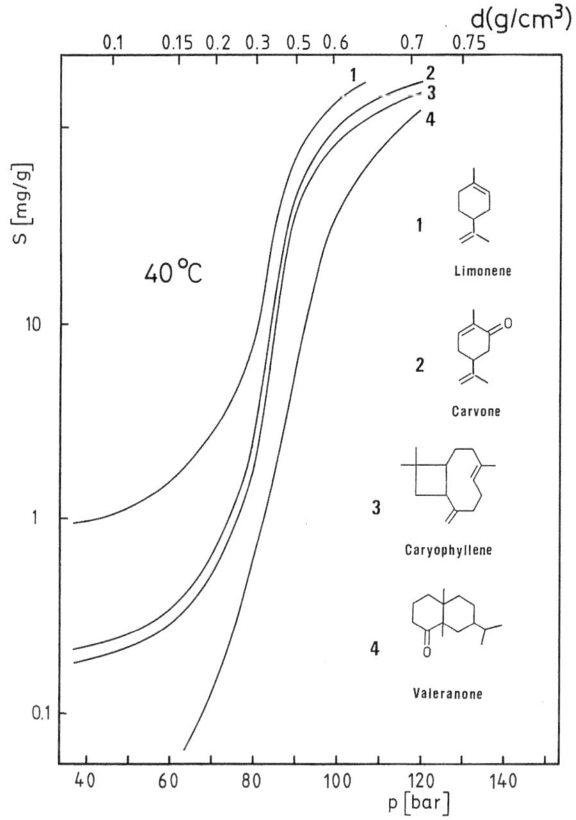

Fig. 56. Solubility isotherms of typical essential oil components in dense carbon dioxide at 40 °C [35]

containing monoterpene, and caryophyllene, a sesquiterpene hydrocarbon, are about equally soluble in supercritical carbon dioxide. Their solubility curves at 40 °C run very close together. On the whole, these two substance groups are rather less soluble than the monoterpene hydrocarbons. The phenylpropane derivatives, anethole and eugenol, show a behaviour similar to carvone. The solubility curve of valeranone in the lower pressure region runs distinctly lower than those of the other components, i.e., the oxygen-containing sesquiterpenes of higher molecular weight are less easily soluble under these conditions. This is due to their lower volatility. As the gas density increases, however, the solubility increases very markedly, so that at pressures above 90 bar at 40 °C the solubility curve of valeranone approaches those of the other essential oil components.

The slope of the solubility isotherms of essential oil components becomes steeper, the lower the temperature is. At 40 °C, solubilities of 50 mg/g CO_2 at about 90–100 bar are attained, and these values can be reached at 30 °C under only 75–80 bar. At lower temperatures the solubility curves climb to complete miscibility at the pressure of liquefaction of the carbon dioxide. The solubility isotherms of the individual substances show differing gradients at temperatures above 50 °C, so that fractionation effects are possible. As the temperature falls, the courses of the curves become more and more similar.

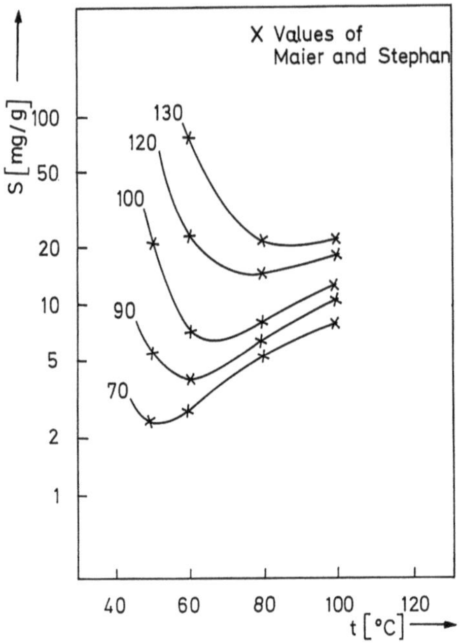

Fig. 57. Solubility isobars of 1-menthol in supercritical carbon dioxide (curves marked with pressure values in bar) [144]

The solubilities just discussed were measured using the analytical dynamic procedure. On the other hand, Maier and Stephan used the analytical static method to investigate the phase behaviour of the monoterpene alcohol, menthol, in supercritical carbon dioxide [144]. Their results are collected together in Fig. 57.

The solubility isobars of menthol pass through a minimum which is displaced to higher temperatures and solubilities as pressure is increased. The high dissolving capacity of carbon dioxide for components of essential oils, mentioned above, is also seen clearly. About 10% menthol by weight dissolves in the supercritical gas at pressures of 100–120 bar and in the temperature range of 40–50 °C. This value is appreciably increased by further increase in pressure. The facts discovered demonstrate good agreement with the values from the analytical dynamic method.

The conditions of separation as well as of extraction play an important part in obtaining essential oils. Whereas poorly volatile compounds (e.g. fatty oils) are obtained quantitatively simply by lowering pressure and passing into the gaseous state of the extraction agent, special conditions are needed for the more volatile substances. These show residual solubilities in gaseous extraction agents, due to their high vapour pressure and to molecular interactions between the gaseous carbon dioxide and the dissolved substances.

The solubility behaviour of characteristic compounds in the low pressure region and at various temperatures was studied with a view to reducing these residual solubilities [34, 145]. The separation isobars and isotherms are given in Fig. 58. The isobars (Fig. 58, left) pass through a minimum and it can be seen that lowering the pressure displaces the separation minima towards lower temperatures and solubilities, since the vapour pressure of the essential oil components becomes smaller at lower temperature. From this it is evident that low pressure and low tem-

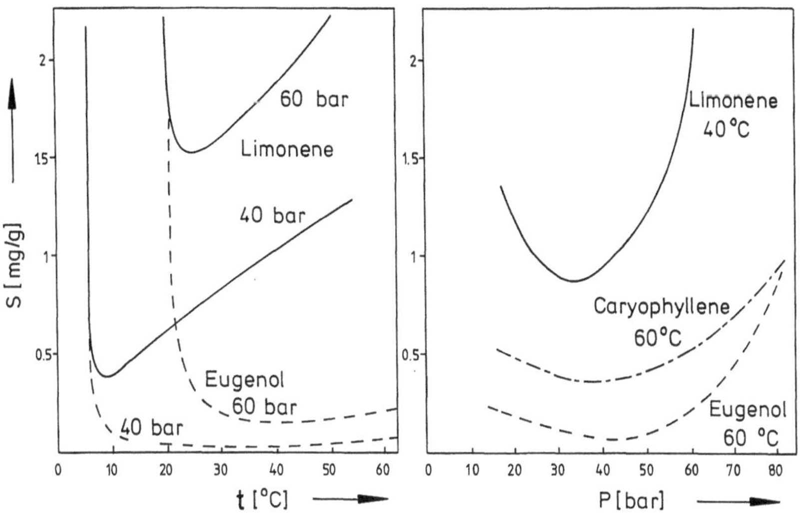

Fig. 58. Separation isobars (left) and isotherms (right) of typical constituents of essential oils, as functions of temperature and pressure [34]

perature favour the separation. Care must be taken, however, that the separation temperature is above the liquefaction temperature of the carbon dioxide at the particular pressure since essential oils are highly soluble in the liquefied gas.

It is also noteworthy that the separation pressure cannot have any desired low value at a given temperature because the isotherms also pass through a minimum (Fig. 58, right). At low pressures the high vapour pressure of the essential oil components effects some solution. As pressure increases, the fractional contribution of the vapour pressure to the pressure in the system falls and the solubility is accordingly reduced. As pressure increases still further, the isotherms traverse a minimum and then climb steeply as a result of increasing gas density, i.e. through solvation effects. The lower solubilities of caryophyllene and eugenol compared to limonene in the pressure and temperature region shown is due to their distinctly lower volatilities.

The procedure parameters of pressure and temperature therefore have to be adjusted in the best way so as to prevent losses, especially of very volatile components. It is less of a problem when the extraction agent is circulated and the gas thereby acquires an equilibrium concentration. After the initial phase of a preparative extraction, during which there are losses of substances due to the residual solubilities, the amount of extracted material which is easily volatile and not separated remains constant in the circulating gas, i.e., all freshly dissolved amounts of substance in this phase of the extraction collect in the separating vessel. A condition for this is, however, that the part of the extract which has not separated out does not separate and concentrate in any part of the plant other than the separating vessel, otherwise the course of the process is impaired. In particular it must be watched that no extract separates in the compressor, the heat exchangers, the flowmeter and the gas storage vessel. The advantage of as quantitative a separation as possible, attainable, for instance, at 30 bar/0 °C, is counteracted by the high energy costs of recompression and cooling; it is thus economically better to effect separation at a higher pressure and a correspondingly higher temperature. In this case, however, somewhat longer times of extraction must be tolerated because the full extracting capacity of the uncharged solvent is not utilised, due to the residual solubilities.

Selective extraction of essential oils with supercritical carbon dioxide is possible at gas densities between 0.4 and 0.6 g/cm^3. No fractionation of components of the oil is effected in this range because they are all very soluble. The solubilities of components which it is desired to fractionate into groups of differing volatility or polarity must show the largest possible differences. The fractionation conditions thus have to be optimised.

The monoterpene hydrocarbons are more soluble in dense carbon dioxide than are the other essential oil constituents as a result of their high vapour pressure, low polarity and small molecular mass. It is evident from the solubility isobars of limonene and carvone (see Fig. 59) that no fractionation is possible at gas densities larger than 0.4 g/cm^3, that is, at temperatures below 35 °C at 80 bar. Each solubility curve passes through a minimum when the temperature is raised isobarically. On the one hand, the dissolving capacity of the gas diminishes on account of the fall in its density, and on the other, the higher temperature leads to increase of vapour pressure of the substances. At higher temperatures the differing volatilities together with the dissolving power of the supercritical carbon dioxide cause the oil components which are to be separated to have distinctly differing solubilities. The solubility curves show an analogous course at higher pressures. Here, however, solubility differences and hence

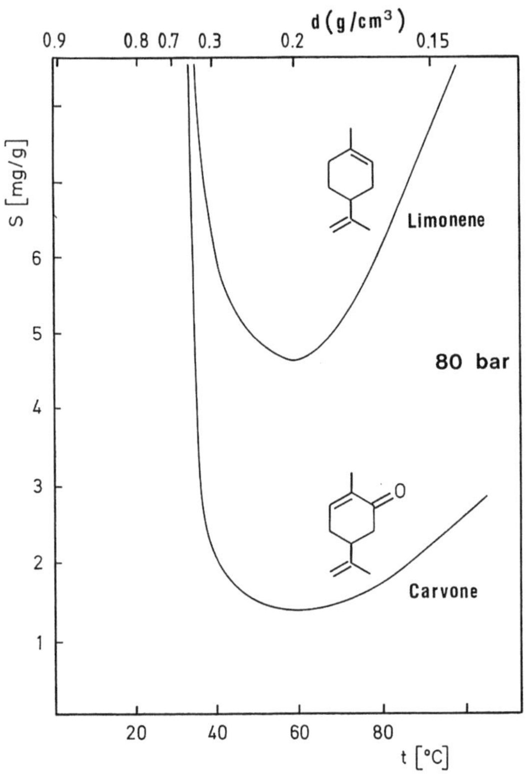

Fig. 59. Solubility isobars of limonene and carvone, as functions of temperature and density of the compressed carbon dioxide [35]

a fractionation effect are brought about only at higher temperatures. For instance, the solubility minima are at 70 °C under a pressure of 90 bar. An advantage here is that the dense gas phase carries more substance.

The oxygen-containing sesquiterpenes are less soluble than the more volatile essential oil components in supercritical carbon dioxide of low density (<0.5 g/cm³). This is clear from Fig. 56. Fractionation is thus possible by selective extraction of the more soluble monoterpenes, their derivatives and the sesquiterpene hydrocarbons with supercritical carbon dioxide, e.g. at pressures of 70–80 bar and at 40 °C. This fractionation has the disadvantage that the concentration in the gas is low. Another possibility is fractionated separation, discussed below with the example of a model mixture of caryophyllene and valeranone (Fig. 60).

If extraction is performed under conditions of high solubility, e.g. at 90 bar and 35 °C, all the components dissolve well. On isobaric increase of temperature, for instance to 50–60 °C, the less soluble sesquiterpene derivatives are preferentially separated because their solubilities fall distinctly more acutely with temperature. The other essential oil components mainly remain in solution and can be separated in a different procedure in a further step.

A greater problem than the separation of the highly volatile monoterpene hydrocarbons or the poorly volatile oxygen-containing sesquiterpenes is the fractionation

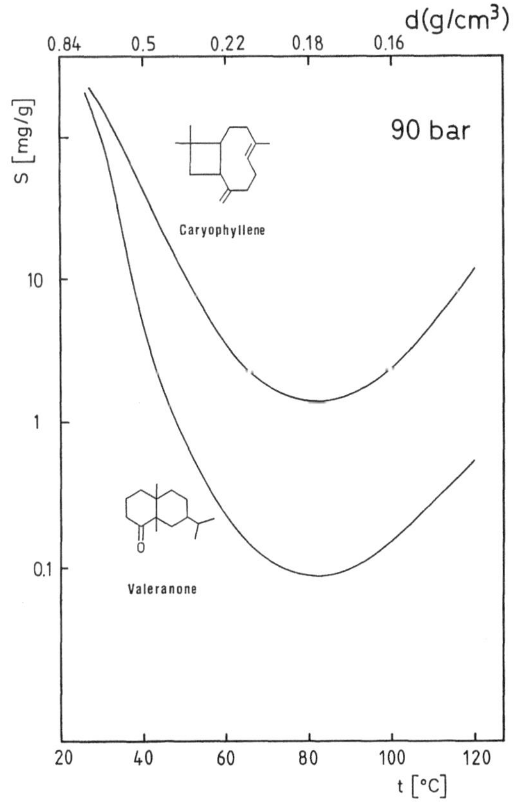

Fig. 60. Solubility isobars of caryophyllene and valeranone in dense carbon dioxide [35]

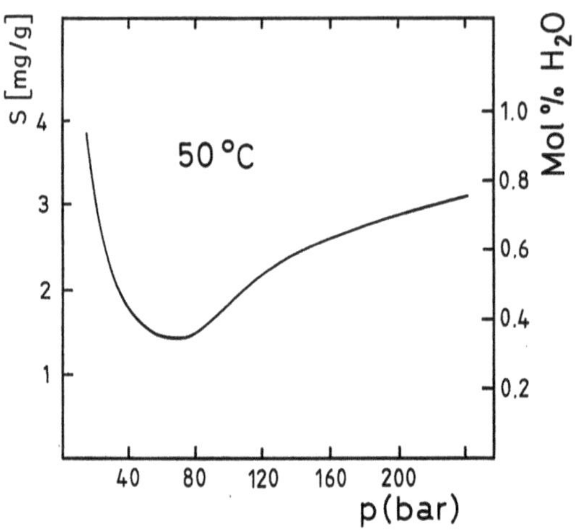

Fig. 61. Saturation concentration of water in dense carbon dioxide, as a function of pressure [146, 147]

of sesquiterpene hydrocarbons and the oxygen-containing monoterpenes. These two compound classes have similar solubilities in dense carbon dioxide: further, their vapour pressures are closely similar. However, they differ distinctly in polarity, so that better selectivity can be expected with an extraction agent of higher polarity. The polarity of carbon dioxide can be increased by addition of a polar component, which is done simply by saturating with water. Water has the advantages that it is subsequently easily separated from the essential oil and that it is safe physiologically.

Fig. 61 shows the saturation concentration of water in dense carbon dioxide at 50 °C. The concentration of the water in the gas is high at low pressures on account of its high vapour pressure. As pressure is increased, the solubility curve traverses a minimum and at high pressures approaches asymptotically a limiting value of about 3 mg water/g CO_2.

There appear to have been no systematic investigations of the dielectric behaviour or the polarity of substance mixtures of a polar component and a supercritical gas. It can, however, be presumed that the dielectric constant of carbon dioxide is increased by saturating with water. Under otherwise constant conditions of pressure and temperature, polar compounds ought therefore to be more soluble in dense carbon dioxide saturated with water.

This is illustrated below with the help of the solubility isotherms of a model mixture of the sesquiterpene hydrocarbon, caryophyllene, and the oxygen-containing phenylpropane derivative, anethole (Fig. 62). The isotherms of caryophyllene (A) and anethole (B) in compressed CO_2 at 40 °C are very close together, so that separation is impossible. Saturating the extraction gas with water increases the solubility of the oxygen-containing component (B*) and lowers that of the hydrocarbon (A*). This is most marked in the steep density-dependent rise of the isotherms at pressures between 80 and 90 bar. For example, anethole is about 8 times as soluble as caryophyllene (arrows joined by dashes) at 85 bar. At gas densities above 0.5 g/cm^3, i.e. at pressures above 90 bar at 40 °C, the solubilities of the compounds increase so rapidly at the expense of selectivity that separation becomes more difficult again. A similar behaviour is shown also at higher temperatures, e.g. 50 or 60 °C. Here,

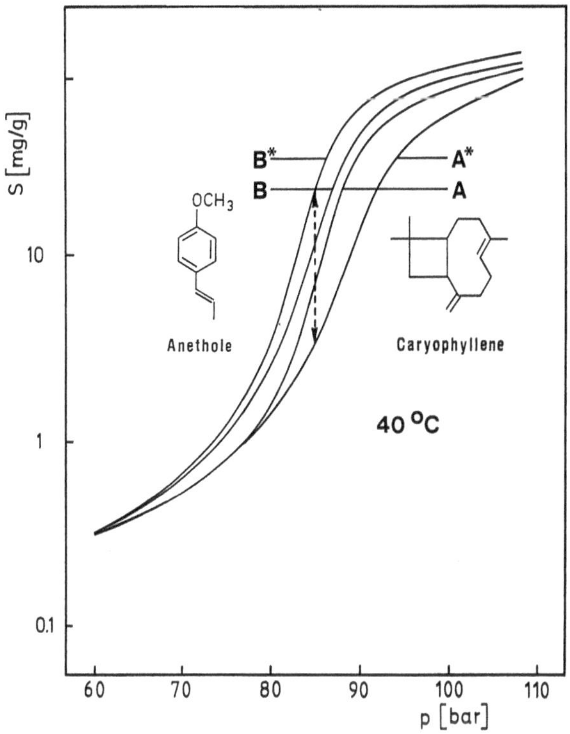

Fig. 62. Influence of saturation with water on the solubilities of anethole and caryophyllene in carbon dioxide [35].
A, B solubilities in pure CO_2; A*, B* solubilities in CO_2 saturated with water

however, the effect of a distinct increase or decrease in solubility is observed only at higher pressures because a definite gas density of about 0.3 to 0.4 g/cm^3 is required to dissolve the substances.

The general statement can be made that the phase region of the dense carbon dioxide which permits fractionation of the essential oils into definite compound classes is characterised by a smaller gas density than that for optimum extraction of the pure oil as a whole (values of 0.4 to 0.6 g/cm^3). Density values between 0.2 and 0.4 g/cm^3 are suitable for fractionation, temperatures of 50 to 80 °C then bringing additional advantage. Under these conditions the differing vapour pressures of the substances are exploited as well as the dissolving properties of the supercritical carbon dioxide. The pressure and temperature values given for extraction and fractionation do not represent optimum conditions in every case, that is, for every essential oil. In border-line cases, a uniform extraction of highly volatile oils can take place, whereas fractionation effects appear with poorly volatile oils which, at least partly, consist of higher molecular compounds. The conditions of extraction and fractionation have thus to be optimised for each problem, i.e., for each essential oil studied. The regions quoted can serve as guides.

The extraction gas must have high selectivity to succeed in fractionating substances of similar solubility properties. Brunner studied the different behaviours of some gases in the extraction of an equimolar mixture of hexadecanol and octadecane [148]. These two compounds are about equally volatile and their solubility behaviour is comparable with that of sesquiterpene derivatives. Table 15 contains the solubilities

Table 15. Solubility of an equimolar mixture of hexadecanol and octadecane in various gases at 150 bar and 70 °C [148]

	C_2H_6	C_2H_4	N_2O	CO_2	CF_3Cl
Solubility of $C_{16}OH/C_{18}$ in % by weight	25.5	6.9	5.5	3.2	2.8

of the substance mixture at 150 bar and 70 °C. The dissolving capacity of the gases investigated rises from monochlorotrifluoromethane via carbon dioxide, dinitrogen oxide and ethene to ethane.

The separation factors of the different gases for the model mixture are drawn in Fig. 63. A definition of the separation factor is that a value exceeding 1 shows that the solubility of octadecane preponderates in the gas phase. It can easily be seen that the separation factors of the gases are of the same order of magnitude. Monochlorotrifluoromethane is an exception, having values of 5 to 6. When pressure is isothermally raised the values generally become smaller as a result of the increasing concentration in the gas phase of the substances to be separated. An exception here is ethene which shows at 150 bar a separation effect comparable with that at 100 bar, although the solubility of the model mixture has risen appreciably.

The solubility behaviour of more or less volatile and soluble substances (essential and fatty oils) has now been discussed thoroughly for individual examples; it is intended to make plain the relationships in Fig. 64 by giving a general survey of the variations of solubility of a substance in contact with a real gas within a large range of pressure. The solubility isotherms of one and the same substance are depicted at two different temperatures. Starting from solubility values which correspond to the vapour pressure of the substance at the given temperature, a concentration minimum of the substance

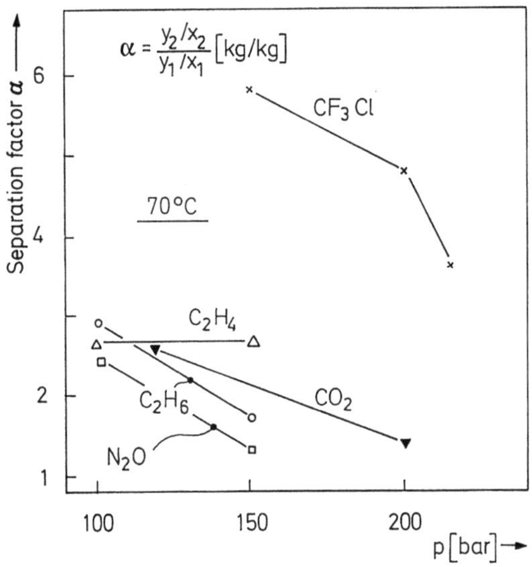

Fig. 63. Separation factors of an equimolar mixture of hexadecanol and octadecane in various gases [148]. 1 hexadecanol; 2 octadecane; Y concentration in the gas phase; X concentration in the condensed phase

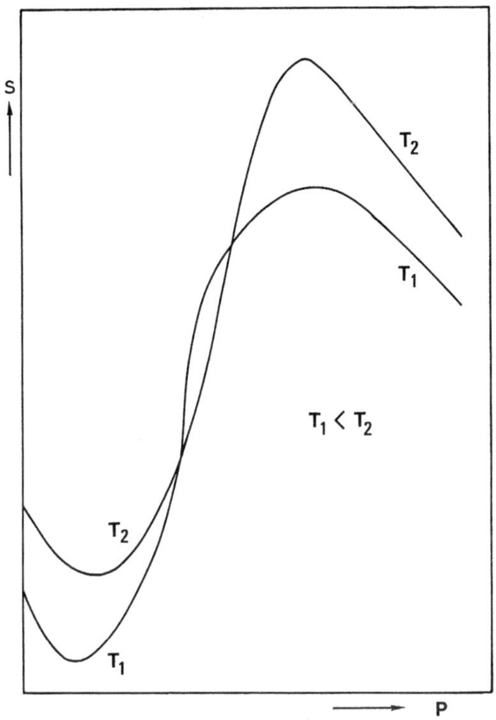

Fig. 64. Schematic course of two solubility isotherms as a function of pressure

under consideration in the binary mixture is first traversed as the pressure of the system increases and this is followed by an exponential rise in the solubility of the condensed phase due to the rapid increase in density of the gas. The solubility finally attains a maximum and then falls slowly. The position of the characteristic points (maxima, minima, intersections) depends on the substance. Increase in temperature displaces generally the minimum to higher, the maximum to lower, pressures. Since the solubilities at the extreme points usually differ by several powers of ten, is is hard to obtain both values for one and the same substance. The solubility maxima of, for example, triglycerides, wax esters and sterols have been experimentally demonstrated, likewise the minima of essential oil components. A double reversal of the dependence of solubility behaviour on temperature results from the two isotherm intersections. Rise in temperature can increase solubility at low as well as at high pressure. In an intermediate, relatively narrow pressure range, there is an opposite behaviour, characterised by fall in solubility with rise in temperature.

IV.3.b Deterpenation

Large amounts of essential oils are used for flavouring food and drink in the domain of foodstuffs. The real aroma donators are the oxygen-containing constituents of these oils, whereas the hydrocarbons, which make up the major part of some oils, contribute little to the total flavour. For example, the very frequently used citrus fruit oils contain up to 95% monoterpene hydrocarbons, mainly limonene. These hydrocarbons are relatively unstable compounds; their decomposition products impair the quality of the taste and odour of the flavoured preparation. There is thus considerable interest in procedures for preparing concentrated essential oils of low hydrocarbon content.

Various procedures are employed at present to obtain oils containing low hydrocarbon amounts; some are combined with others. The most important are:

- fractional distillation in vacuo
- extraction of the more easily soluble oxygen-containing constituents with dilute alcohol
- counter-current partition between a polar (e.g., alcohol) and a non-polar solvent (e.g. pentane)
- chromatography

These procedures have critical disadvantages. Thus organic solvents employed, some of which are even toxic, have to be removed from the products, posing residue problems and accompanied by losses. Further, distillation demands considerable energy and may lead to thermal stress of the flavouring substances, often sensitive. In the chromatographic methods it must be especially borne in mind that powerful adsorption forces can cause chemical reactions (e.g., dehydration, cyclisation) and reduce the value of the oils. Disadvantageous too is the fact that the conventional procedures can be pursued continuously only to some extent.

Extraction with dense gases is in discussion as an alternative to the classical procedures. As seen from the solubility determinations in dense carbon dioxide, it is possible to fractionate the essential oils into substance groups of differing volatilities or polarities.

If the attempt is made under the most favourable conditions to dissolve certain constituents selectively from the mixture, the desired separation is only conditionally attained because the other components are also soluble to some extent and are at least partly co-extracted. Considerable improvement is derived by introducing the liquid mixture to be separated into a high pressure column with internal structure designed to increase the surface area for counter-current extraction with a dense gas; reflux of the product at the column head brings about an additional betterment of the separation performance. The operating principle of the continuous deterpenation of essential oils is seen in Fig. 65 [34]. The terpene-containing oil is continuously fed into the middle of the high pressure column. Mass transfer takes place between dense gas and essential oil. In its climb through the column the gas becomes enriched preferentially with the more soluble and more volatile monoterpene hydrocarbons which separate out as terpene fraction after quitting the column. The oil, of greatly reduced terpene concentration, is drawn off continuously at the foot of the column.

The pressure and temperature of the supercritical carbon dioxide necessary for separating the essential oil components can be derived from the solubility curves. The optimum pressure for extracting the monoterpene hydrocarbons is between 70 and 90 bar. At a lower pressure the concentration of terpenes in the dense gas is uneconomically low. At pressures above 90 bar and temperatures below 80 °C the solubility of the condensed phase increases so markedly at the expense of the selectivity that fractionation is no longer possible. The temperature conditions in the high pressure column are of great importance for a good separation. The temperature at the column head regulates charging of the supercritical carbon dioxide with the mono-

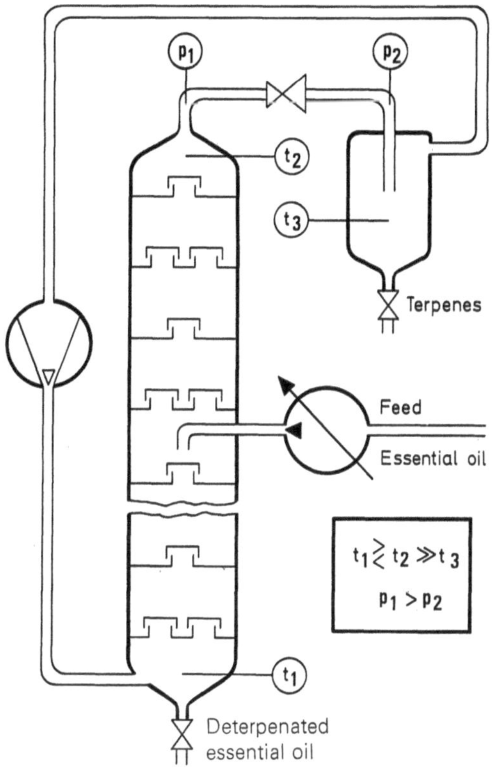

Fig. 65. Principle of an apparatus for continuous removal of terpenes from essential oils

terpene hydrocarbons at the given pressure. A definite reflux of head product is set via the difference in temperature between that of fractionation (in the middle of the column) and that at the head. For this there are two possibilities in principle, which are explained below with the help of Fig. 66 (of also Fig. 59).

Case I: In the temperature region to the right of the minima of the solubility isobars of the terpenes (T) and aroma compounds (A) to be separated, reflux is effected by cooling at the column head, analogous to rectification. To a certain extent this is a type of carrier gas distillation, so that the separation effect under these conditions is based principally on the difference in vapour pressures.

Case II: At temperatures to the left of the solubility minima, reflux of the terpene fraction is obtained by increasing the temperature at the upper end of the column. Fractionation is based largely on polarity differences between the substances in this temperature range at pressures below 80 bar. The influence of vapour pressure differences also becomes greater when pressure is increased, which means that the temperature must be raised also so that the solubility of the condensed phase does not increase too sharply.

At pressures lower than 80 bar (p1) case I yields better results than case II because the differences in vapour pressure of the substances necessary for fractionation are larger. At higher pressures, e.g. at 90 bar (p2), case II has the advantage of higher concentrations in the gas together with the likewise large differences in vapour

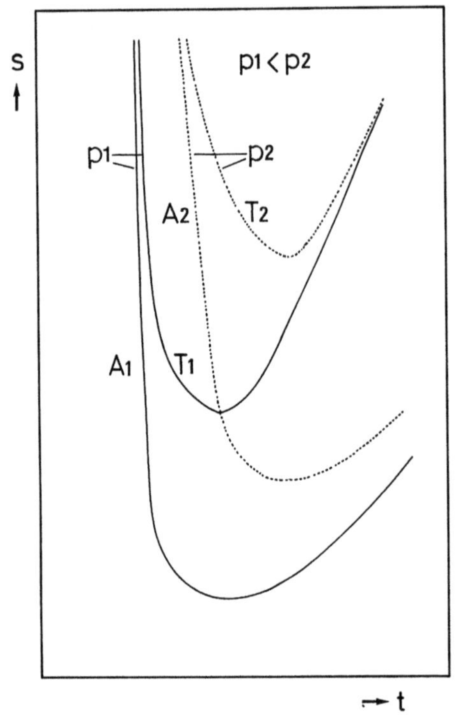

Fig. 66. Schematic portrayal of the solubility isobars of the terpene fraction (T1 at p1, T2 at p2) and the aroma fraction (A1 at p1, A2 at p2) as functions of the temperature, e.g. p1 = 80 bar, p2 = 90 bar

pressure. Still higher pressure, in work to the right of the minima, has the disadvantage that a high temperature (> 100 °C) is required for fractionating.

The influence of the various process parameters on the quality of the separation is discussed below, taking the example of terpene removal from citrus peel oils. The basic studies [34] were carried out in the experimental installation described in Chapter III.4 (see Fig. 31).

Table 16 shows, for instance, the results of counter-current extraction of bitter orange peel oil *(Oleum Aurantii pericarpii)*, a citrus oil obtained by cold pressing and containing 90% of monoterpene hydrocarbons, using supercritical carbon dioxide at 80 bar pressure. The temperature conditions were chosen so that the highest temperature was in the middle of the column and the terpenes were refluxed through cooling of the column head. A lower temperature was set in the lowest third of the high pressure column so as to subject the valuable flavouring materials to the minimum possible temperature stress. The oxygen-containing compounds were exposed only briefly to higher temperature since they flowed down through the column. The temperature at the bottom corresponded to the conditions at the solubility minimum of the oil components at 80 bar pressure. This had the advantage that the compressed gas rising through the column could become more and more enriched with the components to be separated. Under the given conditions at the column exit the supercritical carbon dioxide contained 0.62% by weight of terpenes. The terpene fraction was separated as a completely colourless oil, with less than 1% of oxygen-containing compounds. The essential oil, sensibly free of terpenes, which was withdrawn at the foot of the column, possessed high grade taste and odour qualities and was

Table 16. Process parameters of terpene removal from bitter orange peel oil

Conditions of fractionation of the oil	Gas pressure: 80 bar Temperature in the middle of the column: 85 °C Temperature at the head of the column: 75 °C Temperature at the foot of the column: 60 °C
Conditions of separation of the terpene fraction	Pressure: 30 bar Temperature: 0 °C
Gas flow	2.1 kg/h
Amount of essential oil fed in	16 g/h (terpene content of 90%)
Yield: Terpene fraction Aroma fraction	 13.3 g/h (content of oxygen-containing compounds <1%) 2.7 g/h (hydrocarbon content of 42%)

bright red. This is because all the flavouring and colouring materials (carotinoids) of the oil are present in concentrated form. The high concentration of carotinoids is desirable because the flavouring preparations are thereby coloured at the same time with safe natural colouring matter.

The result of the fractionation described is in Fig. 67, expressed as capillary gas chromatograms of the original bitter orange peel oil (a) and the fraction of low terpene content (b). Comparison of the two shows the marked enrichment of the oxygen-containing components in b (longer retention times) and the fall in content of monoterpene hydrocarbons (shorter retention times).

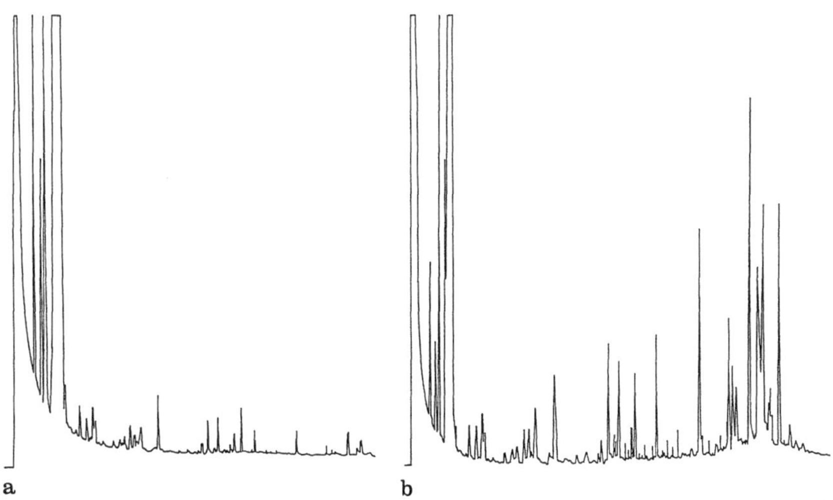

Fig. 67. Capillary gas chromatograms of the bitter orange peel essential oil (a) and the fraction low in terpenes (b) [34]

Two points must be remembered in judging the chromatograms: first, the flavouring constituents of citrus oils, for instance, citral, decompose very easily in the gas chromatograph and are thus not detected; secondly, the flame ionisation detector responds more sensitively to hydrocarbons. The gas chromatographic detection therefore favours the monoterpenes. Quantitative determination of the hydrocarbon content showed that 92% of the terpenes originally present in the oil were separated in the refining. The separation result can be improved further. The degree of enrichment of the flavouring materials or the reduction in hydrocarbon content in the fraction low in terpenes can be regulated via the relative solvent throughput (m_{rel}) under given conditions of pressure and temperature. At a given gas flow the degree of terpene removal can be set via the dosage of essential oil per unit of time.

A rapid and informative method of determining the fall in content of the hydrocarbons is through measurement of the angle of rotation of polarised light. The chief component of the citrus peel essential oils, the monoterpene hydrocarbon (+)-limonene, has a specific rotation (α_D^{20}) of + 126° 8′. The oxygen-containing components have only low values. Orange peel oil *(Oleum Aurantii dulcis*, also an object of the investigations) contains about 90% (+)-limonene, so that the oil studied should have a specific rotation of α_D^{20} of 106° 36′. As terpenes are progressively removed with supercritical carbon dioxide, the specific rotation falls steadily.

The conditions found suitable in the experiments for terpene removal from the oil of the bitter orange can be applied to the oil of the orange, (see Table 17). The concentrated fraction of flavouring materials, sample 1, was obtained by introducing the essential oil at 80 bar and 85 °C into the middle of the high pressure column and extracting it with carbon dioxide in counter-current. Lowering the temperature at the column head to 75 °C yielded a reflux of the terpene fraction in this case also. The working temperature was thus to the right of the solubility minima of the substances to be separated (cf Fig. 66). Under these given conditions the concentration in the gas was 0.84% by weight. At a relative solvent throughput of 109 g/g the terpenes were completely extracted from the orange peel oil. The specific rotation of the concentrated flavour fraction was only 18° 42′.

Another way of separation employing pressure and temperature is to choose the fractionation temperature (i.e. the temperature in the middle part of the high pressure column) on the left side of the solubility minima of the substances to be separated

Table 17. Deterpenation of the essential oil from orange peel, *Oleum Aurantii dulcis*, using supercritical carbon dioxide

	Separation conditions p (bar)/t (°C)	Temperature of column head °C	S_T^* (mg/g)	m_{rel}^{**} (g/g)	α_D^{20}
Original oil					106°36′
Aroma fraction:					
Sample 1	80/85	75	8.4	109	18°42′
Sample 2	90/55	75	10.4	91	55°54′
Sample 3	90/55	75	10.4	72	83°24′

* gas charged with terpenes, ** relative solvent throughput referred to the dosage of essential oil; specific rotation of the terpene fraction: α_D^{20} = 110°18′

(cf Fig. 66). The extraction pressure is raised additionally to 90 bar, so that, at the column head temperature of 75 °C, the gas phase is more highly charged (1.04% by weight); the rate of fractionation is then increased over that of sample 1. Terpenes are then refluxed by raising the temperature at the column head. An increase of gas throughput from 72 g/g as with sample 3 to 91 g/g (sample 2) leads to an enhanced extraction of hydrocarbons, as seen from the lower α_D^{20} value. The yield of aroma fraction is then 80% less than from sample 3.

The examples of terpene removal from the essential oils of citrus peel show that it is possible to carry out a non-drastic and efficient separation of the monoterpene hydrocarbons by means of counter-current extraction with supercritical carbon dioxide. Some uses of the concentrated flavours make it desirable to remove sesquiterpene hydrocarbons and the poorly volatile accompanying substances of the citrus oils (carotinoids, waxes) as well as the monoterpene hydrocarbons. This can also be done using counter-current extraction with CO_2. However, a second separation stage in which the carbon dioxide is best saturated with water, is necessary to obtain pure flavouring components from the essential oils.

The procedure described for continuous removal of terpenes is not restricted to citrus oils alone. For other essential oils of differing chemical composition the influential parameters of pressure and temperature have to be adapted for fractionation. It must be remembered that the ratio of extraction gas throughput per time unit to the dosage stream of the crude product depends on the amount of undesired hydrocarbons and the rate of fractionation.

IV.3.c Extraction of Pharmaceutically Active Components

Extracts of chamomile flowers have been used in large amounts as a result of their anti-inflammatory and antispasmodic activity. Several constituents participate in this function, among which may be mentioned the sesquiterpene lactone, matricin, the sesquiterpene alcohol, (—)-α-bisabolol, the bisaboloxides A and B and the en-ine-dicyclo-ethers. The aim of the extraction is to obtain the active principles and odour constituents without damage to those which are thermolabile.

Extraction experiments on the analytical scale showed that the substances of interest are easily soluble in supercritical carbon dioxide [2]. The chamomile principles of therapeutic importance can be extracted at 40 °C in a pressure range as low as 70 to 100 bar. Subsequent investigations on the preparative scale, using dinitrogen monoxide as well as carbon dioxide for extraction, show that the yield and the content of active principles in the extract depend on the process parameters of pressure, relative throughput and type of gas [28]. There is an upper limit to the extraction temperature, imposed by the thermal instability of the active principles of the chamomile. In particular, matricin decomposes at temperatures above 80 °C. When carbon dioxide is used, at a relative throughput of about 20 g/g, the extraction yield at 40 °C increases rapidly in the pressure range from 90 to 160 bar. Thus at 90 bar a 0.4%, and at 160 bar, a ca. 1.4% extract is obtained, for example. It is interesting that, under otherwise identical conditions, distinctly higher yields are obtained with dinitrogen monoxide than with carbon dioxide. At 160 bar and 40 °C and with a relative solvent throughput of 6 g/g, dinitrogen monoxide dissolves 1.33% of the chamomile active ingredients, and carbon dioxide only 1.06%. The extracts obtained showed good agreement of the contents of active principles, so that the higher total yield with the dinitrogen monoxide results from co-extraction of other material.

The active ingredients of chamomile flowers dissolve even under conditions when the extraction gas has only a small dissolving capacity, e.g., at 40 °C in the pressure region of 80–100 bar. The desired components can thus be extracted selectively. Table 18 compares the contents of active ingredients of various extracts.

Table 18. Contents of active principles from CO_2-extracts of chamomile flowers

p (bar)	t (°C)	Extraction yield	Matricin	α-Bisabolol	En-ine-dicycloethers
300	40	1.65%	7.5%	9.2%	11.7%
90	40	0.4%	17.0%	21.0%	16.7%
80	40	0.12%	3.5%	27%	14%

If work is performed at a high dissolving capacity of the extraction gas, a good yield is obtained of an extract which contains relatively little active material. At a pressure of 90 bar the proportion of extracted ballast falls appreciably, so that the contents of matricin and bisabolol in the extract are more than doubled. At 40 °C and 80 bar fractionation effects within the active principles occur, due to the further fall in gas density. The content of essential oil components in the extract increases further but the content of the less volatile matricin falls.

The composition of the extract can thus be changed by varying the extraction conditions. It is possible also to concentrate the chamomile active ingredients by stepwise precipitation of the high pressure extract. There is no obligation to use only the supercritical gas for extracting chamomile flowers; liquid carbon dioxide also dissolves the active principles easily [24, 149].

Calamus root is mainly used internally as a bitter flavouring agent. The drug and its preparations stimulate production of saliva, gastric juice and bile and are thus appetite stimulants. In addition calamus possesses the property of easing cramp and flatulence. Apart from its employment in popular medicine as "stomach root", calamus root and its oils and extracts are used above all in the liqueur industry in the preparation of bitter liqueurs.

The extraction of calamus root, free from β-asarone, *(Acorus calamus L. var. americanus Wulff)* with supercritical carbon dioxide yields the genuine essential oil without decomposition [154]. Exhaustive extraction at 90 bar and 40 °C gives an 8.3% yield of oil with the typical odour of the drug; steam distillation gives only 6.4% oil. This is because some thermolabile compounds are decomposed and calamus contains relatively involatile sesquiterpene diketones, which, however, dissolve in the CO_2. Analytical comparison of the extracted oil with that from steam distillation shows considerable differences within the group of oxygen-containing sesquiterpenes (see Table 19). A principal component of the steam distillate, 2,6-diepi-shyobunone, is completely missing from the CO_2 extract. Shyobunone itself occurs only in small amounts. Instead of these artefacts of the steam distillation, the CO_2 extract contains the genuine existing, thermolabile acoragermacrone. The chief components of the

Table 19. Influence of the method of preparation on the composition of calamus oil [154]

Component	Oil extracted with supercritical CO_2 (%)	Oil obtained by steam distillation (%)
Hydrocarbons	10	10
(β-Sesquiphellandrene)	1	2
Shyobunone	2	22
Preisocalamenediol	10	12
Acoragermacrone	12	—
Acorenone	6	9
Isocalamendiol	2	3
Acorone	37	8
cis-Isoasarone	—	—

extract are the typical calamus bitter principles, acorone and isoacorone. The sesquiterpene diketones are only slightly volatile in steam but are easily soluble in supercritical carbon dioxide and hence completely present in the extract.

The flowers of *Chrysanthemum cinerariaefolium (Trev.) Visiani* (pyrethrum flowers) contain six esters of the chrysanthemum-mono- or dicarboxylic acids, known under the collective term of pyrethrins (see Fig. 68). These can be biologically degraded and are characterised by powerful activity as contact insecticides and being at the same time of very low toxicity towards warm-blooded creatures. The pyrethrins I are the most active, especially Pyrethrin I; Cinerin I and Jasmolin I (= Pyrethrin I) are slightly less. The II-compounds are more weakly active. The importance of the pyrethrins is underlined by the fact that in 1980 about 100 tonnes of flowers were extracted per day with classical solvents in the cultivated areas of Afrika, especially Kenya.

Stahl and Schütz described in a comprehensive work the extraction of pyrethrum flowers with supercritical and liquefied carbon dioxide [29]. It was shown that the thermolabile active principles could be extracted without decomposition in the temperature range 20–40 °C but that at temperatures exceeding 60 °C first signs of de-

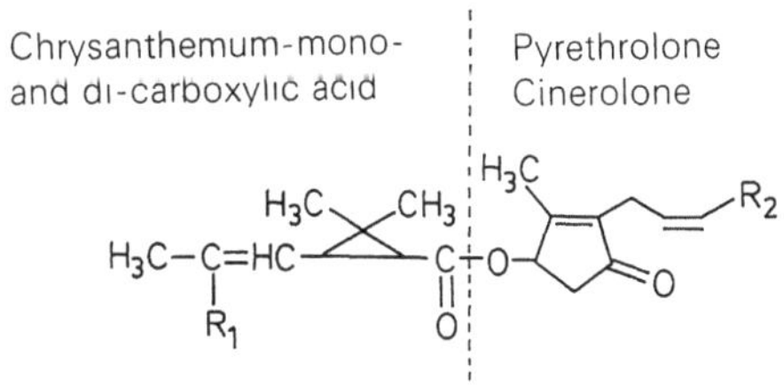

	R_1	R_2
Pyrethrin I	$-CH_3$	$-CH{=}CH_2$
Pyrethrin II	$-COOCH_3$	$-CH{=}CH_2$
Cinerin I	$-CH_3$	$-CH_3$
Cinerin II	$-COOCH_3$	$-CH_3$
Jasmolin I	$-CH_3$	$-CH_2CH_3$
Jasmolin II	$-COOCH_3$	$-CH_2CH_3$

Fig. 68. Chemical structures of the pyrethrins

Table 20. Degree of extraction of pyrethrum active components as a function of the extraction parameters [29]

Extraction parameter				Degree of extraction	
Pressure (bar)	Temp. (°C)	m* g/g	t h	Total Pyrethrin I (%)	Total Pyrethrin II (%)
90	40	32	3	78	50
100	40	32	3	93	86
100	20	32	3	94	95
250	40	50	5	98	98

* amount of extraction gas/unit mass of drug

composition occurred. The yield of active ingredients and the composition of the extracts depend to a great extent on the extraction conditions.

Table 20 shows the dependence on the extraction parameters of the degree of extraction of the pyrethrum active principles. At 90 bar and 40 °C the solubility of the pyrethrins, especially of the somewhat less soluble pyrethrins II, is insufficient for exhaustive extraction in the given time. An extraction pressure of 100 bar under otherwise identical conditions improves the yield distinctly. In liquid carbon dioxide (100 bar, 20 °C) the greater solubilities of the pyrethrins leads to a degree of extraction of 95%. The residual content of active ingredients can be reduced to 2% and less by increasing the extraction pressure or the amount of gas for extraction.

Fig. 69 shows the yield in percent of total pyrethrins (curve 1) and the content of active ingredients in the extract (curve 2), expressed as functions of the time of extraction. A flower sample containing 0.8% of total pyrethrins (I and II) was extracted at 100 bar and 40 °C with altogether 46 g CO_2/g of sample (5 h). A degree of extraction of 97% was reached in the experiment. High contents of pyrethrin were

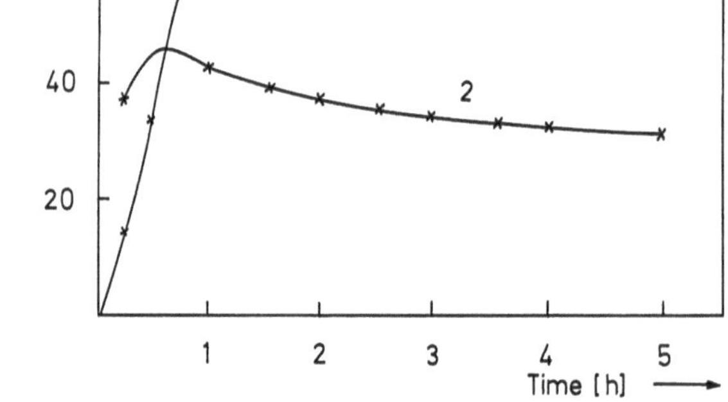

Fig. 69. Course of extraction as a function of time [29].
See text for extraction conditions; 1 yield of total pyrethrins, 100% corresponding to 0.81% in the drug; 2 pyrethrin content in the total extract

Table 21. Content of active principles of various pyrethrum extracts

Conditions of extraction	Yield of extract (% of the drug)	Total pyrethrins (% of the extract)	Colour*
Drug 1 (0.4% total pyrethrins)			
CO_2, 90 bar, 40 °C	0.65	36.2	0.2
CO_2, 100 bar, 40 °C	0.94	29.9	0.4
CO_2, 100 bar, 20 °C	2.06	21.8	0.9
Petrol ether	3.6	11	20
Drug 2 (1.6% total pyrethrins)			
CO_2, 160 bar, 40 °C	2.8	51.3	0.3
Petrol ether	6.2	26	70
Drug 3 (0.8% total pyrethrins)			
CO_2, 100 bar, 40 °C	2.4	30.8	0.7
Petrol ether	6.6	11.8	90

* $E^{1\%}_{1\,cm}$, referred to 1% pyrethrins

attained in the extract at the beginning of extraction (curve 2); after about 30 min the content reached a maximum and then slowly fell as extraction continued, mainly due to increasing co-extraction of undesired accompanying substances.

Table 21 compares contents of active principles of pyrethrum extracts from three different charges. The analytical data from a classical solvent extraction are quoted in each case for comparison.

The extract with liquid carbon dioxide (100 bar, 20 °C) contained a lower percent amount of pyrethrins than with supercritical gas at the same pressure, although the yield of extract was higher. The contents of active ingredient of the CO_2-extracts were 2 to 3 times greater than those using petrol ether. Depending on the drug and conditions of extraction, products from the high pressure procedure contained over 50% of active components, which demonstrates clearly the high selectivity of carbon dioxide.

As a result of their high content of colouring matter (carotinoids, chlorophylls), undesired waxes and resin-like impurities, pyrethrum extracts prepared using the classical procedure have to be refined before use. Particularly important is decolourisation of the extracts, e.g. for sprays in household use. In this connection the carbon dioxide extracts are more valuable than classical crude extracts. In addition to the higher contents of active components they contain practically no undesired colouring material, as shown by the low values in the last column of Table 21. A disadvantage of the simple high pressure extracts is their consistency; they are viscous to solid masses, like the petrol ether extracts, as a result of co-extracted vegetable waxes. It is possible, however, to precipitate these waxy constituents from the real active ingredients using a special prior procedure, so as to yield a fluid and scarcely coloured extract of high pyrethrin content (cf p. 162).

Bunzenberger, Lack and Marr confirmed the above results from extraction of pyrethrum with dense carbon dioxide [109]. The attraction of the high pressure procedure is reflected by a patent application for the extraction of pyrethrum flowers with liquid carbon dioxide [150].

The root of the Zingiberaceae species, *Curcuma xanthorrhiza Roxb.* and *Curcuma domestica Val.* (turmeric) contains substances of pharmacological interest which find application in bile preparations. The essential oil has been shown to possess the property of increasing bile flow whereas the curcuminoids have choleretic properties. The curcuminoids are also important as natural colourants for foodstuffs.

The turmeric essential oils have a high proportion of poorly volatile sesquiterpene derivatives. Their isolation through steam distillation is thus time-consuming and they are also thereby subjected to a temperature stress of long duration. Thermolabile constituents are then decomposed. According to our own studies the essential oils can be obtained from turmeric roots rapidly, quantitatively and without breakdown by using dense carbon dioxide [174]. The turmeric oils are comfortably soluble in the dense CO_2 as shown by the fact that extraction already at 120 bar and 40 °C yields a concentration in the gas of ca. 8 % by weight. Under these conditions 8.5 % oil were extracted from *Curcuma xanthorriza* and 4 % from *Curcuma domestica*.

Table 22 gives the chemical compositions of the CO_2-extracted oils from *Curcuma xanthorrhiza* and *Curcuma domestica* [175]. Components present in amounts less than 1 % of the essential oil are not listed.

With few exceptions both oils contain the same compounds although the quantitative chemical compositions differ. It is noteworthy that tolylmethylcarbinol, a principal

Table 22. Quantitative composition (%) of the essential oils extracted by CO_2 from turmeric root, obtained by GC and preparative HPLC [175]

	Curcuma	
	xanthorrhiza	*domestica*
α-Pinene	1.3	3.1
Camphene	3.1	0.3
Sabinene	2.7	0.1
α-Phellandrene	1.3	1.6
Limonene	1.2	0.5
1.8-Cineol	3.3	5.5
Terpinolene	0.4	1.3
Camphor	5.8	0.7
Caryophyllene	1.5	2.6
allo-Aromadendrene	4.7	—
Humulene	—	3.1
Borneol	6.3	—
Zingiberene	5.5	8.2
β-Bisabolene	1.7	2.4
β-Curcumene	8.5	1.3
β-Sesquiphellandrene	1.9	7.3
ar-Curcumene	11.3	5.2
Isofuranogermacrene	8.3	—
Turmerone	1.2	22.6
Turmerol	0.6	11.4
ar-Turmerone	1.2	16.5
Atlantone	—	3.5
Xanthorrhizol	27.0	—

component from steam distillation, could not be detected in the CO_2-extracts; it must therefore be regarded as an artefact.

The yellow coloured curcuminoids (diferuloylmethane dyes) are not satisfactorily extracted with dense carbon dioxide, even at high pressures (up to 600 bar) and temperatures (up to 80 °C) because they are relatively polar. To obtain an economical high pressure extraction a polar entrainer, such as methanol, must be added to the carbon dioxide.

Valerian preparations are used as natural therapeutic agents on account of their mild sedative action. Both the essential oil and extracts and tinctures from valerian root are recognised as pharmacologically active. Which components are responsible for the sedative action has not yet been finally explained.

Muravlev and Smetanin described the extraction of valerian roots with liquid carbon dioxide at 25 °C and 65 bar [176, 177]. They found that the yield using the same amount of solvent (20 g/g) was sensibly independent of the time of extraction from 35 to 140 min and hence of the rate of flow of the carbon dioxide.

The extraction of Indian valerian *(Valeriana wallichii DC.)* was the object of work by Stahl and Schütz [178]. The principal aim of this investigation was to obtain with great care the valepotriates present in *Valeriana wallichii* in high concentration (ca. 3%). Carbon dioxide and trifluoromethane at 70–200 bar and 30–40 °C were used as extraction gases. It was shown that the valepotriates could be obtained undecomposed in high yield (>90% extraction). The yield of valepotriates in the preparative extraction was little influenced by pressure in the range from 100 to 200 bar; this demonstrated their good solubility even at low pressures, an assumption supported by the observation that 80% of them were extracted at a relative throughput of only 4 g/g. An extract concentration of 1% by weight was attained in trifluoromethane at 70 bar and 30 °C, and of 0.7% by weight in supercritical carbon dioxide at 100 bar and 40 °C. These results show that, at low pressures, trifluoromethane is a better solvent than carbon dioxide for valepotriates. As far as their solubilities in dense gases are concerned, valepotriates are intermediate between the highly soluble components of the essential oil and the other poorly soluble valerian constituents.

Table 23 shows that the valepotriate content of valerian extracts depends on the extraction agent. Supercritical carbon dioxide and trifluoromethane yield 60–70% valepotriate which is clearly more than the conventional solvents. The supercritical gases thus possess greater dissolving selectivity.

At the present time the content of valepotriates in valerian extracts is being subjected to a more critical scrutiny. The latest information indicates that they are no longer without risk, and extracts containing little or no valepotriate are now of interest.

Table 23. Valepotriate contents of valerian root extracts obtained in various ways [178]

Extraction agent	Didrovaltrate (%)	Valtrate (%)	Total extract (%)
Petrol ether	46–47	5.7–5.8	4.5–4.7
Dichloromethane	36–39	4.0–4.7	7.7–7.9
Carbon dioxide, supercritical	55–67	6.2	3.9–4.4
Trifluoromethane, supercritical	59–65	6.6–7.0	3.5–4.4

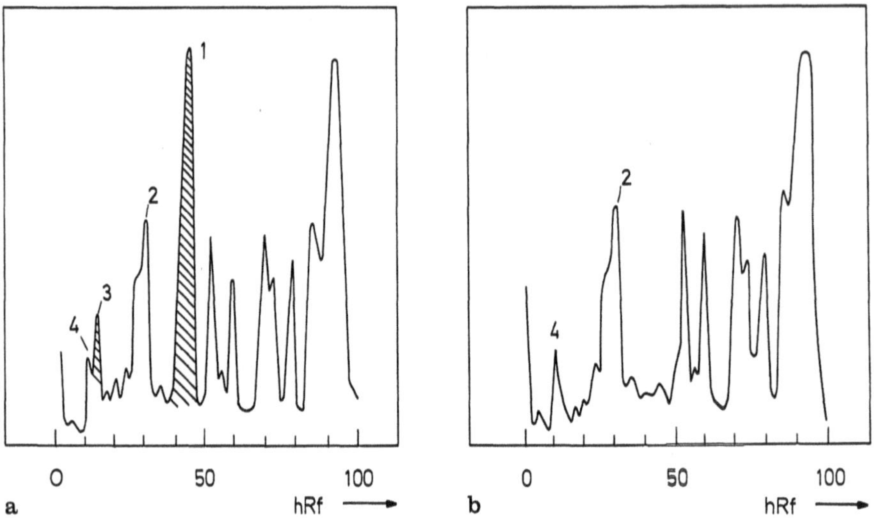

Fig. 70. Thin-layer chromatographic evaluation of various valerian extracts.
a primary extract (with CO_2); **b** valepotriate-free CO_2-extract; 1, 3 valepotriates; 2 valerenic acid; 4 hydroxy-valerenic acid

Such CO_2-extracts of valerian roots, free of valepotriates, can be obtained [179]. Dispensary valerian is extracted with supercritical carbon dioxide to yield first a primary product which is subjected to a careful alcoholysis converting the valepotriates selectively into difficultly soluble derivatives; these remain behind as insoluble residue from a subsequent CO_2-extraction of the product. In this way it was possible to obtain from *Valeriana officinalis* a light yellow extract, free of valepotriate, in about 2% yield. The other components of the extract were not changed by this treatment, as seen clearly in Fig. 70. Confirmation was obtained with the help of capillary gas chromatography and HPLC. HPLC analysis of valerenic acid and its derivatives showed that there was neither a qualitative nor quantitative difference between the two extracts. The former method applied to the primary extract and the CO_2-extract, free of valepotriates, likewise revealed no change in the composition of the essential oil.

Common wormwood *(Artemisia absinthium L.)* is a popular bitter flavouring agent on account of its content of bitter principles and essential oil; it is used as a carminative, cholagogue and stomachic.

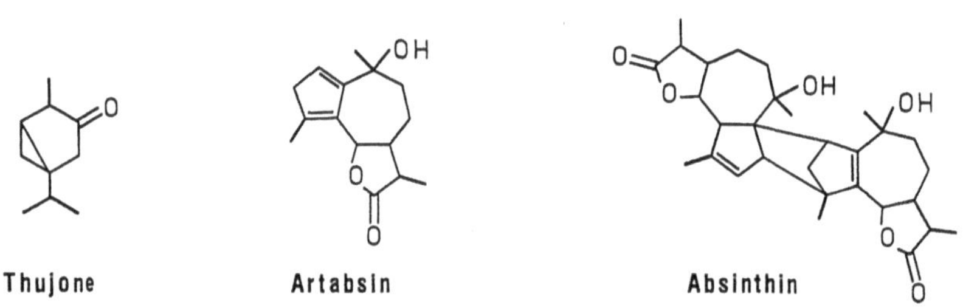

Fig. 71. Structural formulae of the most important constituents of common wormwood

Some 100–150 tonnes are processed yearly in Europe, the main part finding use in spirits. The bitter taste of this drug is due to sesquiterpene lactones. The most important bitter principles of sensory influence are above all absinthin and, rather less, artabsin. The use of alcoholic preparations of wormwood is, however, limited by the β-thujone in the essential oil, which is injurious to health [151].

The poisonous β-thujone cannot be separated from the wormwood using ordinary procedures of separation because the valuable bitter principles are thermally unstable and classical solvents display poor selectivity, so that some of the bitter compounds are co-extracted with the β-thujone. High pressure extraction with carbon dioxide does, however, permit detoxification of the wormwood [32].

To establish the best conditions for extracting with dense carbon dioxide the solubilities of the most important drug components must be known. The chief bitter principle of wormwood, the dimeric sesquiterpene lactone, absinthin, is practically insoluble in supercritical carbon dioxide at 40 °C and pressures up to 200 bar. At 300 bar the solubility is about 22 μg per g gas. Only 4 μg/g are dissolved in liquid carbon dioxide at 80 bar and 20 °C. The solubilities of thujone and artabsin in supercritical CO_2 at 40 °C are shown in Fig. 72. The monoterpene ketone thujone is very soluble. On account of its high volatility, almost 0.6 mg thujone dissolves per g gas in sub-critical carbon dioxide at only 60 bar and 40 °C. Increase of pressure to 90 bar brings about an exponential rise in solubility to values of about 5% by weight. In comparison,

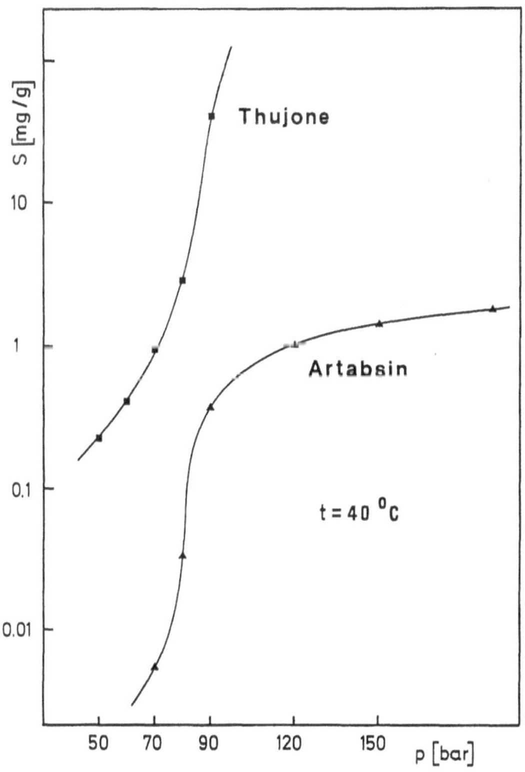

Fig. 72. Solubilities of thujone and artabsin in supercritical carbon dioxide at 40 °C, as a function of pressure [32]

artabsin is about 100 times less soluble; for example, 1 g carbon dioxide at 90 bar and 40 °C dissolves about 0.4 mg substance.

Quantitative experiments on the solubility of these constituents of wormwood in dense carbon dioxide show that the essential oil can be separated together with the thujone. Fig. 73 shows the course of the extraction of the important components of wormwood with supercritical carbon dioxide at 100 bar/40 °C, as a function of the relative throughput of gas.

The toxic thujone can be rapidly and completely extracted from the drug. Almost 95% of it are extracted at a relative gas throughput as low as 4 g/g; the artabsin content is then still about 50% of the original value. The content of absinthin in the drug suffers negligible change. Exhaustive thujone extraction is accompanied by a loss of about 75% of the artabsin. The high bitter value of the drug material of about 50000, due above all to the absinthin, remains practically the same (see Table 24).

The yield of extract with liquid carbon dioxide (80 bar, 20 °C), 2.3%, was slightly greater than that with the supercritical gas, 2.1%. The bitter value of the wormwood was a little lower since artabsin is more easily soluble in liquid CO_2 and was completely co-extracted. These experiments show that the harmful thujone can be completely separated by extraction with carbon dioxide, and without loss of value through a reduced bitter content, thermal stress and the presence of solvent residues.

Extraction of common wormwood with dense carbon dioxide is a clear example of how the high pressure procedure runs with protective care. Depending on the con-

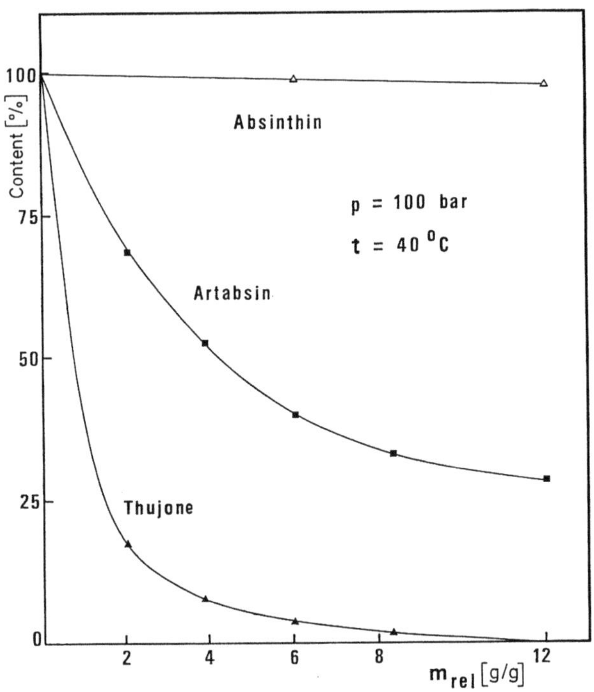

Fig. 73. Course of the extraction of the important constituents of wormwood using supercritical CO_2, expressed as a function of the relative gas throughput m_{rel} [32]

Table 24. Contents of 2 wormwood extracted with CO_2, compared to values in the starting material

Drug	m*	Yield of extract (%)	Bitter value according to DAB 8***	Absinthin content (%)	Artabsin content (%)	Thujone content (%)
*Absinthii herba*** (starting material)			50000	0.42	0.23	0.08
Absinthii herba (extracted with CO_2 at 100 bar and 40 °C)	12	2.1	50000	0.41	0.06	—
Absinthii herba (extracted with CO_2 at 80 bar and 20 °C)	8	2.3	49000	0.42	—	—

$m^* = \frac{\text{extraction gas in g}}{\text{sample of drug in g}}$;

** 0.63% of essential oil in the starting material, 13% of which was thujone

*** Deutsches Arzneibuch (= German Pharmacopoeia)

ditions, steam distillation yields blue, green or even orange-brown essential oils [152]. These colours come from decomposition products of the thermolabile sesquiterpene lactones, artabsin and absinthin. Artabsin can be obtained without breakdown in the extraction with dense carbon dioxide, whereas it rearranges to dihydrochamazulenes under the conditions of the steam distillation [153].

IV.3.d Extraction of Spices

Its taste and odour constituents determine the value of a spice. The industrial application of crude spices for yielding aroma and flavouring foodstuffs is retrograde, especially since treatment with ethylene oxide gas for pest removal from, and sterilisation of, vegetable starting materials has been considerably limited by law. The use of spice oils and oleoresins as an alternative to the crude spices is increasing steadily. These concentrates have a variety of advantages over the starting spice products, such as their sterile nature, longer shelf life, better use of the active principles, easier handling and better means of standardisation.

Oils of spices are obtained in the classical way through steam distillation. Many spices contain also valuable components of low volatility, such as pungent substances or certain resin fractions which determine the taste profile of the crude material. They are generally obtained by extraction with low-boiling solvents. The high pressure extraction with dense carbon dioxide combines the working principles of the two classical separation procedures mentioned, yet without possessing their disadvantages. It thus offers a genuine alternative, furnishing concentrates of high grade sensory quality under conditions of thermal non-aggression which are equivalent to the pure oil or contain any desired large fraction of the less volatile components.

That the high pressure procedure with dense carbon dioxide for preparing extracts of spices is attractive has been known for a long time. A Russian article was published

Table 25. Survey of extracts of spices and herbs prepared with liquid CO_2 in the USSR in the years 1966 to 1972 [27]

Material Extracted	Yield of Extract (%)	Extract amount produced (kg)
Cloves	18.6	14 144
Nutmeg	10.5	8 941
Mace flowers	20.7	3 573
Black pepper	5.1	849
Green pepper	3.9	348
Ginger root	3.6	1 438
Parsley seeds	3.6	1 493
Dill seeds	3.6	1 498
Coriander fruit	1.8	88
Bay leaves (Laurel)	2.5	321
Aniseed fruit	4.3	175
Calamus root	5.1	3 094
Celery root	7.0	380
Chamomile flowers, leaves and seeds	2.9	4 243
Yarrow herb	1.1	403
Hop cones	7.8	5 430
Eucalyptus leaves	1.8	144
St. John's Wort (*Hypericum*)	2.2	776
Carrot roots	2.7	230
Pot marigold flowers (*Calendulae*)	2.3	210

in 1974, describing the extraction of 20 different natural products, principally spices, with liquid CO_2 [27]. Table 25 lists the various starting materials and the yields of extract derived from them. The amounts of extracts produced in this way in the years 1966 to 1972, obtained from amounts exceeding 1000 tonnes of dried crude material, are also given.

The first publication about the use of supercritical gases to extract spices was in 1971 [26]. Vitzthum and Hubert described in this patent the extraction of pepper, cloves, vanilla and cinnamon with supercritical gases at pressures from 280 to 400 bar and temperatures of 50–60 °C.

The patented procedure has two stages. The essential oils are dissolved out of the spices with dry, supercritical carbon dioxide in the first. The flavouring components are then extracted with supercritical carbon dioxide containing water. In numerous cases the second stage is superfluous, as, for example with piperine, the chief pungent principle of pepper; the taste and odour components are dissolved out at the same time is the first stage.

Behr et al. also described a two-stage procedure for preparing spice extracts [157]. They quoted the examples of pepper, ginger, cloves, fennel and sage. The essential oils were removed from the spices with liquid carbon dioxide under mild temperature conditions; this was followed by the second stage of extracting the flavouring components with supercritical CO_2. This procedure did not permit complete separation of essential oil and flavouring materials, however, as the examples given themselves show. For instance, an extract of fennel with liquid carbon dioxide at 130 bar and

30 °C contained only 62 % essential oil. All the components of ginger extractable with carbon dioxide appeared already in the first stage (liquid CO_2, 280 bar, 29 °C), so that no measurable extract was obtained in the second.

The extraction of pure spice oils and aroma components with dense carbon dioxide requires an optimised solvent selectivity because most starting materials containing essential oil have also lipophilic constituents of low volatility, such as fatty oils, waxes, resins and colouring matter. This is not discussed in many examples of use. The extractions are sometimes carried out under drastic conditions, e.g. at pressures of 250–300 bar and temperatures of 40–60 °C, although relatively low pressures (90–150 bar at 40–50 °C) suffice for obtaining essential oils and flavour materials. Some undesired accompanying material is co-extracted even in extraction with liquid carbon dioxide which is an excellent solvent for spice oils.

The basic possibilities for selective isolation of spice oils and aroma components are: fractional extraction on the one hand; and extraction together of both the volatile oils and the poorly volatile components, then following with fractional separation on the other hand. As solubility studies show, concentrations of essential oils in the gas phase of ca. 5% by weight are reached even at 90 bar and 40 °C (gas density of 0.5 g/cm³). Fatty oils and other compounds of low volatility are only poorly soluble under these conditions of pressure and temperature, so that their content in the total extract is low.

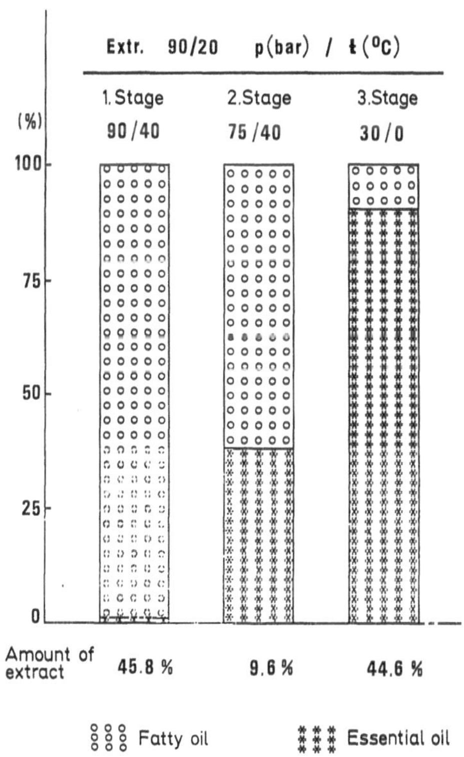

Fig. 74. Schematic representation of the contents of essential and fatty oils in various fractions from a caraway extract in liquid carbon dioxide [155]

Stahl and Gerard described the various ways of selectively obtaining essential oils in the extraction of spices with dense carbon dioxide [155]. They extracted the fruit of anise, star anise and caraway, and also cloves and Ceylon cinnamon. Comparison of the extracting and separating conditions and the yields revealed that a fairly extensive separation of the essential oils from fatty constituents, waxes, resins and similar involatile lipids was possible when the parameters of the procedure were suitably chosen (see Table 26). The essential oil part of the CO_2-extracts was determined by steam distillation but it must then be borne in mind that thermolabile substances may be destroyed under the harder conditions of steam distillation and thus escape determination.

Extraction of anise fruit at 90 bar and 40 °C gave a 2% yield of a low-viscosity, light yellow essential oil. The large fraction, ca. 85%, of this extract which was volatile in steam shows that scarcely any compounds of low volatility were co-extracted under these conditions. Anise fruit pre-extracted in this fashion then yielded almost 10% of fatty oil through extraction with supercritical carbon dioxide at 300 bar and 40 °C. This example shows how essential and fatty oils can be separated by fractional extraction.

Both substance classes are extracted together if carbon dioxide of higher gas density is employed. By lowering this density in stages, the fatty oil can be first separated out and then the essential oil from the extraction gas, as shown in the examples of star anise and caraway oils. Fig. 74 portrays schematically the fractionation of caraway oil by this step-wise precipitation.

The extraction with liquid CO_2 (90 bar/20 °C) was followed by the first separation through isobaric temperature increase to 40 °C; this yielded almost entirely the fatty oil, with only about 1% of essential oil. In the second stage the pressure was reduced from 90 to 75 bar, at constant temperature, which yielded an intermediate fraction of only about 9.6% of the whole, and consisting of both oil types. The main portion of essential oil was separated out in the third stage by lowering pressure and temperature to 30 bar and 0 °C, respectively; this extract contained over 90% of the constituents volatile in steam. The number of separating stages and hence the efficiency of the separation can be improved appreciably by placing a mass transfer

Table 26. Conditions and yields of extractions of spices containing essential oils [155]

Spice	Content of Essential Oil* (%)	Conditions of Extraction p (bar)/T °C	Conditions for Separation p (bar)/T °C Stage 1	Stage 2	Stage 3	Yield of Extract (%) with % Fraction volatile in steam Stage 1	Stage 2	Stage 3
Anise	2.1**	90/40, followed by	75/40	30/0	—	0.7 [80]	1.3** [85]	—
		300/40	75/40	40/5	—	9.7	0.2	—
Star anise	4.2**	90/20	90/50	80/55	30/0	3.4 [5]	0.7** [76]	2.8** [90]
Caraway	4.1**	90/20	90/40	75/40	30/0	3.8 [1]	0.8 [38]	3.7** [90]
Cloves	19.0**	90/40	85/40	30/0	—	2.9 [75]	13.6** [82]	—
		90/0	90/40	70/40	30/0	0.1 [0]	12.6 [81]	4.1 [82]
Ceylon cinnamon	1.5**	100/40	85/40	50/20	30/0	0.1	1.4	0.2
		90/0	85/45	75/40	30/0	0.7	0.6	1.1**

* determined by steam distillation, ** sensory assessment, in Table 27

column after the extraction stage. This enables the essential and the fatty oils to be separated quantitatively into two fractions.

Table 26 shows also that optimum conditions are needed for separating the essential oils completely. The extraction of cinnamon oil is an impressive example (100 bar/ 40 °C). Only about 87% of the essential oil separates from the extraction agent under the usual conditions of 50 bar and 20 °C. By lowering further to 30 bar and 0 °C, the most volatile components of the cinnamon oil, mainly monoterpene hydrocarbons, make up the principal part of the material are separated out.

The essential oils of anise, star anise, caraway, cloves and Ceylon cinnamon obtained by CO_2-extraction show practically the same colour and consistency as when derived by steam distillation; the sensory qualities differ, however (see Table 27). The odour and taste of the various oils were compared with those of a commercially available essential oil of high grade quality (standard). The oils obtained by extracting with dense carbon dioxide were rated more highly than those derived by steam distillation from the same starting material. Marked deviations of the tested oil from the standard are due to differences in chemical composition (e.g. with cloves). In this case also, the CO_2-extract was judged to be better than the steam distillate, which failed to fulfil the conditions.

Table 27. Sensory assessment of essential oils [155]

Drug	Steam distillate	CO_2-extract
Anise (*Anisi fruct.*)	terpene-like, not so full as the standard	high and full quality, comparable with the standard
Star anise (*Anisi stellati fruct.*)	deviating from the standard, not so pure	very sweet, robust, pure anise note
Caraway (*Carvi fruct.*)	pure quality, slightly weaker than the standard	good quality, clear and pure caraway note
Cloves (*Caryophylli flos*)	non-typical, not fulfilling the requirements	usable quality of character, some deviation in type
Ceylon cinnamon (*Cinnamomi ceyl. cortex*)	Scarcely any qualitative difference; samples make a good general impression; some shortage of fullness of aroma and body	

The extraction of pepper with liquid and supercritical carbon dioxide has been described in the literature several times [26, 31, 157, 159, 160, 161]. Depending on the quality, pepper contains 1 to 3.5% essential oil and 5 to 10% pungent principles, chiefly piperine, as constituents of value. Extraction gives very variable yields as a result of large variations in quality. In general it is possible to say, however, that higher yields are obtained with supercritical than with liquid carbon dioxide. More details are given below, based on the results of the studies of Hubert and Vitzthum [26, 160]. Extraction of black pepper with supercritical carbon dioxide (e.g. at 350 bar/60 °C) yielded a yellow, spongy oleoresin with a pure flavour of pepper. Table 28 contains analytical data of the ground pepper before and after extraction,

Table 28. Analytical data on pepper extraction with supercritical carbon dioxide [160]

	Material		CO_2-extract	Degree of extraction	Commercially obtained CH_2Cl_2 extract
	before	after			
	extraction				
Piperine (%)	10.1	0.3	44.0	97.7	41.2
Essential oil	3.5	0.1	12.7	81.1	15.0
Water (%)	9.3	6.2	7.0		0.2
Colour of extract			yellow		olive green

of the CO_2-extract and of an ordinary commercially obtained extract with dichloromethane. It is unsatisfactory that the CO_2-extract contained only 12.7% essential oil. According to our own experience and results of other authors [23] CO_2-extracts of pepper from such high grade starting material (3.5% essential oil) can contain up to 30% spice oil. The degree of extraction of essential oil, quoted as only 81% and derived from the recovery in the extract, is probably due to the fact that the volatile pepper oil could not be fully recovered.

Table 28 shows also that the pepper oleoresin, extracted with supercritical carbon dioxide and separated out in one stage, contained a few percent water which is difficult to remove from the spongy extract and reduces the shelf life of the product. It would be a help to separate the co-extracted water during the procedure. According to Hubert and Vitzthum this is possible by keeping the separation temperature as high as the extraction temperature [26]; the water does not then separate out together with the pepper oleoresin. This has a bad influence on the complete isolation of the essential oil, however. Another way is to pass the gas stream charged with the extracted material through a solid granulated drying agent, such as calcium sulphate, which absorbs only water [162].

Extracts of pepper with dense carbon dioxide have usually a higher content of the pungent substances than those from classical solvent extraction; this is because of the greater selectivity of the CO_2 solvent. Piperine concentrations up to 66% are attained, depending on the quality of the starting material and the extraction conditions [157]. Since, further, the high pressure extraction enables the spice oil to be completely and carefully extracted, a high grade pepper extract may be expected. CO_2-extracts obtained a better assessment in a sensory test than extracts with classical solvents [161]. Odour and taste were characterised as pure and typical. The high quality of the dense gas extract showed up also in a "concentrate mixture", prepared by forming an addition product of the oleoresin and dextrose. Meat products (German sausage sorts) flavoured with this concentrate mixture were judged to have very good taste and odour.

Since the constituents of pepper which determine its value (essential oil, pungent principles) possess widely differing solubilities in dense carbon dioxide, their fractionation is possible. Like all essential oils, that of pepper is easily soluble in dense carbon dioxide. The pungent substances are distinctly less soluble [163]. As seen in Fig. 75, the principal pungent component of pepper, piperine, yields only an approximately 0.5 mg/g solution in liquid CO_2 at 25 °C. The solubility is greater in supercritical

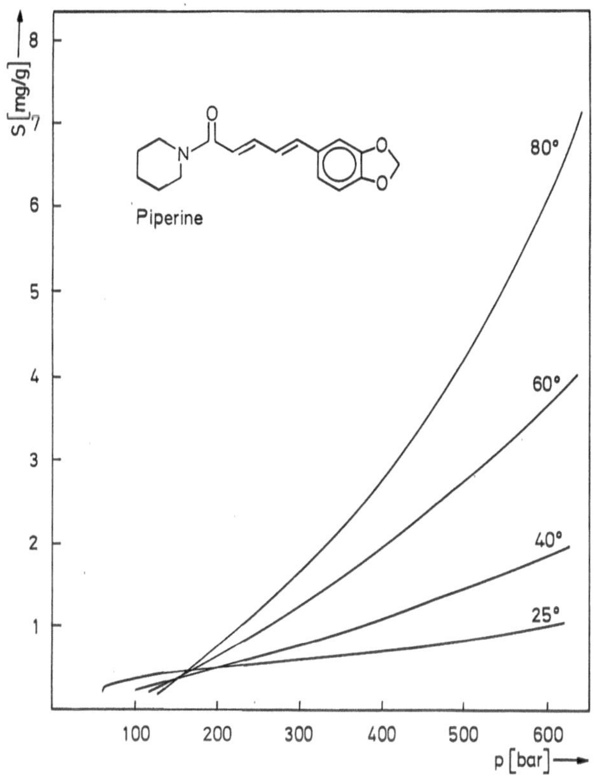

Fig. 75. Solubility isotherms of piperine in dense carbon dioxide, as a function of pressure [163]

carbon dioxide, especially at higher temperatures and pressures. As a result of these markedly differing solubilities, the flavouring and pungent substances can be obtained separately in the CO_2-extraction. This offers in a simple way the possibility of standardising the pepper extracts.

The valued constituents of the capsicum spice (e.g. paprika, chillies) are pungent principles, such as capsaicin and dihydrocapsaicin (see Fig. 76) and also carotinoids which are employed in the food industry as natural colourants.

Hubert and Vitzthum described the extraction of ground chillies with supercritical carbon dioxide [160]. They obtained a red oil, with a degree of extraction, based on capsaicine, of 97%. Coenen and co-workers published work on the extraction of paprika fruits with supercritical carbon dioxide [131, 169]. The authors proposed

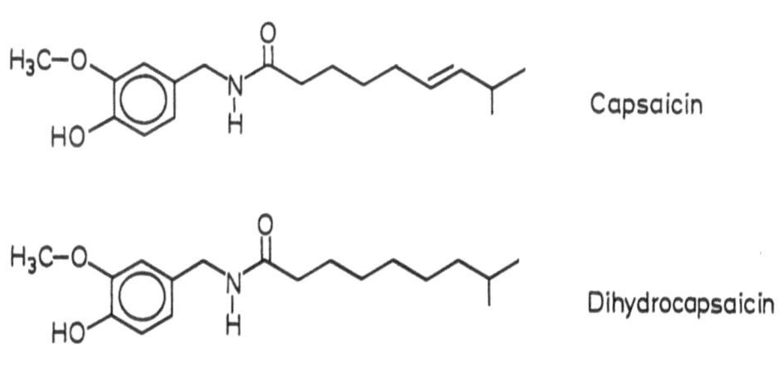

Fig. 76. Principal pungent principles of the capsicum spices

a two-stage extraction to separate the pungent substances from the carotinoid fraction; they employed a large amount of solvent with respect to the amount of starting material. The flavouring and pungent components were extracted at 120 bar and 40 °C and the concentrate of colouring material was then obtained at 320 bar and 40 °C. The central aim of these investigations was to obtain natural vegetable colourants; this is treated more thoroughly in Chapter IV.4.

The extraction of coriander seed with liquid carbon dioxide at temperatures between 0° and 20 °C and the corresponding saturation pressures has been described by Russian authors in several publications [164, 165, 166]. Fatty oil is co-extracted with the essential oil under these conditions. The composition of the CO_2-extract was investigated analytically, the essential oil being separated by steam distillation [165]. It was shown in another work that temperature influenced the composition of the extracts obtained with liquid carbon dioxide also. The highest content of linalool in the CO_2-extract was attained at 0 °C; at higher temperatures considerable amounts of poorly volatile components were co-extracted [166]. The degree of comminution of the seeds also influenced the extraction of the constituents, a particle size of 0.3 mm yielding the best results.

Other Russian authors described the extraction of cardamom fruit [167] and dill seed [168] with liquid carbon dioxide. Components of low volatility were present also in these spice extracts, along with the essential oils really desired. Compared to the steam distillate, the cardamom extract contained a distinctly larger amount of organic acids and phenols. The CO_2-extract of dill seeds was made up of 73.8% of essential oil and 26.2% of fatty oil. These two oil types were distinguished by treatment with cold methanol in which they are differently soluble. A quantitative separation in this way appears doubtful, however.

The extraction of thyme with liquid carbon dioxide on the laboratory scale was described by Bestmann, Erler and Vostrowsky [158]. Dry thyme leaves were treated in a high-pressure soxhlet extractor with carbon dioxide at 5–10 °C and the corresponding liquefaction pressure of 40–45 bar; a yield of 2.1% was obtained, corresponding roughly to that from classical steam distillation. Gas chromatographic and mass spectrometric analysis of the essential oils showed appreciable differences between the CO_2-extract and the steam distillate. Thus the former contained 2.5% p-cymene whereas the latter, from the same starting material, contained 18.9% and in a purchased rectified oil of thyme, as much as 47%. The distilled oils also contained markedly larger amounts of γ-terpinene. These C_{10}-hydrocarbons are artefacts which are formed during the distillation of oxygen-containing oil components (e.g., p-cymene from thymol). The CO_2-extract contained 30% low-volatility products, mainly higher molecular hydrocarbons and free fatty acids, due to co-extraction of leaf waxes.

In the preceding pages the discussion has been of the treatment of dried natural products with dense carbon dioxide to prepare spice extracts. Kitchen herbs, such as dill, estragon or parsley are preferably used in the fresh state because important flavouring constituents are partly or wholly lost during drying. Schütz et al. described in a patent application the preparation of concentrated fresh plant extracts which showed the typical odour and taste of the starting materials, could be stored practically indefinitely and were sensibly sterile [30]. The fragrance and flavour chemicals were completely isolated by a single-stage extraction with carbon dioxide at supercritical pressure (up to 300 bar) and at temperatures preferably between 0 and 31 °C. A temperature of 0–20 °C was suggested to obtain complete recovery, especially of the volatile components, under gentle conditions; care had to be taken that the separation pressure was not so high as the corresponding saturation pressure. The procedure was explained in more detail taking the examples of the extraction of dill herbs, estragon leaves, blackcurrant buds and mimosa flowers which are of interest from the point of view of perfumery. Liquid carbon dioxide yielded extracts smelling strongly of the fresh plant material. The extracted material was sensibly odourless. When extraction of the fresh plant material and recovery of the extract was carried out at 45 °C the scents were qualitatively lower grade and the extracted residue smelt mouldy, presumably due to enzymatic reactions or microbiological processes.

Vanilla is frequently used to flavour foodstuffs. Synthetic vanillin does not fully yield the highly aromatic flavour of the vanilla pods, so that there is considerable interest in natural vanilla extracts which contain the other important aroma principles in addition to vanillin.

A strongly smelling aromatic oil is obtained by extracting vanilla pods with supercritical carbon dioxide at 400 bar and 45 °C [26]. As a result of these drastic conditions an extract prepared in this way contains only 28% of flavouring substances. The content of flavour and fragrance compounds can be increased by using supercritical carbon dioxide under more selective conditions, e.g. at lower gas pressure. Schütz et al. described the isolation of the flavouring components of vanilla pods using liquefied carbon dioxide at pressures between 80 and 350 bar and temperatures between 10 and 30 °C [170]. The flavouring substances, some of which are sensitive, were separated out at a temperature below 20 °C. The authors recommended grinding the vanilla pods finely at a low temperature, between −50 and −30 °C to obtain a quantitative yield. Another possibility is to extract the crudely comminuted pods under the conditions quoted. This pre-extracted material is then subjected to a sudden fall in pressure which brings about the desired disintegration of the cell tissue. The pods then swell up, increasing their volume by about 30%, rendering the constituents more accessible to the extraction agent and enabling then to be extracted quantitatively afterwards. Gas chromatographic analysis showed that the CO_2-extract contained more flavouring substances than an ordinary alcohol extract of the same starting

Table 29. Chemical composition of lemon peel oils [31]

	CO_2-Extract (%)	Cold pressed oil (%)
Monoterpene hydrocarbons	92.4	95.0
Limonene	62.9	66.6
Neral	0.3	1.2
Geranial	0.2	1.15
α-Terpineol	1.2	0.25
Citronellol	—	0.4
Nerol	0.8	0.01
Geraniol	1.3	0.03
Neryl acetate	0.45	0.4
Geranyl acetate	0.45	0.35

material. Ice cream mixtures were treated with each vanilla extract for testing the sensory quality. The ice flavoured with the CO_2-extract was given clear preference in the taste test.

Another example of extraction of flavouring substances is the essential oil from citrus peel. These oils are generally obtained by cold pressing of the fruit peel. The dissolved carotinoids impart intense colour to the oils and the oils also contain several percent of non-volatile material, including cuticular wax; considerable amounts of essential oil remain also in the residue from pressing.

A yield of 0.9% of essential oil was obtained from extracting lemon peel with supercritical carbon dioxide at 300 bar and 40 °C, as described by Calame and Steiner [31]; the oil had a different composition from that derived by cold pressing. The analytical data of both oils are given in Table 29. The principal differences were in the contents of aldehydes and alcohols. The CO_2-extracted oil contained less citral (neral, geranial) but more nerol, geraniol and α-terpineol.

Accompanying matter of low volatility was also co-extracted under conditions of 300 bar and 40 °C. Knowing this, it is better to choose the extraction conditions judiciously so as first to extract selectively the essential oil and then, if desired, the colouring matter and other lipophilic constituents in a second extraction stage. Air-dried, freshly powdered orange peel was extracted with supercritical carbon dioxide at 90 bar and 50 °C to obtain the essential oil, which was precipitated by changing the conditions to 30 bar and 0 °C [171]. After separating water which had been extracted at the same time, this gave a clear, completely colourless essential oil of very good sensory properties and in 5% yield. Carotinoids, waxes and other difficultly volatile, lipophilic compounds were then extracted at 800 bar and 60 °C. This extract (in 1.3% yield) had a waxy consistency and was coloured deep red.

IV.3.e Extraction of Fragrances

High grade natural fragrances have their established position in the perfume and cosmetic industries. Efforts have always been made to isolate and concentrate the scented compounds of nature; great care is needed with sensitive essential oils. Extraction with dense carbon dioxide can be considered here as a gentle procedure for isolating high grade perfume oils of a fragrancy resembling that of the original material.

This is especially clear in the isolation of the fragrance substances from lilac flowers. Oils obtained with classical methods usually lack the typical lilac nuance.

Table 30. Composition of a CO_2-extract of lilac flowers [31]

Benzyl benzoate	32.0%
Elemicin	14.6%
Lilac alcohols	9.8%
Benzyl alcohol	7.3%
Phytol	7.8%
Cinnamyl alcohol	4.1%
Hexadecanol	3.2%

On the other hand dense carbon dioxide at 90 bar and 34 °C yielded an extract of 0.0024% strength which possessed completely the characteristic scent of lilac [31]. This extract was almost solid but by removing the flower waxes (ca. 90% of the extract) could be worked up into an absolute, the composition of which is given in Table 30.

The use of steam distillation to isolate the essential oil from lavender flowers leads to partial hydrolysis of linalyl acetate, one of the chief components of the oil, which contributes to its value; the ester content is thereby reduced. Extraction with liquid carbon dioxide gives lavender oils which furnish the typical scent of the blossoms [172]. The essential oils obtained by extraction contained 30–32% linalool and 47–49% linalyl acetate. The influence of the extraction temperature and time on the yield and the content of essential oil in the extract were also discussed in this investigation.

Moyler pointed out the advantages of extraction of perfume oils with liquid carbon dioxide [33]. He elucidated the best conditions for high pressure extraction to obtain high grade perfume components, based on the examples of juniper berries, ginger root and hops. The extracts with liquid carbon dioxide at low temperatures, between 0 and 10 °C and pressures up to 80 bar were compared with classical steam distillates or solvent extracts; the advantages of the CO_2-extracts were especially evident. Freshly distilled essential oils usually have a "still note" ("off odour") which results from products of decomposition of oil components or constituents of the original material. This "still note" of distilled oils disappears slowly during storage, i.e. when there is a chance of the product maturing. A long storage period is inconvenient, however, and is unnecessarily expensive. CO_2-extracts have no "still note" so that these disadvantages are not encountered.

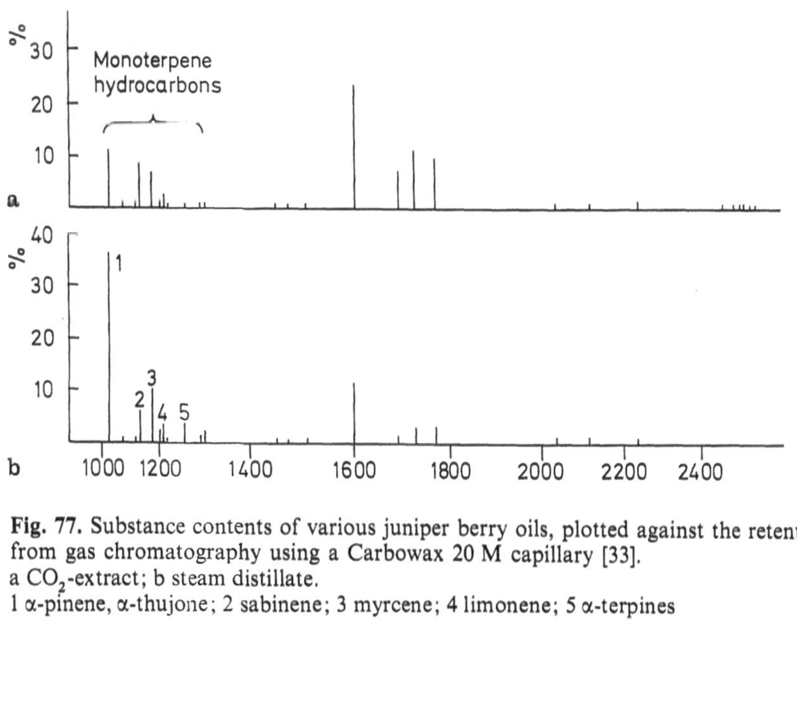

Fig. 77. Substance contents of various juniper berry oils, plotted against the retention indices from gas chromatography using a Carbowax 20 M capillary [33].
a CO_2-extract; b steam distillate.
1 α-pinene, α-thujone; 2 sabinene; 3 myrcene; 4 limonene; 5 α-terpines

There are fewer monoterpene hydrocarbons in the CO_2-extracts than in the steam distillates. An impressive example of this is shown in Fig. 77. A CO_2-extracted juniper berry oil contained only 32% monoterpene hydrocarbons but the steam distillate from the same starting material had 69%. The lower content of hydrocarbons yields a stronger odour and, above all, advantages for dissolving the product in aqueous alcoholic solutions.

According to Moyler, CO_2 extracts have a more sharply defined top note, due principally to the presence in the extract of some volatile components which are missing in the steam distillate. This is proved by the analytical comparison of various hop oils. CO_2-extracts show also more back note. Extraction with "cold" liquid carbon dioxide takes out resinous fractions of greater volatility as well as the essential oil; this gives the extract more fixing properties and a more complete odour impression without impairing the solubility of the product. Extracts using classical solvents likewise have a better fixation of scent than steam distillates. The solvents show poor selectivity, however, so that the extracts contain undesired accompanying matter, such as higher molecular resins, fats and waxes, especially in the preparation of extrait perfumes and scents, which makes then hard to handle.

As a result of their odour, some members of the valerian plant family (Valerianaceae), especially the genera *Valeriana* and *Nardostachys*, are of special importance. Thus, for example, the essential oils from *Valeriana celtica L.* (spike oil) and *Nardostachys jatamansi DC.* (Indian narde oil) are of interest in perfumery and find use in protective cosmetics. The principal constituents of these oils are sesquiterpene hydrocarbons and their oxygen-containing derivatives which are less volatile and also sometimes thermolabile. Classical steam distillation yields only oils of inferior quality from these starting materials. Extraction with dense carbon dioxide enables the genuine original essential oil to be obtained in good yield and excellent quality, at low procedure temperature [173].

A yield of 2.1% of essential oil was obtained selectively in the extraction of the Indian spikenard with supercritical carbon dioxide at 90 bar and 40 °C. About 92% of the extract was volatile in steam. Quantitative extraction of the essential oil demanded a relative solvent throughput of only 4 g CO_2/g drug sample, as a result of the good solubility under these conditions. The main constituents of the CO_2-extract were the sesquiterpene ketone, valeranone, and the sesquiterpene alcohol, nardol, which together make up 54% of the oil. Using a two-stage separation the oxygen-containing compounds could be enriched in the first separator and the sesquiterpene hydrocarbons in the second. The major part (81%) of the extract, dissolved at 90 bar and 40 °C, was brought to separation in the first separator at 90 bar and 60 °C. Only 3% hydrocarbons could be detected in this fraction. The remainder of the extract was recovered in the second stage at 30 bar and 0 °C and contained over 50% of the sesquiterpene hydrocarbons present in the drug.

The essential oil was extracted from the real spike in a way analogous to that used to obtain oil from the Indian spikenard because these are closely related root drugs which, in addition, contain essential oils of similar composition.

The spike oil was obtained by extracting the finely powdered drug with supercritical carbon dioxide at 90 bar and 40 °C and then precipitating the extract in one stage at 30 bar and 0 °C. This yielded 3.4% of a golden yellow oil. It is noteworthy that 6 h steam distillation of the same drug material gave only 1.6% of essential oil. The

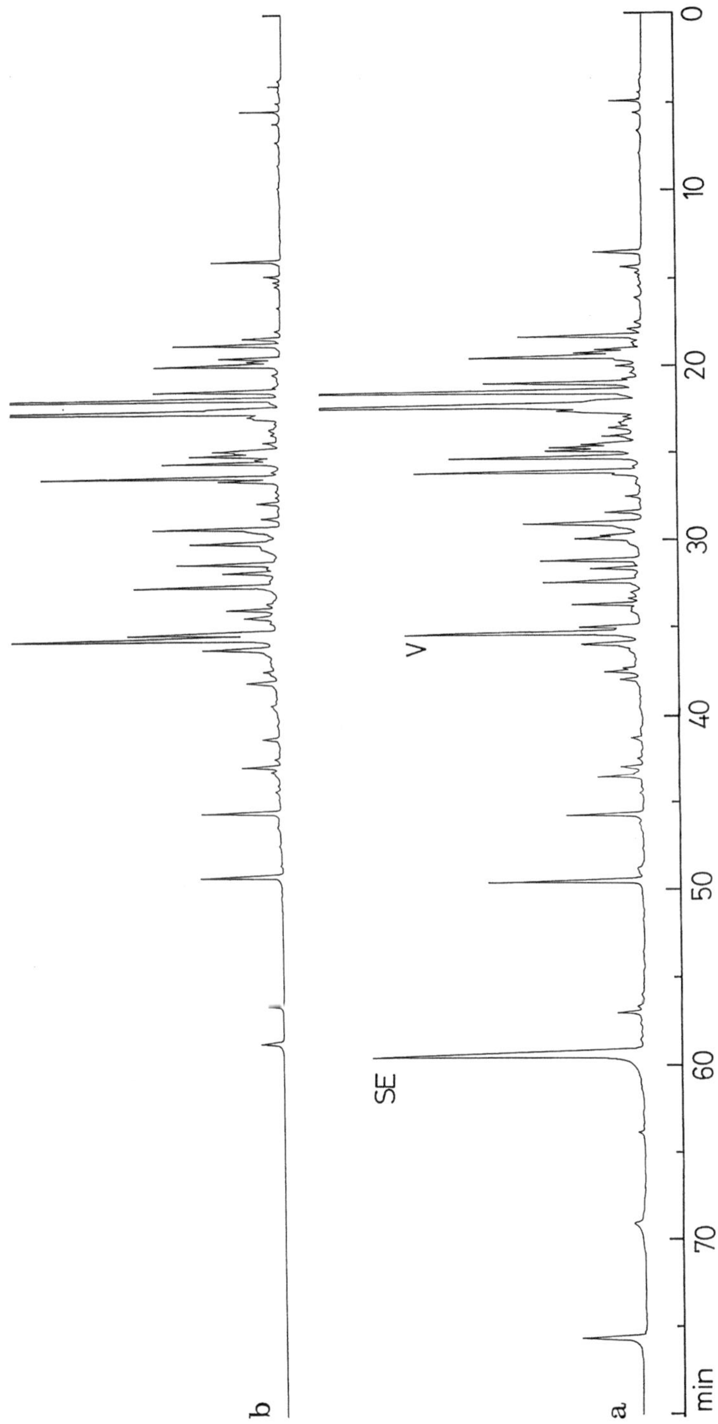

Fig. 78. Capillary gas chromatogram of a CO_2-extract (a) and a steam distillate (b) from *Valeriana celtica* L. [173].
V valeranone; SE sesquiterpene ester

CO_2-extract received a distinctly better sensory assessment than the steam distillate. The quality of the spike oil was further improved by extracting fresh plant material with dense carbon dioxide.

Analytical comparison of the various essential oils from real spike, using capillary gas chromatography, shows well-defined differences, which perhaps explain the vastly superior quality of the CO_2-extract over that of the steam distillate. The gas chromatograms of the CO_2-extract (a) and the steam distillate (b) are in Fig. 78. The oils deviate markedly in the longer retention time region, from about 45 min. Thus more poorly volatile components of the essential oil of the CO_2-extract are detected in higher concentrations at retention times of 59 and 76 min; these are present in the steam distillate (chromatogram b) only in small amounts or not at all. They are probably sesquiterpene esters.

IV.3.f Prior Separation of Vegetable Waxes

Long-chain n-alkanes are widespread in the plant world as leaf or cuticular waxes. They have the function of a protective coating and interfere during the extraction of lipophilic scented- and active-components from flower, herb and leaf materials. They confer mostly a waxy consistency on the extracts although the active components would be expected to give a liquid extract. The wax content of scent extracts which are destined for use in the perfume domain leads to partial insolubility in alcoholic

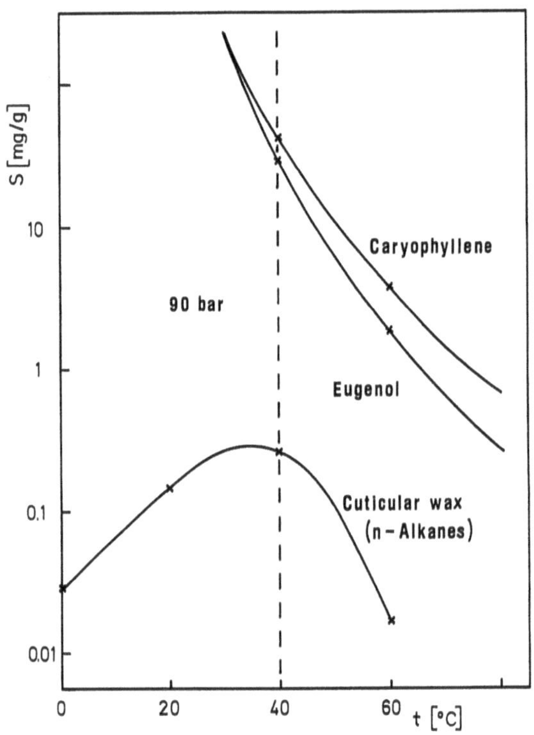

Fig. 79. Solubility isobars of essential oil components and cuticular alkanes in dense carbon dioxide, as a function of temperature [63]

solvents and hence to undesired turbidity of fragrant compositions. It is therefore better to remove the interfering wax components in a separate stage.

Comparative studies of the solubility behaviour of vegetable waxes, particularly of cuticular alkanes, and essential oil constituents, in dense carbon dioxide gave clear differences which can be utilised to separate efficiently the wax components from the fragrance and active components [63]. The density of the carbon dioxide does not alone decide the solubility of a condensed phase. The influence of temperature is also most important and shows large differences from substance to substance. This is seen clearly in Fig. 79. This compares the temperature-dependence of the solubility in dense CO_2 at 90 bar of the n-alkanes of wormwood herb cuticular wax and that of typical essential oil components. The solubilities of caryophyllene and eugenol, as model essential oil, increase markedly with falling temperature, as does the density of the extraction gas, until unlimited miscibility in the liquid carbon dioxide is reached. The long-chain hydrocarbons display different solubility behaviour. At 90 bar the cuticular alkanes have a solubility maximum at 40 °C. The solubility falls when temperature rises, i.e. as the gas density diminishes, or when temperature falls, i.e. with increase in gas density. The lowering of temperature from 40 to 0 °C reduces the alkane content of the dense gas phase to a tenth, for example. The conditions denoted by the dashes, 90 bar and 40 °C, have proved good for extracting essential oils. These conditions lead to solubilities as high as ca. 5 % by weight, under thermally inoffensive conditions; and fewer unwanted accompanying materials, such as fatty oils, colouring matter etc., are extracted in comparison with liquid CO_2 as solvent. Part of the vegetable waxes, of solubility about 0.3 mg/g carbon dioxide, are always co-extracted under these conditions. If the substances dissolved in the supercritical carbon dioxide are separated out, as is usual, by lowering the density, i.e., either by lowering the pressure isothermally or by raising the temperature isobarically, the solubilities of both the essential oil components and of the cuticular wax fall and

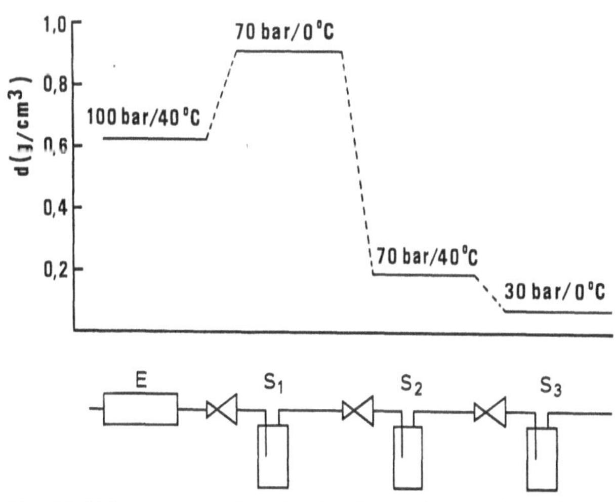

Fig. 80. Scheme portraying the fractionated separation of the CO_2-extract of common wormwood.
E extraction vessel; S1–S3 separation vessels

both classes appear in one fraction. If, however, the temperature is isobarically lowered to 0 °C, for example, a prior separation of the alkanes takes place. Under these conditions the essential oil components become increasingly soluble, so that the wax can be separated in pure form without the remaining dissolved substances.

Fig. 80 shows a tested extraction scheme with which a fractionation of this type can be performed successfully. This was described in the literature for the first time in the case of the extraction of common wormwood with supercritical carbon dioxide [153]. The cuticular wax was extracted along with the monomeric sesquiterpene lactones and the essential oil at 100 bar and 40 °C. The extract is precipitated in three stages:

In the first stage a pure white cuticular wax was obtained by lowering the temperature to 0 °C, which led to an increase in density and liquefaction of the carbon dioxide. The monomeric sesquiterpene lactones and the essential oil were obtained in the second and third fractions; they separate out as a result of stepwise lowering of density. The analysis of the cuticular wax which was recovered at 70 bar and 0 °C, showed 17% wax esters and 83% long-chain n-alkanes, of which 90% were C_{27}, C_{29} and C_{31}-hydrocarbons.

The isolation of active principles from chamomile provides a second example of the differential separation of the interfering wax material from the real active components (cf. also p. 138). The extracts obtained with supercritical carbon dioxide at, for example, 100 bar and 40 °C, are of a wax-like consistency. Use of the separation technique just described enables the cuticular wax to be removed. The wax obtained thus at 0 °C contains over 80% n-alkanes of chain-length from C_{25} to C_{31}. The final separation stage is then performed at 30 bar and 0 °C, leading to an oil of low viscosity and possessing a high concentration of fragrance and active principles. Similar results are obtained in the extraction of pyrethrum flowers (see p. 141). Here, too, the waxy nature of the CO_2-extract, due to the presence of long-chain alkanes, such as n-penta-, n-hepta- and n-nonacosane, can be evaded through use of the prior separation described. This yields wax-free extracts of much better quality and still higher content of active components.

The extraction of oak moss (mousse de chène) with supercritical carbon dioxide and the subsequent fractionation of the extract is an impressive example of the isolation of wax-free fragrances [156]. As a consequence of their high adhesive strength, extracts of oak moss are used for fixing in perfume manufacture in order to upgrade and round off the combination of perfume substances. Classical extracts with petrol ether are intensely coloured green to brown, waxy masses. These resinoids must be freed from the waxes by treatment with alcohol before use. Some of the waxes, present in large amounts, are likewise co-extracted with supercritical carbon dioxide. A prior precipitation of the extract in liquid carbon dioxide (e.g. at 0–5 °C), however, effects an almost quantitative separation of the waxes, so that at the end of the separation a red-brown, oily, strong-smelling extract accumulates.

These examples demonstrate that a judicious choice of the separation conditions enables the wax constituents to be separated selectively from the essential oils and perfume compounds. Extracts are obtained in this way which have a higher content of fragrance and active components, are better soluble in alcoholic solvents and are mostly liquid. This liquid state permits continuous withdrawal of extract in work on the preparative scale, e.g. in the extraction of flowers, herbs or leaves; if the extracts are thermolabile or contain highly volatile components (e.g. essential oils) this is not possible in any other way, e.g. by raising the temperature of the separator. An interesting fact, little noticed up to now, is that certain dissolved substances can be separated by isobaric lowering of the temperature of the extraction agent as well as by isobaric increase of its temperature.

IV.4. Further Isoprenoids

IV.4.a Extraction of Hops

The hops used in brewing have the principal purpose of imparting a bitter flavour and also aroma to the beer. These properties are derived from the resins and essential oils which the glandular hairs (lupulin) of the hop capsules contain. The soft resins comprise the value-determining components of the hop resins. They consist of a mixture of humulones (α-acids) and lupulones (β-acids); chemically they are isoprenoid compounds (cf. Fig. 81).

The typical bitter taste of beer comes through isomerisation of the α-acids during the brewing process. The aroma of the beer is largely due to the essential hop oil.

Dichloromethane is the usual solvent for classical hop extraction; it cannot be removed completely from the spongy hop extract. Dichloromethane extracts thus always contain solvent residues and are dark green to black-green masses.

A highly promising alternative to the classical extraction is to obtain the valuable hop components by extraction with dense carbon dioxide. Hop extracts are in great demand, they have high market value, and the conditions for high pressure extraction (high content of extractable constituents, high gas charging) are favourable; consequently a large number of procedure descriptions have been patented or patents have been applied for [25, 180–190]. Several industrial plants for large scale extraction of hops with carbon dioxide are already in use at the present time in the German Federal Republic, Britain, Australia and USA; in some, supercritical carbon dioxide

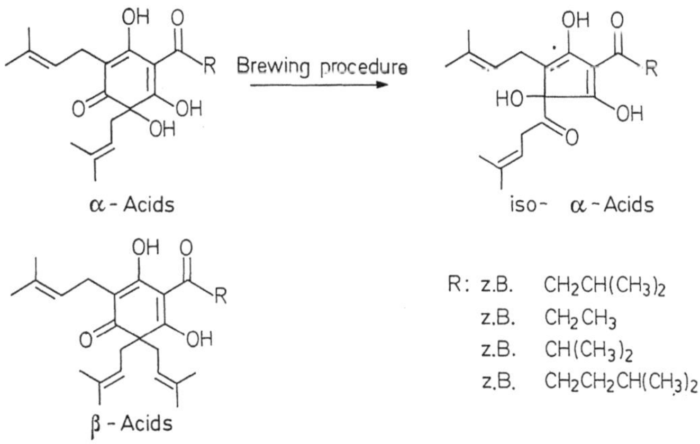

Fig. 81. Chemical structure of the soft resin components of hops

is used, in others, liquid CO_2. The supercritical gas possesses a larger dissolving power and is more flexible as far as the parameters of the procedure and the selectivity thereby obtainable are concerned. However, the supercritical process is dearer than that with subcritical, liquid CO_2 at saturation pressure on account of the higher pressure. Work with supercritical carbon dioxide is rational from the point of view of complete extraction of all the important hop constituents. If hops of reasonable price are available in adequate amounts, extraction need not be quantitative and the cheaper procedure using liquid CO_2 is advantageous.

The extraction of hops with subcritical, liquid carbon dioxide has been described by several authors [27, 191–196]. Liquid carbon dioxide has only limited solvent capacity for the hop resins [193]. As Fig. 82 shows, the temperature of the liquid CO_2 has a decisive influence on the solubility of the soft resin components.

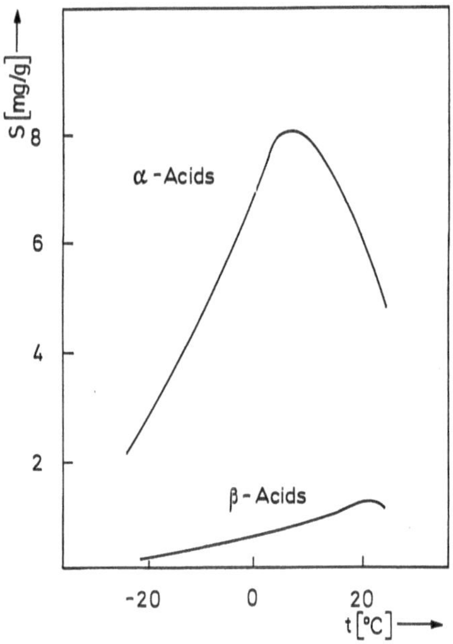

Fig. 82. Solubilities of the soft resin components of hops in liquid CO_2 at various temperatures and subcritical pressure [193]

The solubility of the α-acids attains a maximum value of 0.81% by weight at 7 °C, the value falling sharply whether temperature rises or falls. The β-acids show only about one-tenth of this solubility; these soft resin components reach a maximum solubility of 0.13% by weight at 20 °C.

Hops are extracted industrially with liquid carbon dioxide at ca. 8 °C and 60 bar; the solvent is evaporated, bringing about separation of the extract between 12 and 15 °C (pressure of ca. 50 bar) [194, 195]. Since the value-determining α-acids are only poorly soluble, a throughput of at least 50 kg CO_2 per kg crude material is necessary to reach a degree of extraction of 95% [196]. The volatile hop oil is easily soluble under these conditions and is extracted quantitatively and without danger of decomposition because of the low temperature. The low dissolving power of the liquid carbon dioxide under these conditions has a favourable influence on the quality of the extract. The hop extracts obtained consist almost entirely of soft resin components and the essential oil. They contain neither hard resins nor tannins and only a small amount of fats and waxes [194].

Wilson described the fractionation of a hop extract into an aroma oil and a soft resin fraction [196]. The volatile hop oil in the extract of hops with liquid CO_2 was removed with carbon dioxide in a counter-current column and precipitated separately. This enabled an oil fraction containing ca. 70% essential oil to be isolated. This is of interest to the beer brewer because during brewing a large amount of the aroma substances generally disappears. It is now possible to add the hop aroma directly to the beer while the soft resin fraction of low oil content, is added to the brewing charge.

A problem is created during separation of the extract in the evaporator when semi-crystalline material deposits and interferes with the heat exchange. This is prevented or mitigated by adding 5% ethanol to the evaporator [196]. The presence of ethanol in the evaporator has also another advantage. Most of the alcohol is carried down with the extract but small amounts enter the circulating gas. Alcohol concentrations of 1–2% in the liquid carbon dioxide improve noticeably the solubility of the α-acids without impairing the quality of the extract. If the alcohol concentrations are higher, say 3–4%, unwanted accompanying material, such as chlorophyll, is co-extracted.

The hop resins dissolve better in supercritical than in liquefied carbon dioxide. This means that less supercritical solvent is required and hence a shorter extraction time. The first investigations of extraction of hops with supercritical carbon dioxide were published by Vitzthum and Hubert [25, 160]. Ordinary commercial hop pellets were extracted at 40–60 °C and pressures up to 350 bar to yield olive green, spongy extracts with a strong hop aroma. Some analytical data on the hop extraction are presented in Table 31. The α-acids, the hop bitter principles which give hops their value, were almost quantitatively extracted (degree of extraction of 99%). Unfortunately there is information about neither the extraction of the hop oil nor the content of flavouring substances in the extract.

Krüger compared the dissolving power of carbon dioxide in the liquid and the supercritical states [197]. The investigations were carried out an a hop extract in a phase equilibrium apparatus. Some of the results are given in Table 32.

Table 31. Analytical data on the extraction of hops with supercritical carbon dioxide [160]

	Hop pellets		CO_2-extract	Degree of extraction	Commercial extract
	before	after			
	CO_2-extraction				
Water content (%)	6.0	5.4	7.0		8.0
Total resins (%)	30.3	4.3	90.0	89.9	88.5
Soft resins (%)	26.6	1.3	84.8	96.5	82.0
α-Acids (%)	12.6	0.2	41.2	98.9	39.5
β-Acids (%)	14.0	1.1	43.6	94.4	42.5
Hard resins (%)	3.7	3.0	5.2		6.5

Table 32. Solubility of hop extract in dense CO_2 at 28° and 80 °C [197]

Pressure (bar)	t = 28 °C % by weight	t = 80 °C % by weight
80	0.52	0.026
100	0.67	0.06
150	1.1	0.26
200	1.55	0.74
300	2.5	3.2
400	3.45	9.4

The figures show that better results were obtained with the liquid carbon dioxide in the pressure region below 200 bar. At higher pressures the solubility rises sharply with the temperature. Thus, at 400 bar and 80 °C, 94 mg hop extract dissolved per g CO_2. To attain such a solubility at 28 °C, a pressure of ca. 1000 bar would be needed.

To dissolve good amounts of the hop resins in the extraction with supercritical carbon dioxide, relatively high pressures of up to about 350 bar are needed. Kurzhals and Hubert described a procedure to reduce the expense of the apparatus by carrying out economical extraction at lower pressures [184]. They used a mixture of carbon dioxide and an additional component, e.g. propane, butane or acetone, as solvent. The extraction parameters of pressure and temperature were adjusted so that the mixture of the two components was liquid. Hops were extracted at 82.7 bar and 35 °C, using a liquid mixture of 94 mol% CO_2 and 6 mol% of n-butane. This yielded a light green extract after decompression and evaporation of the mixed extraction agent. Its composition is quoted in Table 33.

A simple way of fractionating the extract was described also in this work. When the pressure of the solvent mixture was reduced to 50 bar, a liquid phase, rich in butane and containing both dissolved and undissolved extract components, and a gas phase, rich in CO_2 but containing no extract, were yielded. The hop resin components were insoluble in the former, liquid phase whereas in particular the hop oil was in solution. After separating the insoluble matter the liquid butane was evaporated, enabling the previously dissolved hop constituents to be separated.

The valuable hop bitter principles, the α-acids, are insoluble in water, hence in

Table 33. Analytical data of hop extraction with a liquid mixture of carbon dioxide and butane [184]

	Initial hop material	Hop residue	Extract
Water (%)	10.5	12.3	3.4
Total resins (%)	23.0	4.1	91.9
Soft resins (%)	20.8	2.0	89.4
α-Acids (%)	10.1	<0.1	46.3
β-Acids (%)	10.7	1.9	41.5
Hard resins (%)	2.2	2.3	2.1
Hop oil (%)	1.0	<0.1	4.9
Tannins (%)	4.3	5.3	0

the beer wort; only after boiling the wort are they converted into the soluble form (iso-α-acids). During preparation of the beer, especially during fermentation, a part of the bitter principles separates out and is thus lost. For this reason there is interest in introducing the bitter quality after fermentation (post-fermentation hop treatment). A condition for this, however, is that the bitter components are in a form that is soluble in the beer.

In this connection Krüger described an interesting procedure of hop extraction with supercritical CO_2 [198]. Alone of the components of hop extracts the humulones (α-acids) possess the property of forming salts with calcium- and magnesium-containing minerals, e.g. bentonite. Through this conversion of the α-acids into the corresponding humulates they evade the extraction into carbon dioxide, making possible their separation from the remaining constituents of the extract. The humulates bound to the bentonite are isomerised by heat treatment (100 °C). This procedure opens up the possibility of combining the extraction of the hop constituents with the isomerisation of the α-acids. Hops are extracted with supercritical carbon dioxide at 50 °C and 300 bar. The miscella is then passed through a reaction vessel filled with bentonite which is kept at 100 °C. The other hop components which are soluble in supercritical CO_2 traverse unscathed the bentonite and are subsequently separated out at 50 bar and 20 °C; the α-acids are retained in the bentonite as humulates and are immediately isomerised there. The bentonite + iso-α-acid salts form a free-flowing powder which can be used immediately for post-fermentation hop treatment; it also contributes to the stability of the beer.

Vollbrecht explained in a publication the advantages of extracting hops with supercritical carbon dioxide and gave information about the quality of the extracts obtained [199]. The CO_2-extracts contain no pesticides which are used in hop plantations at the present time. Even a hop sample which contained the unusally large amount of 527 ppm DTC (dithiocarbamate) yielded an extract with no more than 6.7 ppm of the DTC. The unaggressive temperature of 40 °C and the protective gas atmosphere of the carbon dioxide combine to hinder both oxidation and thermal formation of artefacts from the hop constituents during extraction. This was shown by gas chromatographic analysis of the aroma oil from hop pellets and the CO_2-extract; it is supported also by the fact that the α-acids are present in the extract in their natural state and have not been even partially isomerised. A protective gas atmosphere can be used during drawing off as well as during extraction, so that the extracts are stable over a long period and can be stored for a considerable time without suffering any loss of quality at all.

Extracts of hops with CO_2 have to compete on the market with the classical solvent extracts. Gehrig considered the CO_2-extracts more from the economic side [200]. Compared to the classical procedures, extraction with liquid carbon dioxide at the saturation pressure, or with supercritical CO_2 at 40–60 °C and pressures up to 350 bar, gives smaller yields, and longer times of extraction are required. Under mild conditions carbon dioxide extracts the soft resins and the essential oil but not the so-called hard resins. The author thus proposes a fractionating extraction. First, the soft resins and hop oil are taken out under mild conditions, using liquid or supercritical carbon dioxide. Other hop constituents (e.g. hard resins) are then extracted at higher temperature (120 °C) and a pressure of 350 bar; this increases the yield and all bitter components of technological interest in brewing are obtained.

The procedure described below, for which a patent has been applied, can likewise be regarded from the point of view of augmentation of yield. Müller worked with carbon dioxide as solvent, at subcritical temperatures and supercritical pressures up to a maximum of 500 bar to obtain total resins and flavour components from hops [185]. It was claimed possible under these conditions to extract virtually all the hard resin components in addition to the essential oil and the soft resins.

IV.4.b Steroids

Solubility Behaviour

The class of the steroids displays impressively the close relation between the chemical structure of a compound and its solubility in dense gases. Study of the solubilities of a large number of different steroids in supercritical CO_2 at 40 °C with the aid of high pressure microextraction in direct coupling with thin-layer chromatography (FE-TLC, Chapter III.2) shows, in agreement with the rules of thumb already expressed above (Chapter IV.1), the relationship between solubility and the type and number of substituents in the tetracyclic steroid structure (Fig. 83).

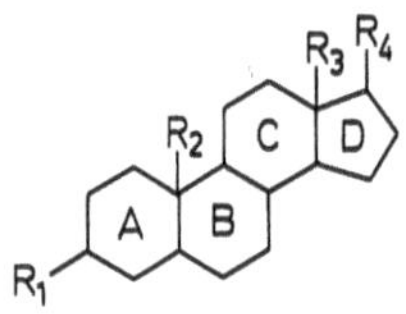

Fig. 83. Skeletal structure of the steroids

Table 34. Relation between the beginning of extraction of steroids and the substituents in the tetracyclic structure (FE-TLC); temperature: 40 °C; throughput: 2 g carbon dioxide [4]

Beginning of extraction (bar)	Functional groups	Example
80	-OH	sitosterol
	2 (=O)	progesterone
90	-OH, =O	androsterone
	-OH, spiroketal ring	diosgenin
	-OH (phenolic), $-C\genfrac{}{}{0pt}{}{C\equiv C}{OH}$	ethinyloestradiol
100	-OH (phenolic), -OH	oestradiol
	-OH (phenolic), =O	oestrone
120	2 (-OH)	pregnanediol
150	3 (-OH)	pregnanetriol
	2 (-OH), lactone ring	digitoxigenin
200	2 (-OH), -COOH	deoxycholic acid
insoluble	3 (-OH), 2 (=O)	hydrocortisone
	3 (-OH), -COOH	cholic acid
	more than 3 (-OH)	glycosides

As Stahl and Glatz [4] showed, steroids with one hydroxyl group (i.e. all sterols) are extracted at pressures from 80 bar upwards, those with 2 hydroxyl groups (e.g. pregnanediol) from 120 bar, and those with 3 hydroxyl groups (e.g. pregnanetriol) above 150 bar (Table 34). Non-polar substituents, such as carbonyl groups, reduce the solubility to a lesser extent. Thus androgens, the male sex hormones, which contain one hydroxyl and one carbonyl group, are extracted at 90 bar. The influence of spiroketal rings in the sapogenins, e.g. diosgenin and solasodin, is similarly small; these dissolve also at 90 bar. The carboxyl group has the strongest effect in lowering solubility. Only deoxycholic acid of the bile acids can be dissolved, at pressures above 200 bar; all compounds with more than three functional groups are practically insoluble.

The subdivision portrayed in Table 34 has been confirmed also by quantitative solubility measurements of some selected steroids in the pressure range up to 200 bar. At 200 bar/40 °C the sterols showed the highest solubilities, of 0.5 mg/g CO_2. A second, less polar substituent, such as the lactone ring in diosgenin and solasodin, sufficed to lower the solubility by a factor of five, a second hydroxyl group by a factor of ten. Compounds with three free hydroxyl groups had solubilities of only 1 μg/g CO_2.

High pressures and temperatures together are required to dissolve in CO_2 practicable amounts of steroids of 1 % by weight and more. This is demonstrated by quantitative measurements of sitosterol at pressures up to 1100 bar and temperatures up to 80 °C (Fig. 84).

A maximum solubility of only 0.3 mg/g gas is shown by sitosterol in liquid carbon dioxide, a low value which is in addition scarcely influenced by pressure. At higher temperatures the influence of increasing pressure becomes more significant and the solubility isotherms become steeper. At 600 bar and 80 °C more than 10 mg of substance dissolve per g gas, already 40 times the maximum amount attainable in liquid CO_2. An intersection of the isotherms is found in the pressure region of 200 bar. The solubility falls with increasing temperature only at lower pressures, otherwise it rises. If the solubility of the substance is referred to the gas density, an increase of temperature always leads to higher solubilities.

The rise in solubility as pressure or gas density is increased does not continue without limit. Solubility maxima are reached at gas densities between 1.03 and 1.06 g/cm^3; their position is displaced slightly towards lower density values as temperature is increased. Further increase in pressure leads to a lowering of solubility. The solubility behaviour of the steroids in dense carbon dioxide therefore shows, on the whole, close similarity to that of the triglycerides and wax esters [201].

Nitrogen, dinitrogen monoxide, ethane, ethylene, sulphur hexafluoride and trifluoromethane have been studied as well as carbon dioxide as potential solvents for steroids [3]. Less than 10 ppb of steroids dissolve in nitrogen and sulphur hexafluoride under conditions of 300 bar and 40 °C, i.e. they are virtually insoluble in these gases. Solubilities could be measured in the other gases showing that the substituent influence follows the same rule. Polar steroids were less soluble than non-polar. However, there were big differences in the absolute solubility values. The solubilities in trifluoromethane in both liquid (20 °C) and supercritical state (40 °C) were appreciably lower than in carbon dioxide. Ethylene and carbon dioxide showed comparable dissolving powers, dinitrogen monoxide and in particular ethane considerably higher powers. At 200 bar and 40 °C, ethane dissolved about ten times as

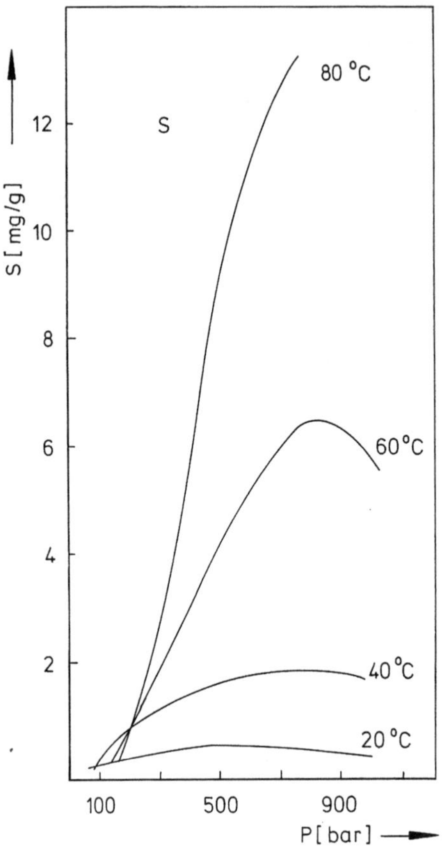

Fig. 84. Solubility isotherms of sitosterol in liquid and supercritical carbon dioxide as a function of pressure [201]

much steroid as carbon dioxide. It is possible, however, to establish any direct correlation between the differing dissolving capacities for steroids of these gases and their physical properties, such as density, dielectric constant or dipole moment.

Saturating the dense carbon dioxide with water lowered the solubility of the steroids. Addition of organic solvents as entrainers, e.g. methanol or hexane in amounts of less than 5% increased the solubility of sitosterol only little. On the other hand, concentrations of up to 20% by weight of solvents in dense carbon dioxide led to a sharp increase but the selectivity and pressure-dependence of the dissolving power of pure CO_2 were lost.

Extraction of Tall Products

Tall products accumulate as residues from the manufacture of paper by the sulphate-pulp process. Lignin and hemicellulose are dissolved out of wood chips by boiling with sulphate-containing alkalies under pressure and high temperature, leaving the solid cellulose as desired product. After evaporating the water, brown, half-solid tall soap is obtained from the alkali solution, consisting mainly of the alkali salts of resin and fatty acids but containing also other lipophilic wood constituents such as sterols. The liquid free acids, the tall oil, are obtained from this by treatment with sulphuric acid. Using vacuum distillation this oil can be separated roughly into resin and fatty acid fractions and the non-distillable residue, the tall pitch, in which the steroids are enriched.

Over a million tonnes of tall oil are acquired globally per year (1980 figures). The resin and fatty acids obtained from it are coveted industrial starting materials; the remaining tall pitch, over 200,000 tonnes, is a waste product although it contains up to 15% phytosterols and is thus an ideal starting material for obtaining these which are needed in large amounts as a basic source for steroid hormones (contraceptive agents, corticoid medicaments). Steroid concentration by classical refining methods requires numerous steps on account of the poor solvent selectivity; in addition considerable expenditure of apparatus, time and energy is involved, especially for solvent recovery. Unavoidable solvent losses increase expenditure and pollute the environment. Stahl and Glatz studied the extraction of tall oil, tall pitch and saponified tall pitch with dense gases as an alternative method for enriching the steroids [3].

The attempts to separate crude tall oil, which contains up to 3.5% phytosterols partly in free state, partly esterified, were made with supercritical carbon dioxide at 80 °C and pressures from 250 to 600 bar. The extraction at 250 bar, which gave a maximum substance concentration in the dense gas of 0.86% by weight, separated the starting material in that the fatty acids were first removed and then the resin acids. The selectivity of the solvent is largely forfeited at 530 bar with a concentration in the gas of 4.5% by weight. However, fractionated separation yielded a rough division of the extract; first a fraction rich in resin acids was obtained and, in the second stage, the more soluble fatty acids were precipitated. The steroids could not be isolated because the free phytosterols show a solubility behaviour similar to that of the resin acids, and the sterol esters, as less soluble compounds of high molecular weight are enriched in the residue which resembles classical tall pitch in its composition. Thus dense gas extraction of tall oil opens up new perspectives only as far as replacement of the ordinary vacuum distillation for separating the acid tall oil constituents is concerned. This is shown especially by experiments in a fractionating column in which a good separation of the resin and fatty acids was possible but the free sterols could not be separated from the resin acids.

Various gases have been tried at pressures of 420–550 bar and temperatures of 80–110 °C for extracting tall pitch. The high temperatures are necessary to convert the material into a sufficiently thinly fluid state to be easily penetrated by the gas. Tall pitch contains over 60% of resin and fatty acids, largely polymers and of high molecular weight; about a half is in the free form, a half as esters. The residual, unsaponifiable portion consists of hydrocarbons, fatty and resin alcohols and phytosterols [202]. The great majority of the sterols in tall pitch (6–15%) are, however, in esterified form [203].

Fig. 85 shows the amounts extractable from tall pitch using various solvents and expressed as a function of the ratio solvent/tall pitch. Considerable amounts of gas are needed for satisfactory extraction in every case when carbon dioxide or ethylene is used. Dinitrogen monoxide and ethane have much better dissolving powers. Ethane dissolves as much as 52% of the tall pitch with nearly 85% of the sterol ester content when the ratio of gas to sample is 10:1. In this way liquid, amber-coloured extracts containing 19–23% of sterols are obtained from a black crude material, half-solid at room temperature with 10.5% sterol content. Comparison with classical

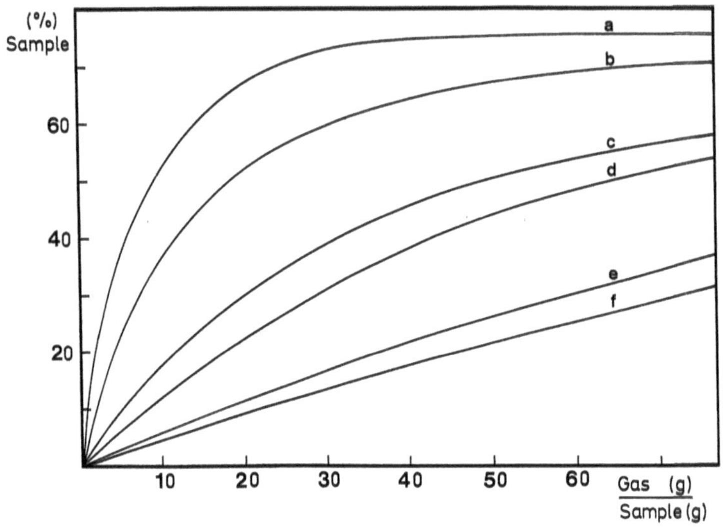

Fig. 85. Yields in extraction of tall pitch as a function of the relative solvent throuhput [3].
a: ethane 480 bar, 110 °C
b: dinitrogen monoxide 480 bar, 110 °C
c: ethylene 480 bar, 110 °C
d: carbon dioxide 550 bar, 110 °C
e: carbon dioxide 480 bar, 110 °C
f: carbon dioxide 440 bar, 90 °C

extracts of tall pitch, made with methanol, toluene or hexane, shows clearly the superior solvent selectivity of supercritical gases. The enrichment of sterols in dense gas extracts can be improved further with a two-step separation. This can be performed by lowering pressure isothermally or lowering temperature isobarically. When the extract is partitioned between the first and the second separator in the ratio of 1:1 to 1:2, the more poorly soluble fractions contain up to 30% sterols, corresponding to 60% sterol esters. The pure sterols can be easily derived from these by saponification in the usual way.

The extraction of tall pitch with dense gases can also be carried out continuously through counter-current extraction in a packed column. The material to be extracted was introduced into the uppermost third of the column, the dense gas led into the lowest third, in an appropriate laboratory installation. The bottom product could be taken off continuously through a cut-off valve; the dissolved top product was recovered in the separator part of the plant and likewise taken off with a cut-off valve. A good mass throughput was possible using ethane, which ensured a large difference in density and rapid separation of the two phases.

The same apparatus is employed for continuous extraction of saponified tall pitch. This is solid and cannot be adequately liquefied by only increasing the temperature; the viscosity has to be reduced by adding a classical solvent. Methanol is especially advantageous because it dissolves easily the alkali hydroxide required for saponification. The mixture from this hydrolysis can then be directly submitted to extraction. Suitable values of the experimental parameters are in Table 35.

Table 35. Experimental parameters of the continuous extraction of saponified tall pitch with dense ethane [3]

Methanol content of the hydrolysis mixture:		15%
Extraction conditions:		500 bar, 110 °C
Mass ratio ethane: saponification mixture		20:1
Separation conditions:	1. Step:	130 bar, 110 °C
	2. Step:	20–40 bar, 80 °C
	3. Step:	20–40 bar, 5–20 °C

About 25% of the saponified tall pitch and over 85% of the sterols contained in it are extracted under these conditions. The extract in the first separator then contains ca. 50% free sterols, the second fraction predominantly the more soluble accompanying substances, and the third fraction the alcohol. Every component can be continuously fed into, and removed from, the process. The crude sterol mixtures obtained are of at least as high quality as those from conventional extraction. There is, however, a noticeably clear reduction in the number of procedure steps when the high pressure method for obtaining sterols from tall products is used instead of the classical methods.

IV.4.c Carotinoids

The carotinoids form an extensive substance class of yellow and red natural colouring materials. Chemically they are highly unsaturated aliphatic or aliphatic-alicyclic hydrocarbons and their oxidation products. Most carotinoids are C_{40}-compounds and belong to the group of tetraterpenes. Examples of compounds of shorter hydrocarbon chains are the apocarotinoids. Some of these vegetable colouring materials are important in the physiology of nutrition, e.g. β-carotene with its provitamin A character. The carotinoids are important in practice above all for colouring foodstuffs and in the pharmaceutical and cosmetic industries. They are also used in large amounts as additives to animal feeds, for example to yield more strongly coloured yolks of hens' eggs.

The first investigations of extraction of carotinoids with supercritical gases were published by Stahl and Schilz [2]. Using microextraction on-line coupled with thin-layer chromatography it was possible to detect the extraction of β-carotene at 40 °C at pressures above 100 bar. These first results stimulated further work on the extraction behaviour of the carotinoids with dense gases [204]. With the microextraction method, carrots, powdered algae, tomato pulp and paprika were extracted with supercritical carbon dioxide at 40 °C and various pressures and the extracts studied with the help of thin-layer chromatography. The results showed that, in addition to the carotinoids, such as β-carotene from carrots and algae, lycopin from tomatoes and capsanthin from paprikas, other lipophilic plant components were extracted. In most cases the carotinoid fractions could be obtained separately from the total extract through differences in solubility.

Supercritical trifluoromethane was compared with supercritical carbon dioxide as an extraction agent for carotinoids and found to possess similar extractive power. It was observed that carotinoids such as β-carotene, esters of apocarotenic acid (C_{32}) and apocarotenal (C_{30}) could be extracted with the trifluoromethane already

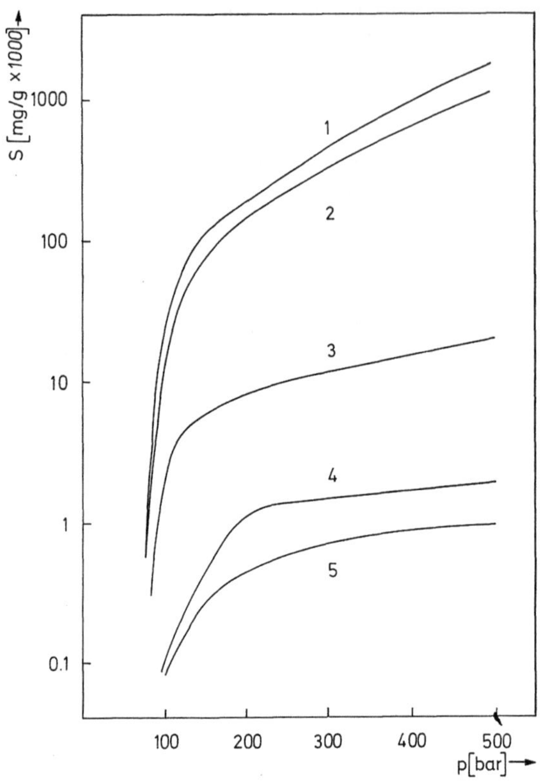

Fig. 86. Solubility isotherms of some carotinoids in supercritical carbon dioxide at 40 °C [204].
1 apocarotenate ester (C_{32}); 2 apocarotenal (C_{30}); 3 β-carotene; 4 canthaxanthin; 5 zeaxanthin

at above 90 bar at 40 °C, and canthaxanthin, a carotene diketone, was shown to be extracted above 200 bar.

Fig. 86 presents carotinoid solubilities in supercritical carbon dioxide. All pure carotinoids are difficultly soluble ($<0.03\%$ by weight at 300 bar/40 °C) even though the compounds of shorter chain length, such as apocarotinoids (e.g. C_{30}) are about 30 times more soluble than the C_{40}-carotinoids (e.g. β-carotene). The solubility is reduced still further by the presence of polar functional groups, such as 2 carbonyl groups in canthaxanthin or 2 hydroxyl groups in zeaxanthin. However, solubility can be appreciably increased by simultaneously raising pressure and temperature. Moreover, carotinoids are extracted in larger amounts from plant materials as a result of the entraining action of lipids (fatty oils, waxes). Thus a solubility of 0.08% by weight of the carotinoid fraction was obtained in a carbon dioxide extraction at 320 bar and 40 °C of paprika, from which the flavour components had been previously removed. This value could be derived from data which were obtained in connection with the paprika extraction mentioned [169, 39].

Entrainers added to the extraction gas increased solubility still more sharply [36]. Various organic solvents were suggested for this, lower alcohols having the best effect. Thus, for example, a 0.18% by weight solution of β-carotene was yielded in carbon dioxide, saturated with ethanol, at 200 bar and 40 °C. This was 200 times the concentration in pure carbon dioxide under the same conditions.

Coenen and co-workers published experimental results on the preparative extraction of paprika fruits [131, 169, 39]. They described the fractional extraction of capsaicin-containing "hot" paprika, the isolation of a pure colouring oleoresin material from special sweet paprika, free of pungent elements and the work up of conventionally obtained paprika extract. The authors proposed a two-stage extraction of "hot" paprika in order to separate the pungent substances from the carotinoid fraction. The flavouring components were first extracted with supercritical carbon dioxide at 120 bar and 40 °C. Since capsaicin, the pungent principle of the paprika, is only comparatively poorly soluble under these conditions, large amounts of solvent were needed for exhaustive extraction, for instance, 130 kg CO_2 per kg paprika. The spongy, water-containing extract obtained in this stage (yield of ca. 15 %) was orange-yellow and of extremely pungent taste. The natural vegetable colouring matter of the paprikas was then extracted at 320 bar and 40 °C, a relative throughput of ca. 50 kg CO_2/kg being necessary to obtain the carotinoids. This second stage of the extraction gave a dark red, liquid concentrate of colouring materials (yield of 2.5 %), sensibly free of pungent substances. The carotinoids and additional lipophilic constituents, especially fatty oil, were obtained by extracting the sweet, red, capsaicin-free paprikas at 300 bar and 40 °C. About 25 g of a water-containing concentrate of colouring material was derived from 300 g paprika with a carbon dioxide amount of 58 kg. The residues from paprika extraction with CO_2 are only pale yellow in colour, almost devoid of smell and taste and are highly suitable as fodder because of their high content of protein and carbohydrate.

Gases other than carbon dioxide are suitable also for extracting the carotinoids from paprika. Better yields and qualities of product, and lower times of extraction characterised extraction with ethane or ethylene [169]. The carotinoids are rather more soluble in these gases, so that a lower gas throughput is required and hence the extraction time can be shortened. A concentrate of colouring matter, largely free of water, could be obtained by extracting with supercritical ethane at 250 bar/45 °C and separating at 46 bar/45 °C.

IV.5. Alkaloids

Alkaloids are nitrogen-containing bases from plants and usually possess a complex chemical structure. With few exceptions they are weak bases and in the plant occur partly combined with vegetable acids or tannins. They can be extracted completely with lipophilic solvents only when present as free bases. This generally means that the starting material must first be made alkaline.

IV.5.a Extraction of Alkaloids Used in Medicine

Stahl and Willing investigated the extractability of many different alkaloids using dense gases at temperatures near the critical values and pressures from 50 to 200 bar [205, 206]. Extraction experiments on the micro scale permitted a comparison of the dissolving powers of the gases, trifluoromethane, bromotrifluoromethane, sulphur hexafluoride, ethylene, carbon dioxide and dinitrogen monoxide. It was found that SF_6 and C_2H_4 were unsuitable for extraction of alkaloids. Only a few compounds,

Table 36. Solubilities of opium alkaloids in dense gases [6]

	p (bar), T (°C)	Solubility in mg/g		
		CO_2	N_2O	CHF_3
Codeine	80/23	0.41	0.43	0.25
	150/40	0.9	1.0	0.56
Thebaine	100/18	0.2		0.73
	150/40	0.2		0.66
Papaverine	100/20	0.09		1.26
	150/40	0.04		0.71
Noscapine	80/21	0.09	0.1	0.49
	150/40	0.09	0.2	0.62

such as caffeine, nicotine and the relatively non-polar indole alkaloids, strychnine and brucine, were reasonably soluble in dense carbon dioxide. Dinitrogen monoxide yielded better results than CO_2. The more polar extraction gases, $CBrF_3$ and CHF_3, dissolve alkaloids better, evidently due to specific interactions. The alkaloids can always be easily extracted when they are not in combined form and are sufficiently lipophilic. Polar substituents in the molecule, such as hydroxyl and carbonyl groups, depress the solubility in the gases studied.

As a sequel to these microanalytical results, a quantitative study was carried out on the dissolving capacity for alkaloids of the gases of interest, namely, carbon dioxide, dinitrogen monoxide and trifluoromethane, in both the supercritical and liquid states and at pressures up to 200 bar [6]. Table 36 contains some solubilities of opium alkaloids.

Solubilities in dinitrogen monoxide are slightly larger than in carbon dioxide within the range of temperature and pressure studied. Various effects are observed in a comparison of the dissolving capacities of CO_2 and CHF_3; these are due to specific interactions between the individual alkaloids and the dense gas. Codeine attains a lower concentration in dense CHF_3 than in CO_2 but the other three alkaloids, thebaine, papaverine and noscapine, are clearly more soluble. Morphine contains two free hydroxyl groups, is thus more polar and is poorly soluble in the gases investigated. It yields a solution of only 5 µg/g in supercritical carbon dioxide at 200 bar and 40 °C. Dense trifluoromethane is the most suitable solvent also for extracting the ipecacuanha alkaloids and colchicine. These compounds attain solubilities of ca. 1 mg/g at 150 bar and 40 °C and much higher values still at higher pressures. The measured concentrations in the gas phase give an incomplete picture as far as preparative extraction is concerned. The times of extraction are decisively influenced by the mass flow of solvent. The higher density of trifluoromethane (20% higher than that of CO_2 at 200 bar and 40 °C, for instance) permits a faster extraction.

It has been established that, in the pressure region up to 120 bar, dinitrogen monoxide and carbon dioxide in the liquid forms extract more of the opium alkaloids than the corresponding supercritical gases. The supercritical gases are, however, better solvents at higher pressures. Higher temperatures at the same time (e.g. 80 °C) increase solubility sharply as a result of the increased volatility of the alkaloids. The solubility isotherms with trifluoromethane intersect in the pressure domain between 150 and 180 bar.

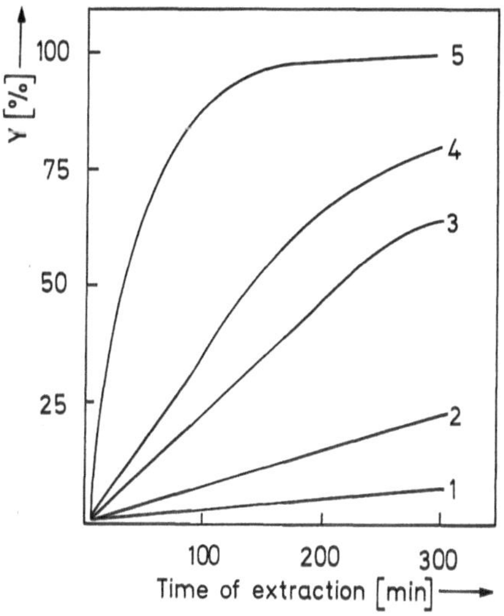

Fig. 87. Course of thebaine extraction with dense carbon dioxide and trifluoromethane under various sets of conditions [5].
Y = Thebaine yield (100% corresponds to a content of 1.8% in the drug).
1) liquid CO_2, 100 bar, 20 °C, 43% water in the drug
2) supercritical CO_2, 200 bar, 40 °C, 31.5% water in the drug
3) liquid CHF_3, 90 bar, 19 °C, 33% water in the drug
4) supercritical CHF_3, 100 bar, 40 °C, 33% water in the drug
5) supercritical CHF_3, 150 bar, 40 °C, 31.5% water in the drug

The extraction of thebaine from *Papaver bracteatum Lindl.* with dense gases was the object of an investigation of Stahl and Willing [5]. This poppy sort contains thebaine and not morphine as principal alkaloid. Thebaine is a spasmodic poison and hence does not appear in the drug scene; it can, however, be relatively easily converted chemically to morphine derivatives of medicinal value. Carbon dioxide and trifluoromethane in the pressure range from 70 to 200 bar at temperatures of 20 and 40 °C are used as gases for extraction. About 95% of the thebaine is bound in the walls of the poppy capsules. It was liberated by moistening the pulverised capsule material, without seeds, with ammonia solution. Adjustment to a water content of 30–40% has been shown to be favourable for the thebaine extraction. The results of extraction are given in Fig. 87.

Below 150 bar supercritical carbon dioxide extracted no significant amount of thebaine; 5 hours at 150 bar and 7 hours at 200 bar extracted, respectively, 7 and 21% of the total thebaine. The extracts with supercritical CO_2 contained up to 90% of fatty oil. Similar results were obtained with liquid carbon dioxide. In contrast, dense trifluoromethane was better suited for the alkaloid extraction. High yields of thebaine were obtained with both liquid and supercritical CHF_3. The maximum extractable amount was relatively rapidly reached at 150 bar and 40 °C (curve 5), yielding extracts which contained about half thebaine and half fatty oil. An interesting fractionation with trifluoromethane at 100 bar and 40 °C was observed. It has much better dissolving power for the alkaloids than for the fatty oil under these conditions, so that extracts containing 80–90% thebaine were obtained.

IV.5.b Decaffeination of Coffee and Tea

Coffee is one of the most popular semi-luxuries and is of great economic importance. Its stimulating action is due to its content of caffeine; crude coffee contains about 0.6 to 3%. Not everyone can tolerate the physiological activity of caffeine, however. It can cause palpitation of the heart and sleeplessness. For this reason the coffee manufacturing industry has also searched for procedures to remove this alkaloid from coffee beans. The classical decaffeinating procedures consist essentially of four principal steps: soaking the crude caffee with water to make it swell up; extraction of the caffeine with a solvent immiscible with water (dichloromethane); evaporating solvent taken up by the beans; and drying the sensibly caffeine-free product. The third step is especially important for the physiological safety of the treated coffee because, after the extraction, part of the solvent remains relatively firmly held in the coffee beans and is freed only when the bean structure is broken up or by evaporation. The caffeine can be removed from coffee and coffee products also with dense gases as an alternative. Patents have been sought for many different process variants during the last two decades [41, 207–227]. No claim of completeness is made here in descriptions of the patents. The firm of HAG in Bremen (Germany) is already working an industrial plant for decaffeination of crude coffee by extraction with dense CO_2.

Before the individual procedures are discussed in more detail, it is advisable to consider the solubility behaviour of caffeine in dense carbon dioxide. Stahl and Schilz used the dynamic flow method to investigate the solubility at pressures up to 200 bar and temperatures between 20° and 60 °C [228]. The solubility isotherms found showed that below 110 bar liquid carbon dioxide (21 °C) dissolved caffeine better

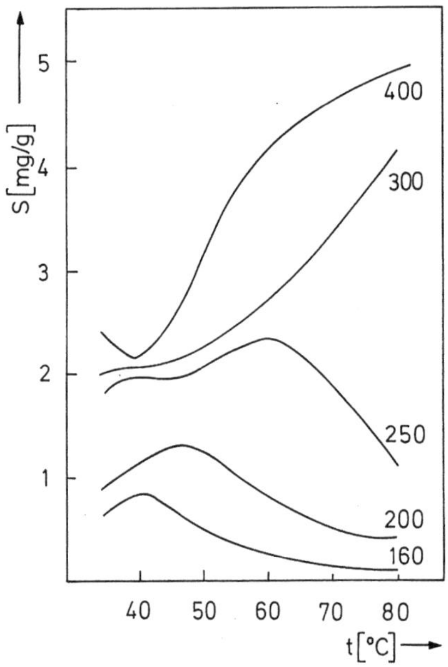

Fig. 88. Solubility isobars of caffeine in supercritical carbon dioxide as a function of the temperature [229]; pressures in bar

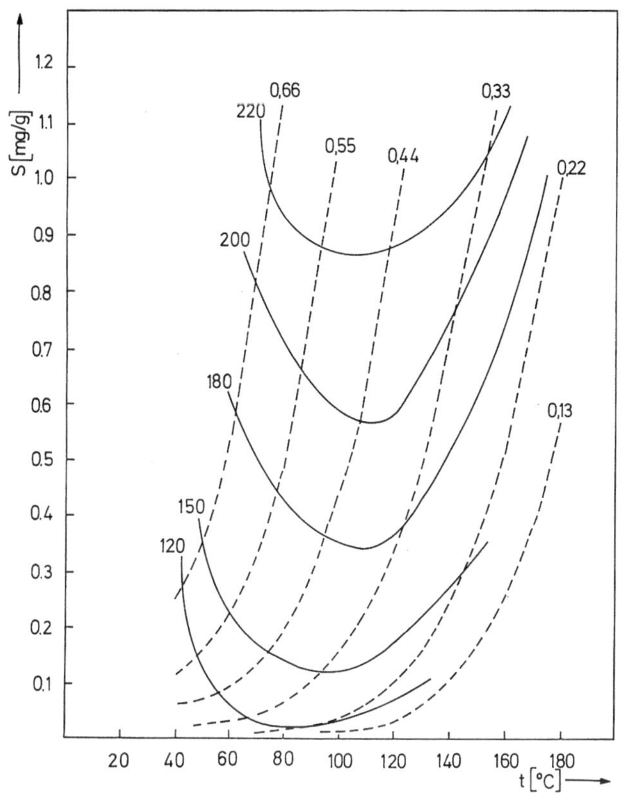

Fig. 89. Solubility isobars (continuous lines) and isochores (dashes) of caffeine in supercritical carbon dioxide, compiled from values of Ebeling and Franck [78]; pressures in bar, densities in g/cm^3

than the supercritical gas. Concentrations of 0.4 mg/g were reached. At higher pressures, supercritical CO_2 is the better solvent (e.g. ca. 1 mg/g at 200 bar/40 °C). Gährs and Marnette, using a phase equilibrium apparatus, worked in a larger pressure range but restricted themselves to supercritical temperatures from 40 to 80 °C [229]. Ebeling and Franck studied the system caffeine/CO_2 in the temperature range of 50 to 160 °C at pressures up to 220 bar [78]. These experiments were also performed with a static measuring method. The absolute values of solubility obtained by the static and by the dynamic procedure were in good agreement.

The influence of the parameters of pressure, temperature and gas density on the solubility of caffeine in carbon dioxide is made plain in Figs. 88 and 89. The course of the solubility isobars varies according to the extraction pressure (Fig. 88). At pressures between 110 and ca. 250 bar the solubility curves first climb with increase in temperature, pass through maxima at 40–60 °C and then fall as a result of the temperature-dependent decrease in density of the carbon dioxide. However, at temperatures of about 100 °C the volatility of the caffeine becomes increasingly significant, so that the solubility isobars rise again steeply as temperature increases (see Fig. 89). At high pressures, such as 300 or 400 bar, the density of the carbon dioxide decreases only slowly with rise in temperature, so that the increasing volatility of the caffeine compensates this effect in any case and brings about a distinct increase in solubility of the caffeine. The alternating dominance of solvation and vapour pressure effects thus influences decisively the course of the solubility curves. The work with caffeine thus also confirms the generally valid statement that increase of temperature at constant gas density can lead to an emphatic rise in solubility of the condensed phase in a dense gas.

The fundamental work on decaffeination of crude coffee was done at the beginning of the seventies by Zosel at the Max-Planck-Institut für Kohleforschung in Mülheim (Germany) [41, 207, 215]. The caffeine is extracted from green coffee beans with the help of moist, supercritical carbon dioxide at pressures between 160 and 220 bar and temperatures between 70 and 90 °C. Carbon dioxide has only a feeble dissolving capacity under these conditions but high selectivity for caffeine; no materials are thus lost which contribute to the coffee aroma during the subsequent roasting. Liquid carbon dioxide has a selectivity inferior to that of the supercritical material at 160 bar/70 °C, so that extraction with it is inexpedient, since not only caffeine but also considerable amounts of desired substances, e.g. fatty oils (content of 10–15%), would be taken out also. An indispensable condition for caffeine extraction is that the crude coffee be soaked up to a water content of ca. 45%.

Zosel described various procedures for removing caffeine from green coffee beans; they differ mainly in the way the caffeine was separated out of the supercritical gas phase [15]. In one variant, supercritical carbon dioxide, saturated with water, was circulated isobarically and isothermally. In the extraction vessel the caffeine was dissolved out of the pre-soaked green coffee beans by the supercritical carbon dioxide, fed to a "washing tower" and washed out of the gas phase there with hot water. The regenerated carbon dioxide was recycled to the extraction vessel. The caffeine content of the crude coffee sank from initial values between 0.7 and 3% to 0.02% after ca. 10 h extraction time. The caffeine was isolated by concentrating the hot, aqueous solution in an air or nitrogen stream and then crystallising out by cooling [207]. The caffeine can also be separated from the circulating supercritical carbon dioxide in an isobaric and isothermal way by adsorption on active charcoal, from which it is then obtained by extraction. A variation of this is especially advantageous: a mixture of pre-soaked coffee beans and active charcoal was fed into a pressure vessel; the grain size of the adsorbent was chosen so that it just filled the hollow spaces between the beans and thus did not change the bulk density of the coffee [215]. About 1 kg charcoal was needed per 3 kg coffee. Supercritical carbon dioxide at 220 bar and 90 °C was then allowed to act in a stationary way on the mixture, so that the caffeine diffused out of the crude coffee directly into the active charcoal; the supercritical solvent accelerated this process markedly. After about 5 h the desired degree of extraction (residual caffeine content 0.02%) was attained. After decompression, the coffee and active charcoal were separated by simple sieving.

The use of supercritical carbon dioxide requires relatively high pressures; this can be circumvented by using another extracting agent. Pressures of only ca. 10 bar suffice when liquid propane or butane are used [222]. However, a solvent/coffee ratio about three times higher than with CO_2 is needed for decaffeinating green coffee beans. Lower working pressures are possible also when carbon dioxide-entrainer mixtures are used [219, 223]. In the procedures described the entrainers were organic solvents, such as acetone or lower alcohols, which had to be evaporated again from the crude coffee after caffeine removal. A high concentration in the dense gas is less the aim than is a highly selective solvent in the caffeine extraction from crude coffee. A low caffeine concentration in the supercritical gas is not a disadvantage as far as the time of extraction is concerned. The diffusion of the alkaloid out of the coffee beans is the rate-determining step, so that a high dissolving capacity or the use of more solvent cannot be exploited. This is shown by the practical experience put to paper in the patents, and also the studies of Brunner [230]. Two extraction curves for the decaffeination of crude coffee with supercritical dinitrogen monoxide are seen in Fig. 90. It is clear that, after a certain time, the rate of extraction no longer depends on the gas flow.

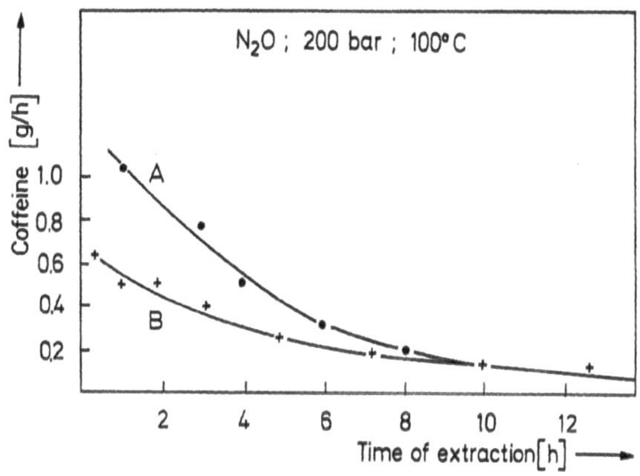

Fig. 90. Extraction curves for decaffeination of crude coffee with supercritical dinitrogen monoxide [230].
A, B: different gas flows, A exceeding B

The water content of the crude coffee is a decisive factor influencing the speed of extraction. The natural openings in the texture of the cell walls are greatly increased in size in coffee beans which have swelled to the maximum extent; diffusion is then facilitated. Transport of the caffeine depends also on the extraction agent, especially on its solubility in the liquid phase of the coffee beans. The dissolved gas lowers the density of the liquid phase and, above all, increases the diffusion coefficients. Detailed studies of the diffusion processes in the decaffeination of crude coffee can also be found in the work of Bichsel [231]. Although extraction with classical solvents is discussed in this work, the results can be at least partly applied to the high pressure extraction.

Van der Stegen and de Witt described another way of removing caffeine from crude coffee [217, 227]. In this, the ground, green coffee beans were extracted with hot water; this water contained no caffeine but so much of the other soluble coffee constituents that their content in the crude coffee was practically unchanged by extraction. The caffeine was finally removed from the aqueous solution by countercurrent extraction with liquid or supercritical carbon dioxide, and separated from this mixture by altering the process conditions or washing with hot water.

It should be possible to extract selectively the caffeine with supercritical carbon dioxide from aqueous extracts of roasted coffee without impairing aroma and flavour of the coffee [218]. The aqueous coffee extract is then brought as a thin film in a packed column into contact with supercritical CO_2 and extracted in counter-current. An aqueous coffee solution containing 4.5% caffeine (total solid content 32%) was treated with a carbon dioxide amount of 47 g/g coffee extract at 200 bar and 50 °C; this reduced the caffeine content to ca. 0.1%. The high relative mass throughput was due to the low partition coefficient of the caffeine in the phase system CO_2/water. A value of 0.054 was calculated by Paulaitis et al. [232] from data in a U.S. Patent [221] for the extraction of caffeine from an aqueous coffee solution using liquid carbon dioxide at 450 bar and 18 °C. Values for supercritical carbon dioxide have not been reported but can be assumed to be of the same order of magnitude. The decaffeination of aqueous coffee extracts with dense carbon dioxide has also been the object of other procedures described which differ from those discussed above only in details [220, 221, 226].

Roselius, Vitzthum and Hubert described a multistage process for preparing a caffeine-free coffee extract [208, 211]. The coffee oil containing the aroma substances was first extracted from roughly ground roasted coffee with dry, supercritical carbon dioxide at 350 bar and temperatures of 40–50 °C. Yields of 10–12% were obtained. The caffeine content of the roasted coffee, 1.1%, remained apparently unchanged during this treatment. After the de-oiled coffee had expanded to a water content of ca. 30%, caffeine was extracted with water-saturated, supercritical CO_2 (250 bar, 60 °C) to lower the residual amount to 0.05–0.1%. The decaffeinated roasted coffee was then finally extracted with hot water. It was shown that a coffee, pre-treated in this way, yielded 4–6% more extract (with water); this was explained by the increase in porosity of the coffee through the pressure treatment with following expansion. Since the fat content had been appreciably reduced, the better wettability may have played a part too. The instant coffee powder obtained by evaporating and freeze-drying the aqueous extract was then aromatised with the coffee oil obtained in the first stage. The dried extract powder contained 0.07% caffeine.

Different coffee oils are extracted from roasted coffee with dense carbon dioxide, depending on the conditions of extraction [209]. The supercritical gas extracts the natural antioxidants of the coffee in addition to the aroma substances and the coffee oil; this increases decisively the storage life of the aroma oil. The temperature conditions during the extraction have an important influence on the quality of the aroma. A "slightly sweet" aroma was obtained at 300 bar and temperatures of 50–70 °C but a "heavy and smoky" aroma of a coffee oil obtained at 300 bar and 130–150 °C. The principles of the coffee aroma and also the antioxidants needed for stability can be completely obtained with supercritical carbon dioxide whereas according to the authors' findings, the poorer dissolving power of liquid carbon dioxide precludes its success. A large number of components of the aroma oils of roasted coffee could be detected and identified through an analytical investigation using capillary gas chromatography coupled with mass spectroscopy [233, 234].

In contrast to the decaffeination of coffee which can be carried out preferably on the crude coffee, tea-caffeine must be separated from a fermented, aromatic material – and the aroma substances must not be lost. Vitzthum and Hubert devoted themselves to this problem and worked out a suitable procedure for preparing caffeine-free tea [42, 304]. The decaffeination has several stages. The aroma was first taken out of the tea with dry, supercritical carbon dioxide. The caffeine was then extracted with water-saturated, supercritical carbon dioxide at 250 bar and 50 °C. After caffeine removal, the moist leaves were dried in vacuo and finally re-aromatised with the extract obtained in the first stage. This was done by placing in a pressure vessel decaffeinated tea leaves and a solution of the aroma substances in supercritical CO_2 at 300 bar and 40 °C and then depressuring. A caffeine-free instant tea is prepared by extracting the decaffeinated leaves with hot water. The product obtained after spray- or freeze-drying is impregnated with the aroma substances as described.

According to information of the authors, caffeine can be extracted from Indian tea only in the presence of water. This implies that the alkaloid is partly in combined form (with tannins) in the leaves. Only the caffeine is separated during the complete three-stage process. The contents of the taste and odour principles and the colouring matter of the tea remain practically unchanged, as can be seen in Table 37. Only a few lipophilic tea flavones are removed together with the caffeine, which agrees with the fact that the caffeine-containing extract yields a yellow powder containing 92% caffeine.

Table 37. Removal of caffeine from Indian tea, using supercritical carbon dioxide [304]

	Starting material	Decaffeinated tea
Moisture (%)	7.26	6.8
Ash, 600 °C (%)	5.15	5.17
Aqueous extract (%)	38.25	35.50
Caffeine (%)	2.99	0.07
Tannins (%)	9.98	10.5
Tea flavones (%)	0.61	0.52
Extractable tea rubigenes (%)	8.90	8.26

IV.5.c Extraction of Nicotine from Tobacco

As in the treatment of coffee and tea the aim of the extraction of tobacco is a refinement of the carrier material. The extract is always of subordinate interest. Nicotine, the relatively easily volatile, low-molecular weight principal alkaloid of tobacco (Fig. 91) is, in its free base form, very soluble in dense carbon dioxide. Concentrations of 7–8% by weight are attained at temperatures of 50–70 °C and pressures of about 150 bar [229]. However, the extractability of a compound is not determined alone by its solubility; its linkage to the plant matrix can make extraction much more difficult. In tobacco nicotine is partly in the free state, partly combined with tobacco constituents. It must be liberated if extraction is to be complete.

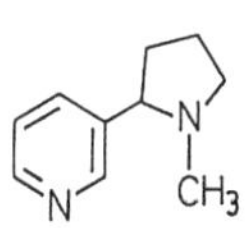

Nicotine

Fig. 91. Structural formula of nicotine

Roselius, Vitzthum and Hubert studied extensively the separation of nicotine from tobacco using supercritical gases [40, 160, 235, 236]. The tobacco is moistened with water before extraction and briefly warmed to 70 °C in an atmosphere of CO_2. This inactivates the browning enzymes so that the tobacco stays light coloured during extraction. This treatment also helps the liberation of combined nicotine. The subsequent extraction is carried out at 300 bar and temperatures from 40 to 70 °C, for example. Separation from the nicotine-containing extract is best done isobarically and isothermally by adsorption. The investigations showed that the nicotine extraction is favoured by a higher water content of the tobacco (cf Table 38). If this water content is low, practically only aroma substances are extracted.

Table 38. Extraction of nicotine from Virginian tobacco (nicotine content of 1.36%) using supercritical carbon dioxide [40]

Conditions of extraction	Moisture content (%)	Amount of solvent (g/g)	Degree of extraction (%)	Residual nicotine content (%)
CO_2, 300 bar, 70 °C	25	5–6	94.1	0.08
CO_2, 1000 bar, 50 °C	15	7–9	91.2	0.12
CO_2 + 5% NH_3, 250 bar, 70 °C	20	5–6	97.8	0.03

Nicotine is separated under milder extraction conditions and using less solvent when the tobacco contains more water. If 5% of ammonia are added to the extraction agent, the procedure runs extremely well; almost 98% of the nicotine was extracted in such an experiment.

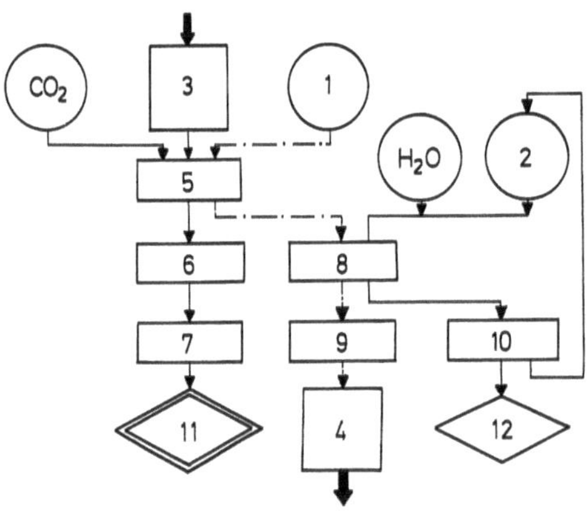

Fig. 92. Multiple-stage procedure for extracting nicotine from tobacco with supercritical carbon dioxide [160].
1 starting tobacco material; 2 adsorbent for nicotine; 3 tobacco from the previous cycle, of reduced nicotine content; 4 tobacco of reduced nicotine content, obtained from 1 in the process; 5 aroma transfer from 1 to 3; 6 dispersion of the aroma; 7 conditioning; 8 removal of nicotine from the charge from 1, already freed from aroma with supercritical CO_2; 9 drying; 10 regeneration of the adsorbent filter; 11 tobacco, largely freed from nicotine and rearomatised; 12 nicotine

The single stage process described leads to a reduction of the aroma content with some tobacco sorts. A three-stage procedure was thus worked out, similar to that for caffeine extraction from tea, which largely protects the tobacco aroma [235]. The dry tobacco is treated with pure supercritical carbon dioxide in the first stage; this takes out the aroma principles but only a negligible amount of the nicotine. The nicotine is then extracted with water-saturated supercritical CO_2. In the third stage the tobacco, sensibly free of nicotine, is treated with the first extract which contains the aroma substances. It is possible to carry out this procedure quasi-continuously in an installation with several vessels. Fig. 92 portrays the three-stage procedure.

The arrow at the top marks the take over of a charge from which the nicotine has been removed and which is to be refurnished with the aroma principles in the present cycle; the lower arrow indicates where the charge produced in the present cycle and freed from nicotine, passes to the next cycle.

The real aim of treating tobacco with supercritical carbon dioxide is to lower its content of nicotine. A desirable side effect of this is that the tobacco leaves expand. The tobacco swells when the pressure of the dense carbon dioxide is reduced and it is removed from the extraction vessel; this increases the effective volume of the tobacco by ca. 10%. The product is scarcely or not distinguishable in appearance from an untreated tobacco.

IV.6. Oils of Low Volatility and Polymers

In the preparation of high molecular synthetic oils homologues are usually formed covering a wide range of molecular mass. Side products are also formed in competing reactions; they are likewise involatile but show differences in other properties. Such mixtures are of low vapour pressure so that thermal methods for removing impurities and isolating uniform fractions are little suited or unusable. In extraction procedures ordinary solvents generally differentiate too little on the basis of the molecular mass; however, supercritical gases show high selectivity in this connection, which can moreover be deliberately varied by changes of pressure and temperature. Advantage is taken of this in the analytical sphere in the use of Supercritical Fluid Chromatography (SFC). For preparative separation of oils of low volatility with dense gases the known methods of fractional extraction, fractional extract recovery and multiplicative separation in columns are available.

Paulaitis et al. described the further fractionation of a mixture of perfluoroalkyl polyethers, obtained as residue from a molecular distillation [232]; no information was given about conditions and experimental details. The fractions were characterised by viscosity measurements because the low volatility precluded the use of gas chromatography. Five fractions were obtained with dynamic viscosities of 0.8–8 Pa · s, the starting oil having the value of 2 Pa · s. It can be concluded from this that the molecular masses of the initial oil ranged over a factor of more than three. A small fraction of low molecular mass could be separated from silicone oil of viscosity 8 Pa · s and average molecular mass of 60000, whereas the principal fractions were of medium molecular mass and another fraction, also small, of very high mass remained behind.

The extraction of solid polymers aims at removing oligomers which have not reacted completely and also unused monomers and other volatile compounds such as solvents or water which could create problems during later use or processing. For example, if the polymers are destined for use as packing material for foodstuffs or in the cosmetic or medicinal domains, their purity is subject to keen scrutiny for physiological reasons, especially as far as compounds of low molecular mass are concerned. The high pressure extraction is ideally suitable for separating such impurities because thermal methods can lead to damage of the polymers and classical solvents yield insuperable problems with residues.

The first investigations of the solubility of polymers were concerned with the phase equilibria of polyethylene (PE) [237–240] and polypropylene (PP) [243] with the respective monomeric gases and were performed some time ago. Further studies have been made in the intermediate time, including work on: the behaviour of polyvinyl chloride (PVC), polymethyl methacrylate (PMMA) and polystyrene (PS) in dense CO_2 [43, 244]; the solubility of PS [245]; the separation of oligomeric cyclic ethers from a polymer product [246]; the extraction of polyacrylamide (PAA), PS, polyoxymethylene, PP-wax and low molecular PE (HPPE) with CO_2, Freons and ethylene [45].

Schröder and Arndt studied in a static system the behaviour of PVC-, PMMA- and PS-sheets towards supercritical CO_2 at temperatures up to 80 °C and pressures up to 1000 bar [43, 244]. The uptake of CO_2 in the body of the sample was followed with measurements of light transmission. The volume of the polymer increased in contact with the CO_2 and it transmitted less light, the effects being larger, the higher the pressure and temperature. For example, PMMA-sheets attained coefficients of expansion of up to 50% at 1000 bar and 40 °C. The CO_2 which penetrated the sample functioned like a plasticiser. After reduction of pressure the body of the samples, first glass-like, became opaquely white. Thin films cut of PS-sheets which were clear at first and without structural elements, showed tiny cavities of 1 µm diameter after treatment for 1 h at 1000 bar/67 °C and then depressurising to the atmospheric value. The desorption of the gas was described in the case of PVC-sheets. Constant weight was obtained after every experiment only by leaving the sheets for 4 h in a drying oven at 60 °C. The amount of gas yielded during this was over 3% of the sample mass after having treated for 2 h with CO_2 at 40 °C and 200 bar; over 4% when 600 bar had been used and nearly 5% when 1000 bar was the pressure. Diffusion coefficients can be calculated from the desorption curves.

The solubilty of the polymer was obtained gravimetrically from the relative loss in mass of the sample after treatment with CO_2. Study of the time-dependence of the dissolving process showed that increase of pressure and especially of temperature accelerated the establishment of equilibrium. A contact time of 1 h sufficed at 400 bar and higher pressure at 40 °C for the PVC-sheets to attain the final solubility value in the static system. The solubility increased with rise in pressure and temperature, the temperature-dependence being especially marked at high pressures. The amount of substance dissolved decreased as the molecular mass increased, as was shown in a series of closely fractionated PVC-samples. The situation was complicated in practice by the fact that the solubility depended also on the relative distribution of molecular masses of the polymer. Further, solvent residues from the preparation method can falsify the experimental results. For example, it is improbable that PVC of molecular mass 110000 can be dissolved in CO_2. The conditions of preparation and of storage of the samples prepared as cast film or compressed sheets also have an influence because they determine the degree of order, the packing density and the vacancy concentration, i.e. the microstructure of the polymer and hence the time of extraction through favouring or obstructing diffusion and mass transport.

Braun investigated the extraction with supercritical CO_2 of water from an acrylamide-copolymer of 30% content solid matter (SM) [45]. A drying degree of over 80% (>70% SM) was attainable in a reasonable time only by extraction at temperatures above 100 °C. The extraction pressures of 150 or 300 bar had no influence on the extraction of water. Even at the high flow rate of 37.5 kg CO_2 per kg PAA per h the degree of drying rose linearly to a definite end value and thereafter stayed practically constant. This fits in with the easy mass movement of water molecules into the supercritical gas phase. Saturation was no longer reached only at flows of over 70 kg CO_2 per kg PAA per hour. The extraction curves then flattened at an early stage and the concentration in the gas fell from about 10 g water per kg CO_2 to 5 g at 150 bar and 120 °C. A more important finding was that monomeric acrylamide could be extracted from ordinary pre-dried PAA (74% SM) with CO_2, C_2H_4, CHF_3 or $CClF_3$ at temperatures below 40 °C. This is of advantage in relation to usual methods because the polymer should not be subjected to a temperature higher than 60 °C, otherwise monomeric acrylamide is re-formed.

In a parallel way the content of monomeric styrene could be reduced from 550 to less than 50 ppm, and also ethyl- and propyl-benzene impurities be removed, by CO_2 extraction of PS at 300 bar/40 °C. Formaldehyde and trioxan/tetroxan present in polyoxymethylene could be separated by high pressure extraction. Polypropylene wax was extracted with CO_2 at 80 °C and 150 bar to give 4.7% of an oily extract, and at 450 bar, a honey-like viscous extract. The average molecular mass of these products was 800 and 1400, respectively. The largest molecules extracted had a molecular weight of 5000, corresponding to a degree of polymerisation of 120 units.

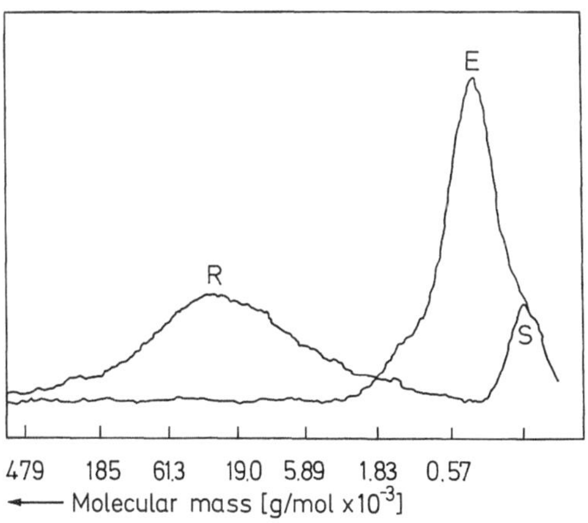

Fig. 93. Gel permeation chromatogram of residue R and extract E in the CO_2-extraction (450 bar/80 °C) of HPPE, S stabiliser [45]

Only 1% extract of components of molecular mass lower than 2000 could be separated from low-molecular PE (HPPE) using CO_2 at 450 bar and 80 °C. Fig. 93 shows the distribution of molecular masses in residue and extract.

IV.7. Extraction of Petroleum Products

The possibilities of applying dense gases in the petroleum branch are manifold; they can, however, be considered under two headings. On the one hand, the "volatility-increasing" action of dense gases is exploited in the separation of substances from tar sands, oil shales, and tar from hard and brown coal, and also for working up residues from petroleum distillation and hydrogenation of coal. The intention here is to separate as large a part as possible of the poorly volatile compounds from a particular solid carrier material or to obtain fractions of different properties from an available liquid mixture of substances. On the other hand, profit is derived from the "viscosity-diminishing" action of dense gases in two ways: either directly when the oil is being pumped, in order to mobilise the viscous tertiary petroleum and hence to utilise the deposits better; or to separate fine solid particles which are present in viscous extracts from coal and oil-containing minerals and cause great problems in subsequent processing.

IV.7.a Deasphaltation of Heavy Oils

The oldest suggestions for using dense gases for extraction were made in the mineral oil sphere. They concern the deasphaltation of heavy hydrocarbon fractions. This separation is based on the fact that the high molecular weight components, termed asphaltenes, are insoluble in dense gases, for example, propane, and hence accumulate as the residue during the extraction. Asphalt removal is important in that the asphaltenes tend to yield coke during the further treatment of heavy oil, and that during their separation undesired metal-containing compounds are also removed; these poison catalysts used for hydrogenation.

As early as 1936 Pilat and Godlewicz described a procedure for deasphaltation of petroleum and for separating high-molecular mineral oil mixtures, for instance with propane at 104 °C and 80 bar [247]. Considerably higher pressures and methane were employed by Messmore who published work in 1943 on obtaining asphalt oil [116]. In 1955 Zhuze and Kapelyushnikov described a method for removing resin from crude petroleum products using supercritical propane at 100 °C and 110 bar [117]. Isothermal lowering of pressure to 40 bar yielded the extract. In analogous procedures, already mentioned, Zhuze et al. described also the isolation of ceresine (earth wax) from ozokerite deposits, and the separation of lanolin from a crude wool fat [248]. In a comprehensive patent Zosel quoted numerous examples of separation of petroleum products through extraction with dense ethylene, ethane, propylene and other gases [249].

Zosel described in another publication the fractionation of a mixture of high-boiling α-olefines (C_{14}–C_{20}) by extraction with supercritical ethane at 45 °C and 60 bar [15]. The gas phase containing the extracted material was led through a fractionating column, at the head of which a "finger" at 85 °C produced a partial reflux of the dissolved substances. During the extraction the pressure was increased slowly from 60 to 110 bar; the individual components dissolved in ascending order of their boiling points and were extracted. They were separated out by lowering the pressure of the charged gas phase to 30 bar.

Another example of use is the continuous deasphaltation of topped crudes from petroleum distillation, using supercritical propane in a 750 litre experimental plant. Zosel recommended a pressure of 130 bar and a temperature of 140 °C, thus differing from the working temperature of 100 °C of Zhuze which is only slightly above the critical point and, together with the oil, can lead to an additional liquid-liquid separation. Under these conditions 75 % of the starting material could be extracted as asphalt-free oil. The insoluble, high-molecular residue contained most of the vanadium compounds. In an experiment using liquid propane at 70 °C and 70 bar, the extract was found to contain higher vanadium concentrations of 2.4 instead of 0.15 ppm, and a higher proportion of weakly volatile components.

Irani and Funk also investigated the removal of asphalt from petroleum fractions, using supercritical propane in comparison with the customary method using liquid propane [8]. The cost of the plant and its running, the selectivity and the solvent/oil ratio needed for separation were evaluated and, apart from obtaining slightly less vanadium in the extract, the supercritical procedure was found to have no particular advantage (Fig. 94).

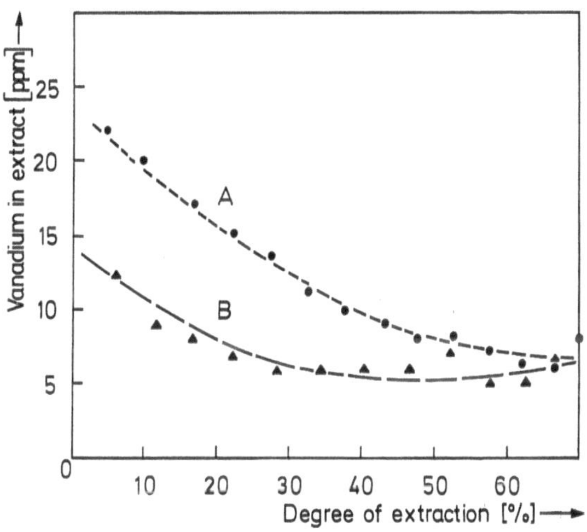

Fig. 94. Vanadium content of the extract from treating heavy mineral oil A with liquid (60 °C/77 bar) and B with supercritical (115 °C/103 bar) propane; the conditions A and B correspond to comparable dissolving capacities of the propane for the mineral oil [6]

The Kerr-McGee Refining Corporation in the USA developed already in the fifties a deasphaltation process termed the ROSE-process (Residual Oil Supercritical Extraction) and pursued it in a pilot plant with a daily capacity of 80 tonnes [250]. The distillation residues to be processed were extracted with sub-critical pentane, so that the asphaltenes remained behind as insoluble components. The pentane containing the extract was heated under a medium pressure to supercritical temperatures, which led to a lowering of the density and hence of the dissolving power. In this way first a resinous and then an oil fraction separated out and the solvent was recovered at the same time.

IV.7.b Particle Separation from Viscous Oils

A modification of the ROSE-process was proposed as a CSD (Critical Solvent Deashing) procedure for separating insoluble particles from viscous extracts and distilla-

residues [251]. The extract containing the solid material is mixed with toluene in the proportion of 2.5:1 and brought to the sub-critical conditions of 290 °C and 55 bar. The bulk of the extract goes into solution whereas the solid particles and the compounds of very high molecular mass precipitate rapidly. Under these conditions the solution has such a low density and viscosity that, even at an upward flow rate of 90 m/h, sedimentation sets in. Recovery of the purified extract is accomplished as in the ROSE procedure by heating to supercritical temperatures, i.e. by reducing markedly the density of the solvent. The CSD process was tested in a pilot plant of 6 tonnes daily production and larger installations have been planned.

Stützer, Brunner and Peter also worked on the separation of very fine particles (in the μm region) from viscous oils [54]. They employed as solvent a hydrocarbon, gaseous under normal conditions (e.g. ethylene or propane), with a liquid entrainer (toluene). The pressures were somewhat higher (50–100 bar) and the temperatures lower (80–180 °C) than in the CSD procedure. The conditions in the extraction vessel were adjusted so that the pseudo-ternary system of oil, gas and entrainer formed a homogeneous supercritical phase, characterised by a lower viscosity and a lower density (0.3 to 0.5 g/cm^3) than that of the solid particles (2 g/cm^3). The rate of sedimentation of the particles depends to a great extent on the concentration of the viscous oil in the solution. For example, a 1:1 (w/w) mixture of propane and toluene was mixed with shale oil containing a 25% dispersion of solid particles of grain size 3–12 μm, at 80 bar and 120 °C. An oil content of 30%, yielded a sedimentation speed of ca. 1 m/h; at values above 50%, the rate of sedimentation fell rapidly to very low values. The purified oil can be separated out of the solution by profiting from a miscibility gap in the quasi-ternary system by increasing the temperature to 170 °C at constant pressure. Residual toluene in the oil must be recovered in a separate distillation process.

According to Henry, who discussed various ways of removing mineral particles from heavy petroleum and coal extracts, the behaviour of the particles is inflenced decisively by their coating of asphaltenes [252]. He carried out measurements in a high pressure-high temperature cell with aluminium windows transparent to X-rays. On the one hand an association of the asphaltenic components combined with an agglomeration of the asphaltene-coated solid particles could be obtained by adding a saturated "anti-solvent"; and on the other hand, dispersion of water in the mineral oil led to a concentration of the solid particles in the aqueous phase when alkalies or detergents were added to improve wetting. Treatment with supercritical solvents (toluene) was investigated as a third way of separating the solid matter; this led to separation out of an asphaltene phase from the heavy oil studied. This phase had a special affinity for the asphaltene-coated solid particles and thus removed them rapidly; the speed was due more to a coalescence effect than genuine sedimentation kinetics.

IV.7.c Extraction of Oil-Containing Minerals and Residues

The kerogen in oil shale and the organic compounds in tar sand provide a reservoir of hydrocarbons more than ten times that of petroleum. In situ methods have been suggested to utilise these sources of raw material similar to those proposed for recovery of tertiary petroleum [253]; they are, however, of problematic value on account of the high viscosity of the bituminous oils. After the oil-containing rock has been broken down there is a possibility of extracting the oil, perhaps in combination with hydrogenation to yield liquid products of higher H/C ratio. Hoffmann et al. studied the treatment of Brasilian Irati oil shale with nitrogen, hydrogen and water at temperatures between 400 and 700 °C and pressures up to 220 bar [254]. High conversion rates of 74% of the organic carbon into liquid, and 9% into gaseous compounds, could be obtained in hydrogen- or hydrogen/water atmospheres at 450 °C/100 bar. A combined process for obtaining liquid products from oil shale was proposed on this basis, in the working cycle of which the necessary hydrogen-, steam- and energy production could also be included.

Martin and Williams were able to show that supercritical toluene at 420–440 °C and 100 bar extracted 88% of the kerogen from Colorado oil shale, whereas ordinary pyrolysis at 600 °C extracted only up to 70% and at a less favourable extract/gas ratio [255]. All organic components down to a residual content of 5% were removed from an Athabasca tar sand (easier to extract) with supercritical toluene at 400 °C and 100 bar or supercritical tetrahydrofuran at 310 °C/100 bar. Using supercritical pentane (230° C/100 bar) as solvent, 13% of high molecular substances remained in the oil sand.

Bott described the extraction of tar sands and peat, stressing in particular the simultaneous separation of the complex constituents into fractions of commercial interest [51]. An active carbon filter was placed after the extraction vessel for this purpose, to adsorb and hold back part of the components taken up by the solvent. The extraction of Athabasca tar sand with sub- or supercritical pentane at 140 to 240 °C and 20–80 bar yielded as first fraction under these conditions 10.6% of deasphaltised, demetallised light oil extract (maltenes). The low molecular asphaltenes were desorbed from the active carbon filter with benzene at 270 °C/60 bar in the second step. The third fraction was a benzene extract of high molecular asphaltenes from the material remaining from the pentane treatment; this had a high metal content. Peat yielded analogously different fractions containing 2.8% light, pre-refined wax, 3.25% low- and medium molecular asphaltenes and 3.8% high molecular asphaltenes.

Eisenbach and Niemann also worked on extracting Athabasca tar sand and isolating material of value from brown coal tar [52]. Supercritical propane at 110 °C and 200 bar extracted a dark red, viscous oil from tar sand; it consisted principally of paraffinic compounds. The yield was 9.5%. During the experiment the extract recovered at 20 °C became more viscous indicating an increasing proportion of higher molecular compounds. Soxhlet extraction with toluene removed a further 4% amount of black solid (asphaltenes) from the residue.

Tar is a source of paraffinic compounds which is still less used. It consists predominantly of two compound classes, the n-paraffins with the related 1-olefines; and oxygen-containing phenols and carboxylic acids, termed creosote. Supercritical ethane at 55 °C and 200 bar gave within a short time an extract from brown coal tar which contained the paraffin- and creosote-fractions at the same time, thus separating them from the unwanted high molecular compounds of more than 40 C-atoms, and also from ash and coke dust. From this total extract, 80% of the crude material used, it was possible to separate the creosote oil, somewhat more polar but also more volatile, quantitatively and selectively in a rectification apparatus, using carbon dioxide as solvent at a reflux ratio of 1.5. Under the chosen fractionation conditions of 135 bar and 55 °C, carbon dioxide has a much lower solvent power than the ethane used in the first extraction step. It is therefore advisable to take away only the smaller creosote oil fraction via the gas phase. n-Paraffins (48%) and 1-olefines (26%) with chain lengths from C_{22} to C_{36} were identified in the paraffin residue. Some overlapping occurred in the separation because the low molecular paraffins, present in only small amounts, were dissolved with the creosote oil. It is interesting that these paraffins can be separated from this mixture by extracting the creosote oil with water-saturated carbon dioxide, but in practice this bother is not necessary. Based on these results a two-stage continuous extraction plant was proposed, which could separate the tar into 45% crude paraffins, 35% creosote oil and 20% residue.

Hydrogenation of brown coal (lignite) gives a poorly volatile fraction from which a high molecular oil can be separated by vacuum distillation. Additional organic compounds can be isolated from the distillation residue by extracting with supercritical solvents. Suitable experiments were carried out with propane, butane and pentane at 200 bar and temperatures in each case ca. 20 °C above the relevant critical value, at 120°, 170° and 210 °C [256]. The solvent power of the supercritical phase increased in the homologous series in the order of ascending molecular mass; this was shown by higher rates of extraction and increases in yield from 20 to 47%. The superiority of pentane survived even when the other solvents were used at the same high temperature. It was shown that pentane dissolved higher molecular, less volatile components (asphaltenes) to some extent, which were insoluble in the propane. The asphaltene content of the propane extract was 11% at the most, that in pentane, 38%. This is expressed also in the softening temperatures of the extracts which were 44° and 62 °C, respectively. If the residue from hydrogenation is extracted immediately, the yields in each solvent are increased by the amounts which would be found in the vacuum distillation. The softening temperature of the extracts then falls to values of 26–31 °C.

Another interesting field of application of dense gas extraction of mineral oil residues is the working up of used or spent oils to furnish re-usable lubricants, as described by Coenen and Rinza [53]. The procedure used hitherto for recovery has been the sulphuric acid-active earth-contact process which has the disadvantage that 1 tonne of residue is obtained from 2.5 tonnes regenerated oil; it consists chiefly of acid resins which pollute the environment and are most expensive to get rid of or dump. As in the classical method, gas oil I (benzine) and water are first removed from the used oil by distillation. The dry oil obtained is treated with carbon dioxide, ethene, ethane or propane under supercritical conditions in an extraction apparatus with multiple-stage separation, as described in detail in Chapter III.4. About 10% of the dry oil then accumulate as a residue of oxidation products, involatile components and solid matter. The extracted substances are separated into three chief fractions by stepwise reduction of the solvent density: 51% heavy residual oil, 31% spindle oil and 8% gas oil II. These products, which still contain residual additives from the used oil, could be employed directly as lubricants. However, at the moment there is market interest only in additive-free secondary raffinates, so that an additional treatment of the dense gas extracts with sulphuric acid and Fuller's earth is necessary. There is, however, a 70% reduction in the amount of acid resin and oil-containing earth compared to the classical method.

A plant with a throughput of 5 tonnes per hour of dry oil was designed on the basis of these experimental results and the outlay and running costs were worked out. The production costs for 1 tonne of secondary raffinate by high-pressure extraction, including the subsequent treatment with sulphuric acid were 182 DM or, profiting from available economising measures, 125 DM. The costs for preparing 1 tonne of secondary raffinate from dry oil, using the sulphuric acid-active earthprocess are about 220–230 DM (1981 prices).

IV.7.d Analytical Investigations

Analytical investigations and methods in the mineral oil sphere have been carried out as well as preparative applications of dense gases. Monge and Prausnitz described an experimental method for determining the solubilities of heavy oil fractions which contain several components and are difficult to analyse, in dense gases at pressures up to 100 bar and temperatures up to 300 °C [257]. The solubility values were obtained from the gas volumes needed to dissolve completely, under dynamic conditions, accurately known masses of the oil sample under study. In connection with gaining oil from deep deposits where high pressure and temperature conditions are found, Simon described a system with a special optical cell which permitted studying the dependence of properties of petroleum (density, viscosity, surface tension, boiling and dew points, separation into phases) on the composition at pressures up to 138 bar and temperatures up to 200 °C [258].

Kuss described experimental methods for studying the viscosity of gas/oil solutions under pressure [259, 260]. Lowering of viscosity by factors up to 1000 occurred when various crude oils were saturated by N_2, H_2, CO_2, CH_4, C_2H_6 and C_3H_8. Killesreiter also investigated the solubilities of various gases, in particular carbon dioxide, in crude oil and the viscosities of the systems formed [261]. The investigations are of interest for oil recovery. The oil usually occurs in porous sand and rock. It contains dissolved gases, mostly methane, which partly separate when the source is tapped, forcing the oil up to the surface. When the gas has been used up, about 70% of the oil deposit usually remain as viscous tertiary oil underground. Successful field experiments have been carried out in the USA to obtain this tertiary oil by injecting gases, usually carbon dioxide, into the oil deposits [262–265]. The use of CO_2 required high pressures but avoided the use of large amounts of expensive hydrocarbons (C_2–C_4).

IV.8. Extraction of Wood, Peat and Lignite

The wood of trees contains 40–50% of cellulose as skeletal material and also 25–30% hemicellulose and incrusted lignin. Cellulose is a carbohydrate made up of long chains of D-glucose units, linked via β(1,4)-glucosidic bonds and with a molecular mass of 200000 to several millions. It is a white material, insoluble in most organic solvents and water. The hemicelluloses are plant gums from a mixture of different polysaccharides which yield D-mannose, D-xylose, D-galactose, D-glucuronic acid and other compounds on hydrolysis. Lignin is a mixture, built up in a complex way, of polymeric phenols from phenylpropane derivatives of the syringyl or guaiacyl types, e.g. coniferyl alcohol. In addition to these constituents, various compounds can be isolated from the resinous juice of the wood. The resin of conifers yields, for example, volatile turpentine oil, consisting largely of terpene hydrocarbons such as α- and βpinenes, and a non-volatile part of chiefly resin acids (colophonium), which are in the main tricyclic diterpene acids, such as abietic acid. The bark contains fats and surface waxes in particular.

As a consequence of this complex composition wood finds varied uses as a raw material. Extraction with dense gases under mild conditions can separate terpenes, resins, waxes, fats and other substances. On the other hand, more drastic extraction conditions, linked with defined thermal degradation of the biopolymers can be made to yield phenols from lignin, or sugars and other compounds from hemicelluloses and cellulose. For this, the usual organic solvents, which are liquid under normal conditions (such as pentane, acetone, dioxan, ethanol) are employed at high temperature as supercritical gases. Complex mixtures are obtained as products; these are separated or modified chemically, depending on the prospective use. Such extraction methods make possible a sensible utilisation of wood as a source of valuable compounds, already in structurally organised forms and with interesting economic possibilities for the future. In contrast, combustion, gasification or carbonisation only with the aim of obtaining energy, does not represent an optimum and profitable way of utilising the biomass.

The non-aggressive extraction of wood with dense gases as a substitute for petrochemical solvents is of less significance and rarely mentioned in the literature. Fremont described in a patent the extraction of turpentine and tall oil from small pieces of the wood of conifers [266]. By stepwise pressure reduction the extract was separated into various fractions of definite composition; first the resin acids, then the fatty acids and finally the terpenes separated out. Such a procedure appears possible in principle but the information given in the patent on how to attain this goal is questionable. The wood extraction mentioned is not carried out only to obtain extracts of value; it facilitates also subsequent work with the wood fragments remaining. For instance, they require fewer chemical additives than untreated wood in the preparation of cellulose pulp; the formation of soap from the tall acids is also reduced. This means that the plant is less soiled, there is less waste water and the environment thus suffers less.

McDonald et al. investigated the extraction of resin- and fatty acids from pine wood and of wax from Douglas fir bark using supercritical carbon dioxide, dinitrogen monoxide, propane and ethylene [49]. Some of the acids were extracted from pine wood at 40 °C using a relative gas amount of 8 g CO_2 per g dry wood substance and studying the influence of pressure on the amount of extract. A yield of 2.1 % by weight (based on dry substance) was found at 210 bar; this rose to 5.6 % at 620 bar, as a result of the increasing solvent capacity. The other gases were tested at 210 bar and at 9 °C above the critical temperature of each gas. Ethylene (15 °C) gave slightly better yields than CO_2; dinitrogen monoxide (45 °C) gave three times as much and propane (105 °C) nearly nine times. With larger amounts of gas of about 70 g/g wood, it was possible at 620 bar to obtain 15 % extract with CO_2 and even 25 % with N_2O. At the beginning of an experiment, the extract contained over 90 % of the resin- and fatty acids; at the end this figure was less than 50 %. A typical composition of the initial fraction is shown in Table 39.

Ethylene was a poorer solvent than carbon dioxide for extracting wax from Douglas fir bark (temperature effect). Yields of only 4 % were attained, rather more with propane. The light yellow wax contained six main components when extracted with pro-

Table 39. Typical composition of an extract from pine wood (initial fraction) in non-aggressive treatment with supercritical CO_2, N_2O, C_2H_4 or C_3H_8 [49]

Acid	Content (%)	Acid	Content (%)
Abietic acid	31	Pimaric acid	4.0
Levopimaric and palustrinic acids	26	Sandaracopimaric acid	1.1
Neoabietic acid	14	Linoleic acid	0.22
Isopimaric acid	8.3	Oleic and linolenic acids	0.14
Dehydroabietic acid	4.8	unknown	ca. 10

pane but only five otherwise. The compounds identified were 1-octadecanol,2-methyl-1-hexadecanol, a saturated C_{24}- and C_{30}-fatty acid. The saponification value of the extract was 110, the melting range, 40–70 °C. Compared to conventional extraction with petrol ether, the extraction of pine wood and fir bark using supercritical gases was faster, and gave better yields with comparable amounts of solvent.

The same authors also tried to obtain the largest possible proportion of liquid products from cedar wood through pyrolysis/extraction with supercritical acetone or methanol at 260–350 °C and 100–280 bar. Up to 74% of the starting material was obtained as extract, acetone yielding slightly more than methanol. The residue was only negligibly carbonised. The results of the elementary analysis of the extract were independent of the extraction temperature. The extract and residue from methanol treatment showed twice as high a methoxyl-group content as the acetone-treated samples, which indicated that the wood had reacted with the methanol. It was possible to identify 2.3% of substituted guaiacols and 2% laevoglucoses (1,6-anhydro-β-D-glucopyranose) in the extracts. The acetone extract was able to replace 30% of the phenol components for preparing phenol-formaldehyde resins, used to stick wafer-boards together, without impairing the adhesive strength of the binding shown in a dry test.

Calimli and Olcay carried out an early investigation of the liquefaction of wood by treatment with supercritical gases, using a discontinuous method [47]. The largest yield of liquid products in the static extraction of spruce wood was with supercritical dioxan at 330 °C and 90 bar in a pressure vessel. The amount and composition of this extract were compared with those of a distillate obtained by pyrolysis at 340 °C under exclusion of oxygen [267]. For this, the decomposition products were first separated into three fractions by extracting with hexane and benzene and then further separated by adsorption chromatography. The volatile components were characterised by GC-MS, the insoluble residue by ^1H-NMR spectrometry.

Thermal degradation led in both methods to about 23% of liquid products, 30% of the wood remained as residue and the rest disappeared as gas and volatile products. The dioxan extract contained more components soluble in hexane and benzene and fewer insoluble components than the dry distillate. The extract contained moreover about 50% of phenolic compounds, twice as many as in the distillate; this means that the phenols were broken down or polymerised during the pyrolysis unless they had been "diluted" by the supercritical solvent. The identification of numerous compounds showed that, qualitatively, the extract and the distillate contained similar components. The most striking difference was that the substances were markedly demethoxylated under the conditions of the classical pyrolysis, whereas this was not observed during the extraction with supercritical dioxan.

Koll et al. obtained better results in the thermal degradation of wood and other biomass products when they used a dynamic procedure which functioned less aggressively and almost entirely avoided the formation of secondary products [48]. Their flow apparatus was constructed largely of HPLC components and a GC-oven. The conditions given here were considered to be the optimum for degradation of wood: a 3 g sample of the material in small pieces was placed in the reactor (HPLC column) and the previously thermostated solvent at 250–280 °C and 100 bar was passed through it for 1 hour at 1 ml/min. The products of decomposition were taken up by the solvent and, after a short interval, exactly defined by the flow rate, removed from the hot reaction zone. This is a basic condition for controlled breakdown to products as uniform as possible. Further, in comparison with a wholly or partially static contact between the wood and the solvent, the flow method led to fewer gaseous and highly volatile products (0–20 %), which are lost when the solvent is evaporated and are regarded as losses in the balance sheet of the materials. Depending on the extraction agent, the breakdown can be purely thermal or thermal-solvolytical, i.e., there is partial reaction of the solvent.

The extraction of birch wood with different organic solvents under the standard conditions given, showed that alkanes and ethers degraded almost only polysaccharides. Esters and ketones led in addition to a moderate breakdown of lignin, and alcohols displayed considerable delignification but little degradation of carbohydrates. About 20–40 % of the wood was broken down by treatment for one hour with the

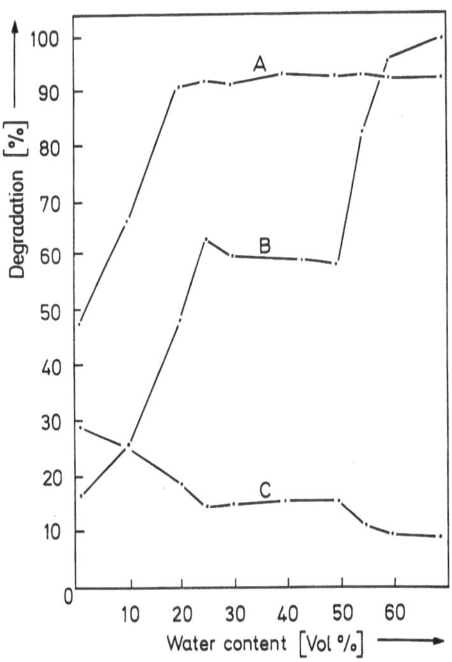

Fig. 95. Degradation of birch wood with ethanol/water mixtures, as a function of water content (V/V) [46].
Reaction time: 1 h at 100 bar and 250 °C. A loss in amount of lignin, B loss in carbohydrate amount, C ratio of A:B × 10

pure organic solvents. If water was added to the alcohols, degradation was much more extensive. Methods for delignifying wood with ethanol-water mixtures were suggested also by Kleinert [268].

Fig. 95 shows the dependence of the degradation of lignin and non-lignin substances (hemicellulose and cellulose) on the composition of ethanol/water mixtures. Almost the whole of the lignin is removed when the water content of the ethanol is 25–50% by volume. The hemicellulose also goes completely into solution. A very light-coloured, fibrous cellulose remains which has the degree of crystallinity of the natural cellulose but a lower degree of polymerisation of only 350 units and is therefore more suitable for chemical purposes than for cellulose processing.

The degradation products of the hemicelluloses can be separated from the lignin with water. They are interesting as a substrate for microorganisms or they serve as starting material for obtaining xylose and xylitol. The hemicelluloses can also be obtained prior to the delignification by extracting with water at 180 °C (pre-hydrolysis). The lignin obtained has an average molecular mass of only 940, corresponding to 4–5 phenylpropane units, and can be employed as the phenolic component for making polyurethanes and phenoplasts, or it can be decomposed further to monomeric and dimeric phenols.

If an ethanol/water mixture with 70% water is used, the whole wood can be dissolved in a short time. The extract has a tar-like consistency; 40% are water-soluble, 60%, chloroform-soluble constituents. The water-soluble part contains 70% of the polysaccharides as monomeric sugars, most of which can be fermented. The chloroform-soluble part, rich in aromatic compounds, can be processed further using petrochemical methods.

Pure microcrystalline cellulose and also chitin which has an analogous structure and is found in the shells of crustaceans, in the skeletal material of insects and in fungus cellulose, are likewise almost completely broken down with supercritical acetone as solvent, at 250 bar and in the temperature range of 250–340 °C (Fig. 96) [46]. Chitin yielded for the first time the primary product of the thermal degradation, 2-acetamido-1,6-anhydro-2-deoxy-β-D-glucopyranose, an amino-sugar which is important in the synthesis of physiologically active oligosaccharides. The principal product from cellulose was laevoglucosan, in yields up to 40%, likewise of interest as a starting material for chemical syntheses.

Peat is another biopolymeric material which offers promise in its extractive degradation with supercritical solvents. Degradations of 80–90% of young peat could be achieved with an ethanol-water mixture (3 + 7, V/V) at 150 bar and 265–275 °C; only 4% of volatile compounds were formed [269]. The extracts were solid and 55% of them

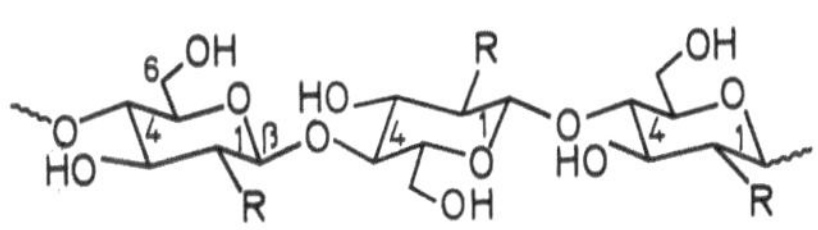

(1)

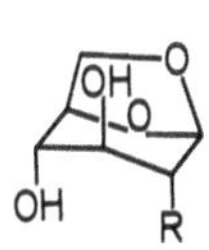

(2)

Fig. 96. Biopolymers (1): R = OH: cellulose, R = NH(Ac): chitin and degradation products (2): R′ = OH: 1,6-anhydro-β-D-gluco-pyranose (laevoglucosan), R′ = NH(Ac): 2-acetamido-1,6-anhydro-2-deoxy-β-D-glucopyranose

were water-soluble. A small chloroform-soluble fraction was also obtained and 20–30% of a product which could be dissolved only in dilute alkali solution. Under the same conditions, only 30% of old peat was broken down. At higher temperatures (295 °C) and with a higher proportion of alcohol in the mixture (40% by volume) up to 60% could be degraded, although considerably larger amounts of easily volatile products were formed during the decomposition of old peat, so that the extracted part amounted to 40% at the best.

Besides degrading biomass material the flow apparatus could be used also for preparing unsaturated carbohydrates. This was done by thermal cis-elimination (ester pyrolysis) of completely acetylated aldo- and ketopyranoses [270]. A further example is the performing of intermolecular thermal-pericyclic ene-reactions under high pressure and temperature; this is not possible under normal conditions [271].

Studies of the liquefaction of lignite form a transition from wood to coal extraction. Lignite is a soft, comparatively young brown coal with the wood structure still visible. Inexpensive solvents, such as methanol or water were suggested for supercritical extraction, and hexane used for purposes of comparison [272]. Methanol gave the best results with yields of 5.5–9%, at 120–220 bar and moderate temperatures of 250–300 °C, well below the normal pyrolysis temperatures. As temperature was increased the yield was lower at 120 bar but increased at 200 bar.

Yields of up to 30% oil were obtained from extraction of lignite with supercritical dioxan and toluene in another study [273]. The yield increased with increase in pressure (although very little above 150 bar), in temperature and in time of extraction. In choosing the temperature it must be remembered that higher temperatures augment the thermal breakdown of the polymers and the volatility of the primary products, but can lead to decomposition and re-polymerisation. The optimum temperature is generally between 300 and 400 °C; this value and its coordination with the time of extraction are of decisive importance. The composition of the extract depended on the solvent. Larger yields of hydrocarbons, principally aromatic, were found in the toluene extracts whereas dioxan extracts yielded larger amounts of the oxygen-containing compounds, mainly phenols.

Bartle et al. investigated the composition of the alkanes in extracts obtained from various Turkish lignites, using supercritical toluene and tetrahydrofuran [274]. They were able to identify n-C_{13} to n-C_{33}-hydrocarbons, acyclic isoprenoids with 18, 19 and 20 C-atoms, 2 pentacyclic triterpenes (C_{31}) and one bicyclic C_{15}-cycloalkane. Since n-alkanes are thermally more labile than acyclic isoprenoids and cycloalkanes, the pattern of the n-alkanes in the extract depended more markedly on the conditions of extraction than that of the other hydrocarbons. At 350 °C comparatively high yields of paraffin hydrocarbons were obtained, which showed no signs of thermolysis; their pattern can thus be regarded as typical for the lignite sort and as a geochemical labeller.

IV.9. Extraction of Coal

Liquid hydrocarbons are used as starting materials for preparing plastics and other products; they are also used in large amounts as fuels, especially for internal combustion motors. At present, coal liquefaction is not yet capable of competing economically with petroleum production but the prospect of exhaustion of petroleum reserves, energy crises and striving after greater independence have encouraged vigorous research on the liquefaction of hard coal. The aim of this work is to overcome the shortage of hydrogen in the coal, to increase its low H/C-ratio. Energy is required for this, which must come from the coal itself or from other sources. Many suggestions have been made of technical procedures, for example, conversion in carbon monoxide/water systems; reaction with H-donor solvents; direct and indirect methods of catalytic hydrogenation (Bergius-, Pott/Broche-, SRC-, Exxon-, Donor Solvent-, Conoco $ZnCl_2$-, Dow-liquefaction procedures). An important criterion of the hydrogenation is economic utilisation of the costly hydrogen; this is accomplished when primarily liquid products are formed and only small amounts of gas and volatile hydrocarbons.

A pre-condition for the conversion is a thermal degradation of the highly aggregated components. The primary decomposition products, mostly radicals, can repolymerise, especially at high pyrolysis temperatures, to yield gases and molecule aggregates which can no longer be converted to useful products. In order to prevent this, pyrolysis is usually carried out in the presence of a high-boiling solvent which dilutes and stabilises the initial products, and also serves to transport the powdered coal and the heat energy supplied, e.g. the LSE(Liquid Solvent Extraction)-process of the NCB (National Coal Board) in Britain. The solvent can be distilled off after the reaction and the coal extract hydrogenated in a second stage under better conditions than in the direct procedure. Difficulties are created, however, by having first to separate undissolved material of coal, and finely divided solid mineral particles from the viscous coal extract.

Extraction/pyrolysis with supercritical solvents offers good chances of overcoming the problems. The degradation products are rapidly dissolved and separated from the coal matrix as a result of the better mass transport properties; and solid particles settle out and can be separated from the extract because the supercritical solution is of low viscosity. The mild thermal breakdown of the coal is performed at temperatures of 400–500 °C, so that prospective solvents must have critical temperatures below this working value. The solvent should also be stable under the conditions of the procedure, so that it can be repeatedly used. Many substances ranging from water to alkanes have been tested with varied success. Aromatic compounds have proved to be suitable; these appear in adequate amounts as end-products of the coal liquefaction and are available at reasonable prices. Addition of a solvent with H-donor properties can stabilise the radical fragments formed in the thermolysis and improve the yield, but regeneration is then needed in a separate step. The usual extraction pressure is in the region of 100–300 bar. This is sufficient to obtain high gas densities, as needed to dissolve larger molecules, if the working temperature is only slightly above the critical temperature of the solvent. The advantages of supercritical solvents in coal extraction have been discussed by, for example, Gangoli and Thodos [275].

The first work on supercritical extraction of coal was carried out by Whitehead, Williams and other co-workers of the NCB (276–278). They developed a SGE-procedure (Supercritical Gas Extraction) for treating the coal, on the basis of which a large scale technical plant was proposed in British-American studies of coal extraction [279]. In this, the coal was to move in a solid bed in counter-current to the extracting supercritical toluene at 100 bar and 400 °C. Yields of no more than 30% are to be expected from hard coal under these conditions. Finn et al. obtained better extraction results in the further development and testing of the SGE-procedure in a pilot plant for throughput amounts of up to 10 kg/h [55]. The plant enabled the influence of the most important procedure parameters on the yield of extract and its composition to be studied in a continuous operation. A mixture of an aromatic and a hydrogenated aromatic compound was used as solvent, which could be yielded as a fraction from hydrogenation of the coal extract. This vague description of the solvent diminishes the value of the results described below.

The extraction pressure was varied from 50 to 200 bar, the temperature from 420 to 480 °C. The temperature had little influence, which suggests that the yield passes through a maximum within the narrow temperature range used. The amount of substance extracted increased continuously from ca. 35% to almost 60% (based on water- and ash-free coal) as pressure was increased and the gas density rose from 0.4 to 0.8 g/cm^3. At a high gas density a ratio of 3:1 for solvent stream/coal was sufficient to take out all extractable components. The mean molecular mass of the extract increased with rise in pressure and gas density from 450 to 700, but it fell with increasing temperature. In comparison with coal the extracts contained less sulphur and oxygen and had a better H/C-ratio, which is a positive feature with regard to the subsequent hydrogenation. The carbon content of the extract increased slightly with increase in temperature at constant pressure; increase in pressure at constant temperature had no influence on the carbon content, however. The amount of gas formed during the conversion rose with temperature from 3 to 7%, the CO_2-content of the gas falling from 50 to 30% and the carbon content rising from 30 to 50%.

Fong et al. investigated the influence of pressure, temperature and solvent on the extraction of coal (particle size of 0.7–1.2 mm) [280]. The degree of extraction with toluene rose from 5 to 18% at 140 bar when the temperature was increased from 275 to 385 °C. The increase continued steadily even when the critical temperature of 319 °C was exceeded, and levelled out above 380 °C. The effect of pressure was less marked. An increase of pressure from 30 to 140 bar (p_c = 41 bar) at 365 °C improved the degree of extraction only from 14 to 17%.

The study of various solvents at 110 bar and 400 °C yielded values between 17 and 25% for alcohols and aliphatic and aromatic hydrocarbons; tetralin led to extensive decomposition of the coal. It was shown that, within a homologous series, the larger solvent molecules possessed better dissolving power, but, beyond a certain size, the extraction result deteriorated again because of the hindrance to diffusion of the solvent within the microporous coal structure. Measurements of pore size of coal showed that 30% of the pore volume are made up of cavities of diameter of less than 12 Å, and 53% are in the region between 12 and 300 Å [281]. In comparison the naphthalene molecule measures 7 Å and the length of the 1-octanol chain is 11 Å. As a consequence, the proposal was made of extracting coal with a supercritical solvent made up of large and small molecules in order to improve the yield.

Kershaw and Jezko made a notable contribution to the extraction of South African coal with supercritical solvents [282]. They used two principal methods: the coal was extracted in an autoclave with stirring at 200 bar in a constant stream of solvent, heating so that temperature rose linearly from 150 to 450 °C (Method A); or the

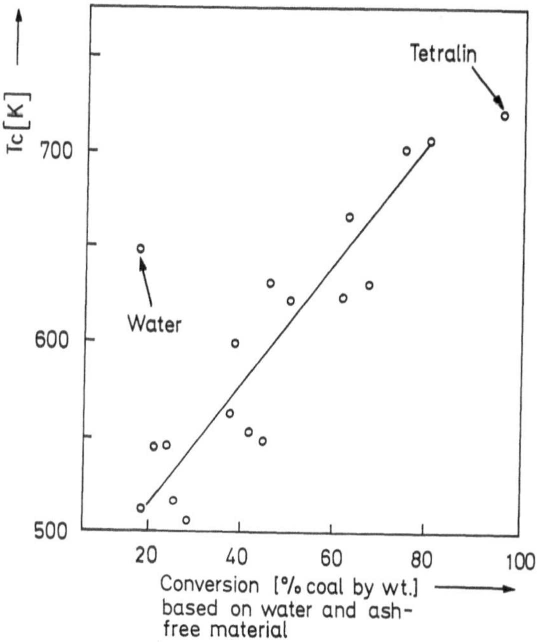

Fig. 97. Degree of conversion of coal at 200 bar and 450 °C, as a function of the critical temperature T_c of the solvent [282].
Degree of conversion (% by weight) $= 0.297 \cdot T_c - 129.3$, correlation. Coefficient $r = 0.94$

coal was mixed with sand to prevent agglomeration, then heated in a reactor so that its temperature rose at 200°/min, and extracted with toluene as solvent (Method B).

The extraction of 14 different sorts of coal showed that the degree of extraction increased as the H/C-ratio of the coal sort rose. The use of Method A with 18 different solvents under supercritical conditions at 200 bar and 450 °C showed a remarkably good correlation between the critical temperature of the solvent and the degree of conversion of coal (water excepted) (Fig. 97) although the solvents were of highly varied chemical nature. Water displayed an inferior result than expected and tetralin a better one, as a consequence of its H-donor ability.

Using Method B the influence of pressure and temperature on the extraction of coal with toluene was studied. The curve relating degree of extraction of the coal with pressure passed through a maximum at 300 bar at every temperature investigated. The maximum at 450 °C was higher than those at 350 °C and 550 °C. There is therefore no advantage of a better yield of extract by using pyrolysis temperatures above 450 °C.

The oils extracted from the coal were separated and analysed and characterised in various ways. The distribution of the n-alkanes in the extract led to the conclusion that there was practically no thermal breakdown of the coal up to 350 °C and that the supercritical solvent extracted only molecules which were held in the pores of the coal. This agrees with Vahrman's molecular sieve theory of coal [283]. Pyrolysis products of the toluene, notably 1,2-diphenylmethane, could, however, be detected with various analytical methods; these are to be expected already at 350 °C [284].

A two-stage procedure for supercritical extraction of hard coal was worked out by Wilhelm and Hedden [285, 286]. Information about the individual processes in coal extraction was derived from the results of extraction in a solid bed at constant

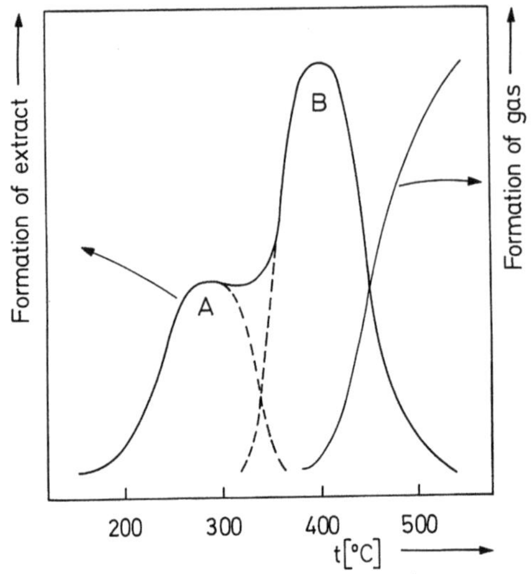

Fig. 98. Extract formation related to rise of solvent temperature during the extraction of hard coal [286]

pressure and with a constant solvent flow while heating to yield a linear rise in temperature from 150 to 550 °C. Pressure, rate of heating, particle size of the coal, nature and flow of solvent and other parameters were varied in many experiments. The amounts of extract and of gas leaving the extractor were measured at short time intervals. This yielded extraction diagrams with curve forms shown schematically in Fig. 98.

The extraction begins at 200 °C while the solvent is still liquid and reaches a plateau at 300 °C which is largely independent of the nature of the solvent. Extract formation then increases and attains a maximum in the supercritical state at 250 to 400 °C; the height of the maximum increases with the density of the solvent in the order benzene, toluene, xylene, mesitylene. The curve then falls steeply and extraction virtually ceases at 550 °C. Gaseous products are formed only above 400 °C. The least amounts of gas (ca. 100 cm^3/g coal) and the best yields of extract were given by mononuclear aromatic compounds among all the solvents tested. The degree of conversion was calculated from the loss in mass of the Saarland hard coal used and attained 30–35% by weight under optimised conditions. When a hydrogen carrier solvent, e.g. tetralin, was used, higher values (up to 70%) could be realised.

As already mentioned, these results can be interpreted by the hypothesis that coal consists of an easily extractable fraction of monomeric substances, a fraction of polymerised compounds and a fraction of components which cannot be extracted and can at the best be broken down into gas and coke. The non-polymerised substances are easily dissolved and removed completely (Curve A in Fig. 98). Higher temperature first causes thermal degradation of the polymeric part, which plastifies the coal. Depending on the conditions, a larger or smaller fraction of the plastic mass is dissolved and extracted (Curve B). The residue yields volatile compounds and coke as a result of secondary reactions. Coal of particle size less than 1 mm enables a better transport of dissolved substances and yields more extract. Adding an H-donor such as tetralin increases markedly the plasticity of the coal and prevents secondary reactions of repolymerisation by hydrogenating the fragmentation radicals formed in the thermal breakdown.

A procedure has been proposed for the supercritical extraction of coal, based on this interpretation of the course of extraction; it is termed the DSSE-process (Donor-Solvent-Supercritical-Extraction-Process). Powdered coal (particle size 0.2–0.5 mm) is suspended in tetralin as H-donor solvent and depolymerised in a first stage in a flow tube at 100 bar and 400 °C. This gives a homogeneous, viscous mass which is extracted in a second stage using supercritical toluene as solvent at 100 bar/380 °C. The viscous mass is most efficiently extracted, without re-polymerisation, in the turbulent shearing field of a jet stream. Neither stage requires a high pressure vessel of large volume. Any solid particles present in the low-viscosity, supercritical extract can be separated without difficulty. The solvent and the dehydrogenated H-donor (aromatics with two and three nuclei) are separated from the extract by distillation. The H-donor components are re-hydrogenated catalytically and fed back into the process. In this way a degree of conversion of the coal of 70 % is said to be obtained from hard coal, the products consisting of 45 % ash-free extract (of softening temperature of 140 °C), 25 % liquid hydrocarbons and 1 % gas. The coal extract can then be further processed to liquid hydrocarbons by catalytic hydro-cracking. A part of the extraction residue is gasified with oxygen and steam according to known procedures to give the hydrogen needed for this; another part of the residual coke is employed to yield the energy for carrying out the process.

Mention may be made finally of the extraction of coal with a mixture of supercritical water and carbon monoxide, carried out at temperatures around 400 °C and pressures below 100 bar. The advantages of the aqueous extraction system are, according to Penninger, the easy separation of solvent and extract after treatment and the powerful effect of the supercritical water in increasing the volatility of the high molecular hydrocarbons in the extraction; this comes about because 50 % of the oil obtained has a boiling point higher than the extraction temperature [287]. Mixtures of water with carbon dioxide, nitrogen, hydrogen and carbon monoxide were tested but only the water/carbon monoxide system gave a coal conversion of 50 % with a 40 % yield of extract. At a higher extraction pressure, extract was formed in preference to gas.

Ross and Nguyen worked extensively with the H_2O/CO-system for coal extraction. They were able to show that when strong alkalies (KOH) or certain soluble salts (potassium formate) were added to the water, the method was much more efficient [288]. The extraction products are poorer in oxygen and richer in hydrogen than the original coal, having a composition similar to that of the products obtained by extracting coal with H-donor solvents (tetralin). Discussion of the chemical processes involved indicates that an important role is played by reactions catalysed with hydroxyl ion.

The chemical compositions of the products isolated in the extraction of coal with supercritical solvents have been studied in a series of further publications [289–292].

IV.10. Miscellaneous

IV.10.a Extraction of Aqueous Solutions

The preparation of organic compounds with the help of biotechnology has been arousing increasing interest recently and is now at the focus of international research activity. Generally known examples of biotechnological procedures are fermentations (preparation of ethanol and acetic acid) and the extensive domain of cell cultures for obtaining certain metabolic products of plants, such as fatty or essential oils, alkaloids, etc. A restricting feature of biotechnological procedures is that usually only low concentrations of the desired compounds are obtained in the water. Separation of the organic compounds from the dilute aqueous solutions by classical extraction and/or distillation demands considerable energy. In addition, there are limits to the use of thermal procedures to concentrate labile natural products derived from cell cultures.

A gentle and energy-saving isolation of products prepared biotechnologically is possible by extraction with dense gases, especially when the partition coefficients of the organic compounds between the dense gas and water are sufficiently large. On the preparative scale the dilute aqueous solution is fed continuously into the head of a mass exchange column. The liquid phase moves downwards as a result of its higher density and is extracted in counter-current by a dense gas. The extraction agent containing the organic compounds and some water is drawn off at the column head and returned to the cycle after the extracted material has been precipitated. The extracted aqueous phase is drawn off at the foot of the column.

The separation of organic compounds from water with supercritical ethylene at 15 °C was the object of work by Elgin and Weinstock [293]. For example, they studied the ternary system methyl ethyl ketone (MEK)/water/ethylene in detail, showing that the partial solubility of the MEK in water was conspicuously lowered by adding dense ethylene. Snedeker investigated the separation of acetic acid/water mixtures with supercritical carbon dioxide and concluded that the extraction of the acetic acid was uneconomic because of the unfavourable partition coefficients in the system carbon dioxide-water [294].

An application which has been well studied is the extraction of ethanol with supercritical gases. Paulaitis, Gilbert and Nash described the phase behaviour of ethanol-water mixtures in combination with supercritical carbon dioxide and supercritical ethylene [59]. They investigated the effect of temperature, pressure and the extraction agent on the selectivity. The solubility of ethanol in supercritical carbon dioxide increases with increase in gas density. The selectivity of CO_2 for ethanol increases at constant pressure as the temperature falls. The authors used the values found to compare the efficiencies of a counter-current extraction and the normal distillation. An amount of energy of about 3.5 MJ/kg extract is required to concentrate a ca. 10% solution by weight of ethanol to ca. 80% by weight using the counter-current extraction with supercritical CO_2 at 35 °C and 170 bar. This is only a half of the energy needed for a distillation under normal pressure. The extraction with supercritical ethylene is less favourable than that with CO_2 since there is a lower concentration in the gas so that more solvent is required. The concentration of the alcohol cannot be increased by dense gas extraction to a value beyond the 95.5% by weight attainable as the alcohol/water azeotrope in distillation.

The separation of alcohol-water mixtures was studied also by Kuk and Montagna [295, 296]. They measured the solubilities of ethanol and isopropanol in supercritical carbon dioxide as functions of the temperature, pressure and the alcohol concentration chosen (Tables 40 and 41). As the alcohol concentration in the starting mixture was increased, the solubility in supercritical carbon dioxide rose sharply, whereas

Table 40. Solubility of ethanol in supercritical carbon dioxide [296]

Initial ethanol conc. (% by wt.)	Pressure (bar)	Temperature (°C)	Solubility (% by wt.)	Selectivity*
15	100	40	1.3	25
	100	60	1.0	15
	210	40	1.8	26
	210	60	2.2	20
30	100	40	2.8	18
	100	60	2.3	13
	210	40	3.6	22
	210	60	4.0	16
50	100	40	5.0	6
	100	60	3.0	2
	210	40	7.7	8
	210	60	8.6	8
70	100	40	8.7	4.0
	100	60	4.8	1.5
	210	40	12.5	4.2
	210	60	15.1	3

$$* \text{ Selectivity} = \frac{(\text{ethanol/water}) \text{ in supercritical gas phase}}{(\text{ethanol/water}) \text{ in liquid phase}}$$

the selectivity of CO_2 for ethanol decreased. At pressures below a certain inversion point (170 bar according to Paulaitis [59]), lower solubilities were attained at the higher temperature; at higher pressures, solubilities were observed to increase with rise in temperature. As seen from the data in Table 40, CO_2 is more selective for ethanol at 40 °C than at 60 °C. The solubilities of isopropanol and ethanol in supercritical carbon dioxide are compared in Table 41. Isopropanol has a more favourable partition coefficient so that a higher concentration in the gas is attained. Extraction with supercritical carbon dioxide concentrated isopropanol up to 86% by weight, only just less than that in the isopropanol-water azeotrope.

According to McHugh, Mallet and Kohn, who studied the ternary system ethanol-water-ethane, the solubility of ethanol in ethane at 40 °C and 50 or 80 bar was independent of the composition of the starting mixture and amounted to 2.15 or 3.1% by

Table 41. Comparison of the solubilities of isopropanol and ethanol in supercritical carbon dioxide at 100 bar and 40 °C [296]

Initial alcohol concentration (% by weight)	Solubility in % by weight	
	Isopropanol	Ethanol
30	8.5	2.8
50	15.6	5.0
70	17.4	8.7

weight, respectively [297]. Large amounts of supercritical ethane were required because of the low solubility of ethanol. Here, too, the alcohol concentration of the ethanol-water azeotrope could not be exceeded.

Brunner and Kreim also worked on the separation of alcohols from aqueous solutions with the help of carbon dioxide and ethane [62]. The separation factors of the CO_2-extraction were distinctly higher than in the distillation under atmospheric pressure at lower ethanol starting concentrations. As this concentration in the liquid phase increased, the differences became smaller. The data obtained enabled an estimate to be made of the cost of separation. This came to about 0.1 U.S. dollar (1984 value)

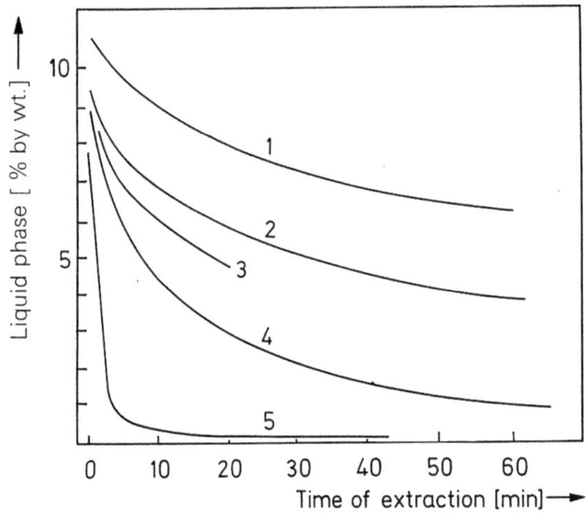

Fig. 99. Extraction of organic compounds from aqueous solutions using supercritical carbon dioxide [62].
1 ethylene glycol/water; 2 acetic acid/water; 3 glycerol/water; 4 ethanol/water; 5 n-butanol/water and isobutanol/water.
Extraction conditions: 150 bar, 60 °C, with exception of 150 bar, 95 °C for ethanol

Table 42. Partition coefficients of some alcohols and esters between water and liquid carbon dioxide at 16 °C [1]

Compound	Partition coefficient
Methanol	0.4
n-Propanol	0.66
n-Butanol	1.8
n-Pentanol	4.5
n-Hexanol	15.0
n-Heptanol	31.0
i-Propanol	0.35
i-Butanol	1.7
tert.-Butanol	0.82
sec.-Butanol	1.23
i-Pentanol	5.0
Ethyl acetate	42
Isopropyl acetate	80
Methyl butyrate	120
Butyl acetate	350
Propyl propionate	370
Isoamyl acetate	850

for the isolation of 1 kg 90 % ethanol. In addition to extraction of ethanol they extracted other compounds from aqueous solutions using supercritical carbon dioxide at 150 bar and 60 °C. As seen in Fig. 99, separation of ethylene glycol was the slowest, followed by acetic acid, glycerol and ethanol in that order of increasing speed. The extraction of higher alcohols was the fastest, for example the butanols. These alcohols are relatively lipophilic and are extracted much faster on account of their higher partition coefficients.

Many organic substances, e.g. ethanol and acetic acid, are very soluble in dense carbon dioxide but their strong affinity for water is an obstacle to their extraction from aqueous solutions. Table 42 contains the partition coefficients of some alcohols and esters between water and liquid carbon dioxide at 16 °C [1].

Higher alcohols, esters and flavouring components have large partition coefficients, making these compounds more attractive examples for extraction from aqueous solutions using dense gases.

IV.10.b Adsorbents

Active charcoal is used on a large scale industrially, for example to remove organic compounds from waste water. For economic reasons the adsorbent must be regenerated. The methods used most frequently for this are thermal desorption with steam at high temperatures up to 900 °C and classical solvent extraction. An alternative way is extractive regeneration with supercritical gases which is advantageous on account of the low energy requirement.

Comprehensive research and development work in the USA have confirmed that it is possible to desorb organic compounds from active charcoal relatively quickly and efficiently with supercritical carbon dioxide [58, 298, 299]. The procedure was tested on various chemicals and pesticides. The draft of an installation with appropriate estimates of costs has shown that both the plant and the running costs are clearly lower than in the classical thermal regeneration processes.

A basic investigation of the problems of adsorption/desorption of organic compounds in the system active charcoal/carbon dioxide was made by Paulaitis and Kander [300]. They measured the adsorption isotherms for phenol on active charcoal from supercritical carbon dioxide at 36 °C/140 bar or 60 °C/175 bar, and, for purposes of comparison, from aqueous solutions at 36 °C and 60 °C. In both cases the charcoal becomes more charged as temperature is lowered at constant phenol concentration in the supercritical or aqueous phases; the increase in the charge was much larger for adsorption from the supercritical carbon dioxide. Table 43 gives the adsorption values for phenol on active charcoal from supercritical CO_2.

The charge of the active charcoal rises steeply with increase in the phenol concentration in the supercritical gas phase. In contrast, desorption from charged active charcoal can take place in pure supercritical gas. A dense gas of high dissolving power has thus the advantage of increasing the rate of desorption. The solubility of phenol in supercritical carbon dioxide at 60 °C increases by a factor of 10 when the pressure is raised from 100 to 300 bar [70].

It was ascertained during the investigation of the adsorption/desorption cycles of phenol in the system active charcoal/carbon dioxide that the adsorption capacity of the charcoal diminishes noticeably after the first adsorption; the complete capacity

Table 43. Adsorption of phenol on active charcoal from supercritical carbon dioxide [300]

36 °C/140 bar		60 °C/175 bar	
Conc. of phenol in supercr. CO_2 (mg/g)	Charge of charcoal (mg/g)	Conc. of phenol in supercr. CO_2 (mg/g)	Charge of charcoal (mg/g)
7.21	611	9.97	568
6.68	558	0.974	355
0.65	512	0.04	108
0.04	241	0.004	43
0.004	142		

cannot be regenerated with supercritical carbon dioxide. For instance, the capacity of 280 mg phenol per g fresh active charcoal fell after the first cycle to 132 mg/g. This capacity then changed only little following further cycles of adsorption and desorption. Similar results were described by de Filippi for the adsorption/desorption of alachlor on active charcoal [60]. The adsorbent regenerated after the first cycle then possessed a charge capacity of only 60% of that of the fresh charcoal. These results show that supercritical carbon dioxide is not wholly suitable for desorbing adsorbed compounds, such as phenol. However, high pressure extraction offers an alternative to conventional procedures for regenerating active charcoal charged with more weakly held substances.

The use of carbon-containing adsorbents is limited by the irreversible adsorption of some compounds. Fetzer et al. investigated the use of supercritical pentane for extracting firmly held polychlorinated organic compounds [301]. It could be demonstrated that pentane in the supercritical state extracted other substances from the adsorbent than only these halides. For example, alkanes, dienes and small amounts of alkyl-substituted benzenes were extracted from active charcoal.

In this connection it may be mentioned that the adsorption on active charcoal can be circumvented by extracting waste waters, contaminated with organic substances, directly with dense carbon dioxide.

IV.10.c Decontamination of Drugs

Plant protection agents (insecticides, herbicides, fungicides) find general use in agriculture to ensure good harvests. Misuse and abuse of these agents can lead repeatedly to inadmissably high residues of the substances and their metabolites in drug plants. It is legally forbidden to marked and use such contaminated products. They are either destroyed, blended with less polluted charges or subjected to decontamination.

It is scarcely possible to decontaminate such tainted drugs with classical solvents since these are not selective enough and leave inevitably hazardous residues in the plant material. After such treatment a possible morphological change of the plant material may have to be taken into consideration.

In some cases, however, dense gases can be used successfully to decontaminate drugs. Stahl and Rau were able to show that lipophilic pesticides could be easily

Table 44. Pesticide amounts on senna leaves before and after dense gas extraction with CO_2 [61]

Material/method	Pesticide	Content of pesticide before extn. (ppb)	Content of pesticide after extn. (ppb)	Maximum legally (German) permitted content (ppb)
Origin A	HCH-isomers	245	8	200
380 bar, 60 °C	γ-HCH	105	4	500
11 hours	Aldrin	3	trace	100
49 g CO_2/g drug	DDT (total)	80	5	1000
Origin B	HCB	765	below detect. limit	100
350 bar, 60 °C	γ-HCH	10	below detect. limit	500
160 min	d-endosulfan	1860	110	30000
175 g CO_2/g drug	PCB	280	below detect. limit	10
	DDT (total)	111	13	1000
Origin B	HCB	765	below detect. limit	100
113–215 bar, 60 °C	γ-HCH	10	below detect. limit	500
150 min	d-endosulfan	1860	42	30000
106 g CO_2/g drug	PCB	280	below detect. limit	10
	DDT (total)	111	4	1000

extracted with supercritical carbon dioxide at 40 °C and pressures above 80–100 bar [61]. Notable examples were the halides difficult to degrade, e.g. hexachlorocyclohexane (HCH), hexachlorobenzene (HCB), aldrin, DDT, heptachlor and endosulfan (some are not permitted in the German Federal Republic but allowed in other countries). The example of contaminated senna leaves is described here; the active principles, the polar sennosides, are insoluble in CO_2. The results of the extractions are given in Table 44.

The contamination of the senna leaves with lipophilic pesticides could be reduced by amounts ranging from 88 to 99% through dense gas extraction. The concentration of the residues of some of the active substances was in some cases below the limit of detection (GC with electron capture detector). Since the anthraglycosides, which determine the value of the drug, are so polar as to be insoluble in supercritical carbon dioxide, their content in the extracted drug remained unchanged. Drugs which could not be used because of their excessive contamination became commercially available in the eyes of the law after a dense gas extraction. The extraction also removed lipophilic cuticular waxes. Hydrophilic plant constituents were then liberated more readily in the infusion of the senna leaves used; this is a desirable side-effect of the extraction.

In another example it was possible to show that partly polar urea herbicides, e.g. isoproturon, could be largely removed from thorn apple herb by dense gas extraction without loss of content of the effective tropane alkaloids [302].

This procedure has a limitation in that predominantly only plants with hydrophilic constituents of value can be treated; lipophilic components, such as essential or fatty oils, would be co-extracted unless they occur in intact seeds, from the surface of which the contaminants can be separated. A further restriction is the fact that good results are obtained only with lipophilic pesticides; more polar substances are less soluble and require a large throughput of carbon dioxide.

The pesticide problem must be regarded from a diametrically opposite side when the extract and not the plant material is the desired product for further use. In connection with hop extraction, Bett studied the solubilities in liquid carbon dioxide of the plant protection agents, endosulfan, mephosfolan, dicofol, dinocap and triadimefon, used in hop cultivation: for comparison solubilities in the customary solvents hexane and dichloromethane were also measured [193]. All the pesticides except triadimefon were sometimes distinctly less soluble in liquid CO_2 at 10 °C. This is, however, unimportant in practice because exhaustive extraction of hops with liquid CO_2 demands large volumes of solvent (at least 20 kg CO_2/kg hops), which must always suffice for dissolving the pesticides completely. A concentration of pesticides in the hop extracts can be avoided as always only by an appropriate and controlled use of the plant protection agents during the hop cultivation. According to Vollbrecht, more polar pesticides such as dithiocarbamates (DTCs) and their metabolites are present only in traces in CO_2-extracts of hops, even when the plant material has been unusually liberally treated [199].

As long ago as 1973, Gjavotchanoff investigated the separation of traces of insecticides from aqueous drug extracts with the help of carbon dioxide and pentane in liquid-liquid partition under pressure [303].

A special application may be mentioned incidentally, namely, the removal of undesired polychlorinated biphenyls (PCBs) from transformer oil [60]. Up to some years ago, PCBs were employed for cooling in transformers. After their hazardous nature had been recognised, they were replaced by silicone oils, mineral oils, phthalate esters, etc. These oils contain nevertheless sometimes undesirably high contents of PCBs from mixing with residual oil. Carbon dioxide at 100 °C and pressures up to 200 bar, for example, was shown to be an adequately selective solvent for removing PCBs from the particular oils. The PCBs from silicone oil and phthalate esters were concentrated in the extract; those from mineral oil, in the residue.

References

1. Schultz, W. G., Randall, J. M.: Food Technol. *24*, 1282 (1970)
2. Stahl, E., Schilz, W.: Chem. Ing. Tech. *48*, 773 (1976)
3. Glatz, A.: Thesis, Saarbrücken, 1985
4. Stahl, E., Glatz, A.: Fette, Seifen, Anstrichm. *86*, 346 (1984)
5. Stahl, E., Willing, E.: Pharm. Ind. *42*, 1136 (1980)
6. Stahl, E., Willing, E.: Mikrochim. Acta *1980/II*, 465
7. Brunner, G., Peter, S.: Chem. Ing. Tech. *53*, 529 (1981)
8. Irani, C. A., Funk, E. W.: Recent Developments in Separation Science, Vol. III, part A, p. 171, CRC Press, Palm Beach, Florida (1977)
9. Tsekhanskaya, Y. V., Iomtev, M. B., Mushkina, E. V.: Zh. Fiz. Khim. *38*, 2166 (1964)
10. Zhuse, T. P., Jushkevic, G. N., Gekker, J. E.: Maslob.-Zhir. Prom. *24*, 34 (1958)
11. Vitzthum, O., Hubert, P.: Germ. pat. appl. 21 27 596 (HAG AG), filed 1971
12. Zosel, K.: Germ. pat. appl. 23 32 038 (Studiengesellschaft Kohle mbH), filed 1973
13. Zosel, K.: Germ pat. 24 41 152 (Studiengesellschaft Kohle mbH), filed 1974
14. Peter, S., Brunner, G., Riha, R.: Fette, Seifen, Anstrichm. *78*, 45 (1976)
15. Zosel, K.: Angew. Chem. *90*, 748 (1978)

16. Panzner, F., Ellis, S. R. M., Bott, T. R.: Inst. Chem. Eng. Symp. Ser. *54*, 165 (1978); — see also Chem. Ind. *1980*, 228
17. Stahl, E., Schütz, E., Mangold, H. K.: J. Agric. Food Chem. *28*, 1153 (1980)
18. Stahl, E., Quirin, K.-W.: Germ. pat. appl. 32 07 914 (SKW Trostberg AG), filed 1982
19. Quirin, K.-W.: Fette, Seifen, Anstrichm. *84*, 460 (1982)
20. Kaufmann, W., Biernoth, G., Frede, E., Merk, W., Precht, D., Timmen, H.: Milchwissenschaft *37*, 92 (1982)
21. List, G. R., Friedrich, J. P., Pominski, J.: J. Amer. Oil Chem. Soc. *61*, 1847 (1984)
22. Braun, G., Schmidt, H.: Ber. Bunsenges. Phys. Chem. *88*, 891 (1984)
23. Stahl, E., Quirin, K.-W.: Fette, Seifen, Anstrichm. *87*, 219 (1985)
24. Pekhov, A. V., Ponomarenko, I. Ya.: Maslozhir. Prom. *34*, 25 (1968)
25. Vitzthum, O., Hubert, P., Sirtl, W.: Germ. pat. 21 27 618 (HAG AG), filed 1971
26. Vitzthum, O., Hubert, P.: Germ. pat. 21 27 611 (Studiengesellschaft Kohle mbH), filed 1971
27. Prokopcuk, A. F.: Izv. Piscev. Technol. *3*, 7 (1974)
28. Stahl, E., Schütz, E.: Arch. Pharm. *311*, 992 (1978)
29. Stahl, E., Schütz, E.: Planta med. *40*, 12 (1980)
30. Schütz, E. et al.: Germ. pat. appl. 31 19 454 (SKW Trostberg AG), filed 1981
31. Calame, J. P., Steiner, R.: Chem. Ind. *1982*, 399
32. Stahl, E., Gerard, D.: Z. Lebensm. Unters. Forsch. *176*, 1 (1983)
33. Moyler, D. A.: Perf. Flavor. *9*, 109 (1984)
34. Gerard, D.: Chem. Ing. Tech. *56*, 794 (1984)
35. Stahl, E., Gerard, D.: Perf. Flavor. *10(2)*, 29 (1985)
36. Best, W. et al.: Germ. pat. appl. 29 43 267 (BASF AG), filed 1979
37. Gerard, D.: Diploma thesis, Saarbrücken, 1980
38. Fremont, H. A.: US Patent 4 308 200 (Champion International Corp.), filed 1980
39. Coenen, H., Hagen, R.: Gordian *83*, 164 (1983)
40. Roselius, W., Vitzthum, O., Hubert, P.: Germ. pat. 20 43 537 (Studienges. Kohle mbH), filed 1970
41. Zosel, K.: Germ. Pat. 20 05 293 (Studiengesellschaft Kohle mbH), filed 1970
42. British Patent Specification 1 333 362 (HAG AG), filed 1972
43. Schröder, E., Arndt, K.-F.: Z. Polym. Forsch. *27*, 135 and 141 (1976)
44. King, J. W.: Polym. Mat. Sci. Eng. *51*, 707 (1984)
45. Braun, G.: Chem. Ing. Tech. *56*, 856 (1984)
46. Köll, P., Metzger, J.: Angew. Chem. *90*, 802 (1978)
47. Calimli, A., Olcay, A.: Holzforschung *32*, 7 (1978)
48. Köll, P., Brönstrup, B., Metzger, J.: Holzforschung *33*, 112 (1979)
49. McDonald, E. C., Howard, J., Bennett, B.: Fluid Phase Equil. *10*, 337 (1983)
50. Bartle, K. D., Martin, T. G., Williams, D. F.: Fuel *54*, 226 (1975)
51. Bott, T. R.: Chem. Ind. *1980*, 228
52. Eisenbach, W., Niemann, K.: Erdöl u. Kohle *34*, 296 (1981)
53. Coenen, H., Rinza, P.: Chem. Ing. Tech. *54*, 386 (1982)
54. Stützer, D., Brunner, G., Peter, S.: Fluid Phase Equil. *10*, 345 (1983)
55. Finn, M. J., Bower, C. J., Hughes, R. D.: Fluid Phase Equil. *10*, 327 (1983)
56. Cottle, J. E.: US Patent 3 119 771 (Phillips Petroleum Comp.), filed 1960
57. Arnaud, M., Wuhrmann, J. J.: Brit. Patent 1 446 140 (Nestle S.A.), filed 1973
58. Modell, M.: US Patent 4 124 528 (Arthur D. Little Inc.), filed 1974
59. Paulaitis, M. E., Gilbert, M. C., Nash, C. A.: Paper presented at the 2nd World Congr. of Chem. Engng., Montreal 1981
60. De Filippi, R. P.: Chem. Ind. *1982*, 390
61. Stahl, E., Rau, G.: Planta med. *1984*, 171
62. Brunner, G., Kreim, K.: Chem. Ing. Tech. *57*, 550 (1985)
63. Stahl, E., Quirin, K.-W., Gerard, D.: Fette, Seifen, Anstrichm. *85*, 458 (1983)
64. Kurnik, R. T., Reid, R. C.: Fluid Phase Equil. *8*, 93 (1982)
65. van Gunst, C. A.: Thesis, Delft, 1950
66. Anon: Europa Chemie *32*, 562 (1981)
67. Pardun, H.: Analyse der Nahrungsfette, Verlag Paul Parey, Berlin, 1976

68. Schwab, A.: Jojoba – Ein hochwertiges Pflanzenöl aus der Wüste, Ledermann-Verlag, Bad Wörishofen, 1981
69. Brogle, H.: Chem. Ind. *1982*, 385
70. Van Leer, R. A., Paulaitis, M. E.: J. Chem. Eng. Data *25*, 257 (1980)
71. Eggers, R., Sievers, U., Stein, W.: J. Amer. Oil Chem. Soc. *62*, 1222 (1985)
72. Hederer, H.: Thesis, Erlangen–Nuremberg 1981
73. Brunner, G., Peter, S.: Sep. Sci. Technol. *17*, 199 (1982)
74. Prausnitz, J. M.: Molecular Thermodynamics of Fluid-Phase Equilibria, Prentice Hall, 1969
75. Perry, C. E. S., Weber, W. H., Daubert, B. F.: J. Amer. Chem. Soc. *71*, 3720 (1949)
76. Peter, S., Brunner, G.: Angew. Chem. *90*, 794 (1978)
77. Peter, S., Brunner, G., Riha, R.: Ger. Chem. Eng. *1*, 26 (1978)
78. Ebeling, H., Franck, E. U.: Ber. Bunsenges. Phys. Chem. *88*, 862 (1984)
79. Snyder, J. M., Friedrich, J. P., Christianson, D. D.: J. Amer. Oil Chem. Soc. *61*, 1851 (1984)
80. Brannolte, H. D., Mangold, H. K., Stahl, E.: Chem. Phys. Lipids *33*, 297 (1983)
81. Leiner, S.: Thesis, Saarbrücken 1986
82. Magne, F. C., Skau, E. L.: J. Amer. Oil Chem. Soc. *30*, 288 (1953)
83. Rao, R. K., Arnold, L. K.: J. Amer. Oil Chem. Soc. *33*, 82 und 389 (1956)
84. Aquilera, J. M.: Alcohol-Based Process for Oilseed Extraction, Paper presented at 42nd Annual Meeting of IFT, Las Vegas, NV, June 1982
85. Friedrich, J. P.: US-Patent Appl. No. 364290 (US Secretary of Agriculture), filed 1982
86. Brunner, G.: DSc Thesis, Erlangen–Nuremberg 1978
87. King, M. B., Alderson, D. A., Fallah, F. H., Kassim, D. M., Kassim, K. M., Sheldon, J. R., Mahmud, R. S.: p. 31 in *Chemical Engineering at Supercritical Conditions*, eds. Paulaitis, M. E., Penninger, J. M. L., Gray, R. D., Davidson, P., Ann Arbor Science, 1983
88. Chrastil, J.: J. Phys. Chem. *86*, 3016 (1982)
89. Czubryt, J. J., Myers, M. N., Giddings, J. C.: J. Phys. Chem. *74*, 4260 (1970)
90. Dickinson, J. T.: US-Patent 2660590 (1947)
91. Palmer, G. H., Fanwood, N. J.: US-Patent 2658907 (1950)
92. Groll, H. P. A.: Germ. pat. appl. 1079636 (1953)
93. Zosel, K.: Germ. pat. appl. 1493190 (Studienges. Kohle mbH), filed 1964
94. Friedrich, J. P., List, G. R., Heakin, A. J.: J. Amer. Oil Chem. Soc. *59*, 288 (1982)
95. U.S. Department of Agriculture, Economic Research Service, Fats and Oils outlook and situation, FOS 310, February 1983
96. Christianson, D. D., Friedrich, J. P., List, G. R., Warner, K., Bagley, E. B., Stringfellow, A. C., Inglett, G. E.: J. Food Sci. *49*, 229 (1984)
97. Quirin, K.-W.: Thesis, Saarbrücken 1984
98. List, G. R., Friedrich, J. P., Christianson, D. D.: J. Amer. Oil Chem. Soc. *61*, 1849 (1984)
99. Anon.: J. Amer. Oil Chem. Soc. *62*, 1166 (1985)
100. God, F. E., El-Zalaki, E. M.: Fette, Seifen, Anstrichm. *82*, 450 (1980) and J. Amer. Oil Chem. Soc. *59*, 488 (1982)
101. List, G. R., Friedrich, J. P., Pominski, J.: J. Amer. Oil Chem. Soc. *61*, 1847 (1984)
102. Fleetwood, J. G., Hudson, B. J. F.: J. Food. Technol. *17*, 11 (1982)
103. Stahl, E., Quirin, K.-W., Mangold, H. K.: Fette, Seifen, Anstrichm. *83*, 472 (1981)
104. Hatzold, T., Groß, R., Elmadfa, I.: Fette, Seifen, Anstrichm. *84*, 59 (1982)
105. Blaicher, F. M., Nolte, R., Mukherjee, K. D.: J. Amer. Oil Chem. Soc. *58*, 761 (1981)
106. Edwards, D. E., Clarke, F. H., Douglas, B.: Can. J. Chem. *32*, 235 (1954)
107. Kolthoff, J. M., Bosch, W.: Rec. Trav. Chim. *47*, (1928)
108. Gross, R., van Baer, E.: Z. Ernährungsw. *14*, 224 (1975)
109. Bunzenberger, G., Lack, E., Marr, R.: Chem. Ing. Tech. *55*, 320 (1983)
110. Eggers, R., Stein, W.: Fette, Seifen, Anstrichm. *86*, 10 (1984)
111. Kramer, J. K. G., Sauer, F. D., Pigden, W. J.: High and Low Erucic Acid Rapeseed Oils, Academic Press, Toronto, 1983
112. Bulley, N. R., Fattori, M., Meisen, A., Moyls, L.: J. Amer. Oil Chem. Soc. *61*, 1362 (1984)
113. Stahl, E., Quirin, K.-W., Mangold, H. K.: Chem. Phys. Lipids *31*, 319 (1982)

114. Anon.: Food Develop. Aug. 1981 p. 34 and Food Eng. Intl., Oct. 1981 p. 45
115. Schneider, F., Sirtl, W.: U.S. Patent 4280961 (Haussener, E.), filed 1980
116. Messmore, H. E.: U.S.-Patent 2420185 (Phillips Petroleum Co.), 1943
117. Zhuze, T. P., Kapelyushnikov, M. A.: Russ. Patent 113325 (1955)
118. Zhuze, T. P., Safranova, T. P., Jushkevic, G. I.: Izvest. Akad. Nauk SSSR *11*, 123 (1958)
119. Panzner, F., Ellis, S. R. M., Bott, T. R.: Proc. Int. Solvent Extraction Conf., vol. 2685, Toronto, 1977; see also: Chem. Ind. *1980*, 228
120. Valteris, R. L.: Birmingham University Chem. Eng. *17*, 38 (1966)
121. Metzner, J., Bekemeier, H., Paintz, M., Schneidewind, E.: Pharmazie *34*, 97 (1979)
122. Pepeljnjak, S., Jalsenjak, I., Maysinger, D.: Pharmazie *37*, 864 (1982)
123. Quirin, K.-W., Gerard, D.: Private communication of FLAVEX Naturextrakte GmbH
124. Verbiscar, A. J., Banigan, T. F., Weber, C. W., Reid, B. L., Swingle, R. S., Trei, J. E., Nelson, E. A.: J. Agric. Food Chem. *29*, 296 (1981)
125. Quirin, K.-W., Gerard, D.: Private communication of FLAVEX Naturextrakte GmbH
126. Stahl, E., Quirin, K.-W., Blagrove, R. J.: J. Agric. Food Chem. *32*, 938 (1984)
127. Weder, J. K. P.: Z. Lebensm. Unters. Forsch. *171*, 95 (1980)
128. Weder, J. K. P.: Food Chem. *15*, 175 (1984)
129. Friedrich, J. P.: US-Patent Appl. No. 534015 (US Secretary of Agriculture), filed 1983
130. Coenen, H., Kriegel, E.: Germ. pat. appl. 2843920 (Fried. Krupp GmbH) filed 1978
131. Coenen, H., Kriegel, E.: Chem. Ing. Tech. *55*, 890 (1983)
132. Eisenbach, W.: Ber. Bunsenges. Phys. Chem. *88*, 882 (1984)
133. Heigel, W., Hüschens, R.: Germ. pat. appl. 3011185 (Kali-Chemie Pharma GmbH) filed 1980
134. Coenen, H., Hagen, R.: Germ. pat. appl. 3229041 (Fried. Krupp GmbH) filed 1982
135. Stahl, E., Quirin, K.-W., Hübgen, A.: Europ. Pat. 0137214 (SKW Trostberg AG) filed 1983
136. Peter, S., Schneider, M., Weidner, E., Ziegelitz, R.: Chem. Ing. Tech. *58*, 148 (1986)
137. Weidner, E.: Thesis, Erlangen 1985
138. Treibs, W.: in Gildemeister, E., Hoffmann, F.: Die ätherischen Öle, Akademie-Verlag, Berlin 1956
139. Weast, R. C.: Handbook of Chemistry and Physics, 55th Ed., CRS Press, Inc., Cleveland USA 1974/75
140. Poynting, J. H.: Philos. Mag. (4) *12*, 32 (1882)
141. Brand, K., Preuss, R. E. G.: Pharmaz. Zentralhalle Deutschland *80*, 441 (1939)
142. Francis, A. W.: J. Phys. Chem. *58*, 1099 (1954)
143. Schultz, W. G.: Liquid CO_2 for selective aroma extraction, Paper presented at the 26th Annual Meeting of the Institute of Food Technologists, Portland Oregon, May 1966
144. Maier, M., Stephan, K.: Chem. Ing. Tech. *56*, 222 (1984)
145. Stahl, E., Quirin, K.-W.: Fluid Phase Equil. *10*, 269 (1983)
146. Wiebe, R., Gaddy, V. L.: Chem. Rev. *29*, 475 (1941)
147. Won, K. W.: Fluid Phase Equil. *10*, 191 (1983)
148. Brunner, G.: Fluid Phase Equil. *10*, 289 (1983)
149. Behr, N., v. Ettingshausen, O., van der Mei, H., Wüst, R.: Europ. Pat. Appl. 0058365 (Henkel KGaA) filed 1982
150. Sims, M.: Europ. pat. appl. 0026559, filed 1980
151. Gesetz über den Verkehr mit Absinth v. 27. 04. 1923 (RGBI I S. 257) i. d. l. F. vom 02. 03. 1974 (BGBL I S. 449, 550) (Law dealing with absinth)
152. Stahl, E.: Naturwissenschaften *39*, 571 (1952)
153. Stahl, E., Gerard, D.: Parfuem. Kosmet. *64*, 237 (1983)
154. Stahl, E., Keller, K.: Planta med. *47*, 75 (1983)
155. Stahl, E., Gerard, D.: Parfuem. Kosmet. *63*, 117 (1982)
156. Gerard, D., Quirin, K.-W.: Private communication of FLAVEX Naturextrakte GmbH
157. Behr, N., van der Mei, H., Sirtl, W., Schnegelberger, H., v. Ettingshausen, O.: Europ. pat. appl. 0023680 (Henkel KGaA), filed 1980
158. Bestmann, H. J., Erler, J., Vostrowsky, O.: Z. Lebensm. Unters. Forsch. *180*, 491 (1985)
159. Katyuzhanskaya, A. N.: Tr. Krasnodar. Nauch.-Issled. Inst. Pishch. Prom. *4*, 177 (1967)
160. Hubert, P., Vitzthum, O. G.: Angew. Chem. *90*, 756 (1978)

161. Weber, H., Gährs, H. J.: Fleischwirtsch. *63*, 1747 (1983)
162. Vollbrecht, R., Schütz, E., Sandner, K.: Germ. pat. appl. 31 18 160 (SKW Trostberg AG), filed 1981
163. Quirin, K.-W., Gerard, D., Kraus, J.: Gordian *86*, 156 (1986)
164. Meerov, Y. S., Bykova, S. F.: Maslo-Zhir. Prom. *37*, 22 (1971)
165. Bykova, S. F., Propova, S. A., Pekhov, A. V.: Maslo-Zhir. Prom. *37*, 33 (1971)
166. Bykova, S. F., Aleksandrov, L. G., Meerov, Y. S., Anoshin, I. M.: Ser. Tekh. Nauk *2*, 18 (1974)
167. Meerov, Y. S., Popova, S. A., Ponomarenko, I. Y.: Tr. Krasnodar Nauch.-Issled Inst. Pishch. Prom. *5*, 203 (1969)
168. Bykova, S. F., Popova, S. A., Pekhov, A. V.: Maslo-Zhir. Prom. *37*, 25 (1971)
169. Coenen, H., Hagen, R., Knuth, M.: Germ. patent 31 14 593 (Fried. Krupp GmbH) filed 1981
170. Schütz, E., Vollbrecht, R., Sandner, K., Sand, T., Mühlnickel, P.: Germ. pat. appl. 31 37 230 (SKW Trostberg AG, Haarmann & Reimer GmbH), filed 1981
171. Gerard, D.: Chem. Ing. Tech. MS 1280/84 (1984)
172. Roslyakova, T. K. et al.: Tekh. Nauki *1983*, 84
173. Gerard, D.: Thesis, Saarbrücken 1985
174. Gerard, D., Quirin, K.-W.: Private communication of FLAVEX Naturextrakte GmbH
175. Setijadi, T.: Private communication
176. Muravlev, I. A., Smetanin, Y. I.: Aktualnye voprosy farmacii *2*, 244 (1974)
177. Smetanin, Y. I.: Aktualnye voprosy farmacii *2*, 247 (1974)
178. Stahl, E., Schütz, E.: Planta med. *40*, 262 (1980)
179. Quirin, K.-W., Gerard, D.: Private communication of FLAVEX Naturextrakte GmbH
180. Russian patent 167 798
181. British Patent Specification 1 388 581 (HAG AG), filed 1972
182. Canadian Patent 98 72 50 (Studienges. Kohle mbH), filed 1972
183. US-Patent 4 104 409 (Studienges. Kohle mbH), filed 1976
184. Kurzhals, H.-A., Hubert, P.: Germ. pat. appl. 28 44 781 (HAG AG), filed 1978
185. Müller, A.: Germ. pat. appl. 28 27 002, filed 1980
186. Japanese patent application 44 864
187. Europ. pat. appl. 0 020 086, filed 1980
188. British Patent Specification 1 576 729 (Brewing Patents Ltd.)
189. Laws, D. R. J.: Germ. pat. appl. 28 01 843
190. Müller, A.: Germ. pat. appl. 29 20 765
191. Laws, D. R. J., Bath, N. A., Pickett, J. A.: J. Am. Soc. Brewing Chemists *35*, 187 (1977)
192. Sharpe, F. R., Grimmett, C., Laws, D. R. J., Bett, G.: J. Inst. Brewing *86*, 234 (1980)
193. McRae, J. B., Wheldon, A. G., Bett, G.: J. Inst. Brewing *86*, 296 (1980)
194. Laws, D. R. J.: J. Inst. Brewing *87*, 24 (1981)
195. Gardner, D. S.: Chem. Ind. *1982*, 402
196. Wilson, R. J. H.: p. 275 in High Pressure Chemical Engineering, GVC-VDI, Düsseldorf, 1984
197. Krüger, E.: Mschr. Brauerei *33*, 104 (1980)
198. Krüger, E.: Lebensmittel-Wissenschaft und Technologie LWT-Ed. *1983*, 7 (Prog. Food Eng.) 217
199. Vollbrecht, R.: Chem. Ind. *1982*, 397
200. Gehrig, M.: p. 281 in High Pressure Chemical Engineering, GVC-VDI, Düsseldorf, 1984
201. Stahl, E., Quirin, K.-W., Glatz, A., Gerard, D., Rau, G.: Ber. Bunsenges. Phys. Chem. *88*, 900 (1984)
202. Hohnbom, B.: J. Amer. Oil Chem. Soc. *55*, 876 (1978)
203. Azarnoff, D. L., Tucker, D. R.: Biochim. Biophys. Acta *70*, 586 (1963)
204. Gerard, D.: Diploma Thesis, Saarbrücken 1980
205. Stahl, E., Schilz, W., Schütz, E., Willing, E.: Angew. Chem. *90*, 778 (1978)
206. Willing, E.: Thesis, Saarbrücken 1981
207. Zosel, K.: Germ. pat. appl. 22 21 560 (Studienges. Kohle mbH), filed 1972

208. Roselius, W., Vitzthum, O., Hubert, P.: Germ. patent 21 19 678 (Studienges. Kohle mbH), filed 1971
209. Roselius, W., Vitzthum, O., Hubert, P.: Germ. patent 21 06 133, (Studienges. Kohle mbH), filed 1971
210. Zosel, K.: US-Patent 4 260 639 (Studienges. Kohle mbH), filed 1973
211. Roselius. W., Vitzthum, O., Hubert, P.: US-Patent 3 843 824
212. British Patent 2 014 425 (1973)
213. British Patent 1 336 276 (1973)
214. British Patent 1 346 134 (1974)
215. Zosel, K.: Germ. patent 27 32 103 (Studienges. Kohle mbH), filed 1977
216. US-Patent 4 247 570 (Studienges. Kohle mbH)
217. van der Stegen, G. H.: Germ. pat. appl. 26 38 383 (DEJ International Research Co, NL), filed 1976
218. Zosel, K.: US-Patent 4 348 422 (Studienges. Kohle mbH), filed 1978
219. Brunner, G., Peter, S.: Germ. pat. appl. 27 37 793 (1979)
220. Prasad, R. et al.: US-Patent 4 246 291 (General Foods Corporation), filed 1981
221. US-Patent 4 251 559 (Societé d'Assistance Technique pour Produits Nestlé S.A.) filed 1981
222. Katz, S. N., Monsey, N. Y., Gottesman, M.: US-Patent 4 276 315 (General Foods Corporation), filed 1981
223. Roselius, W., Kurzhals, H.-A., Hubert, P.: US-Patent 4 255 458 (HAG AG) filed 1981
224. Jasovsky, G. A., Gottesman, M.: US-Patent 4 255 461 (General Foods Corporation), filed 1981
225. Sirtl, W.: US-Patent 4 344 974 (1982)
226. Prasad, R., Scarella, R. A.: US-Patent 4 341 804 (General Foods Corporation) 1982
227. De Witt, H. G., van der Stegen, G. H.: US-Patent 4 364 965 (1982)
228. Stahl, E., Schilz, W.: Talanta *26*, 675 (1979)
229. Gährs, H. J., Marnette, E.: Verfahrenstechnik *17*, 554 (1983)
230. Brunner, G.: Ber. Bunsenges. Phys. Chem. *88*, 887 (1984)
231. Bichsel, B.: Thesis, Bern 1972
232. Paulaitis, M. E., Krukonis, V. J., Kurnik, R. T., Reid, R. C.: Rev. Chem. Eng. *1*, 179 (1983)
233. Vitzthum, O., Werkhoff, P.: J. Food Sci. *39*, 1210 (1974)
234. Vitzthum, O., Werkhoff, P.: Z. Lebensm. Unters. Forsch. *156*, 300 (1974)
235. Roselius, W., Vitzthum, O., Hubert, P.: German patent 21 42 205 (Studienges. Kohle mbH), filed 1971
236. Roselius, W., Vitzthum, O., Hubert, P.: British Patent 1 357 645 (Studienges. Kohle mbH), 1974
237. Hunter, E., Richards, R. B.: US-Patent 2 457 238 (1948)
238. Young, J. T., Doty, P.: J. Am. Chem. Soc. *79*, 761 (1957)
239. Ehrlich, P., Graham, E.: J. Polymer. Sci. *45*, 246 (1960)
240. Koningsveld, R., Diepen, G. A. M., Chermin, H. A. G.: Rec. Trav. Chim. *85*, 504 (1966)
241. Patel, D. J.: Macromolecules *3*, 448 (1970)
242. Bonner, D. C., Maloney, D. P., Prausnitz, J. M.: Ind. Engng. Chem. *13*, 91 (1974)
243. Fujishige, S.: J. Polymer. Sci. *10*, 256 (1972)
244. Schröder, E., Schimmel, K.-H., Arndt, K.-F., Pätz, R.: Plaste und Kautschuk *22*, 310 (1975)
245. Albihn, P., Hedman, K., Kubat, J.: J. Appl. Polym. Sci. *23*, 2829 (1979)
246. Copelin, H. B.: US-Patent 4 306 053 (1981)
247. Pilat, S., Godlewicz, M.: US-Patent 2 188 012 and 2 188 013 (Shell), 1936
248. Zhuze, T. P.: Petroleum *23*, 298 (1960)
249. Zosel, K.: GDR-Patent 41 362 (Studienges., Kohle mbH)
250. Gearhard, J. A., Garwin, L.: Hyd. Proc. *1976*, 125
251. Adams, R. M., Knebel, A. H., Rhodes, D. E.: Chem. Engng. Prog. *75* (6), 44 (1979)
252. Henry, D. J.: p. 207 in High Pressure Chemical Engineering, GVC-VDI, Düsseldorf, 1984

253. Anderson, H. R.: Proc. 16th Intersoc. Energy Conv. Eng. Conf., p. 1202, Am. Soc. Mech. Eng., New York, 1981
254. Hoffmann, R., Künstle, K., Brunner, G.: p. 155 in High Pressure Chemical Engineering, GVC-VDI, Düsseldorf, 1984
255. Martin, T. G., Williams, D. F.: UK-Patent 1 495 722 (1977)
256. Eisenbach, W. O., Göttsch, P. J., Sonnentag, R. J.: Chem. Ing. Tech. *55*, 655 (1983)
257. Monge, A., Prausnitz, J. M.: p. 159 in Chemical Engineering at Supercritical Conditions, eds. Paulaitis, M. E., Penninger, J. M. L., Gray, R. D., Davidson, P., Ann Arbor Science, 1983
258. Simon, R.: p. 173 ibid
259. Kuss, E.: p. 193 in High Pressure Chemical Engineering, GVC-VDI, Düsseldorf, 1984
260. Kuss, E.: High Temperatures-High Pressures *15*, 93 (1983)
261. Killesreiter, H.: Ber. Bunsenges. Phys. Chem. *88*, 838 (1984)
262. Holm, L. W.: JPT Jan., 76 (1976)
263. Matheny Jr., S. J.: Oil Gas J. March, 79 (1980)
264. Simlote, V. N., Withjack, E. M.: J. Pet. Tech. *1981*, 808
265. Doscher, T. M., El-Arabi, M.: Oil Gas J. Apr., 144 (1982)
266. Fremont, H. A.: US-Patent 4 308 200 (Champion International Corporation), 1981
267. Calimli, A., Olcay, A.: Sep. Sci. Technol. *17*, 183 (1982)
268. Kleinert, T. N.: Papier (Darmstadt) *30*, 18 (1976)
269. Köll, P., Metzger, J., Brönstrup, B.: in Recent Technologies in the Use of Peat, ed. Lüttig, G. W., E. Schweizerbart'sche Verlagsbuchhandlung, Stuttgart 1983
270. Köll, P., Steinweg, E., Meyer, B., Metzger, J.: Liebigs Ann. Chem. *1982*, 1039 and 1052
271. Metzger, J., Köll, P.: Angew. Chem. *91*, 74 and 75 (1979)
272. Scarrah, W. P.: p. 395 in Chemical Engineering at Supercritical Conditions, eds. Paulaitis, M. E., Penninger, J. M. L., Gray, R. D., Davidson, P., Ann Arbor Science, 1983
273. Olcay, A., Tugrul, T., Calimli, A., p. 409 ibid
274. Bartle, K. D., Jones, D. W., Pakdel, H.: Sep. Sci. Technol. *17*, 167 (1982)
275. Gangoli, N., Thodos, G.: Ind. Eng. Chem. Prod. Res. Dev. *16*, 208 (1977)
276. Whitehead, J. C., Williams, D. F.: J. Inst. Fuel *48*, 182 und 397 (1975)
277. Whitehead, J. C., Williams, D. F.: US-Patent 3 970 541 (1976)
278. Whitehead, J. C.: Tech. Energ. *43*, 61 (1980)
279. Maddocks, R. R., Gibson, J., Williams, D. F.: Chem. Eng. Prog. *75* (6), 49 (1979)
280. Fong, W. S., Chan, P. C. F., Pichaichanarong, P., Cororan, W. H., Lawson, D. D.: p. 377 in Chemical Engineering at Supercritical Conditions, eds. Paulaitis, M. E., Penninger, J. M. L., Gray, R. D., Davidson, P., Ann Arbor Science 1983
281. Gan, H., Nandi, S. P., Walker, P. L.: Fuel *51*, 272 (1972)
282. Kershaw, J. R., Jezko, J.: Sep. Sci. Technol. *17*, 151 (1982)
283. Vahrmann, M.: Fuel *49*, 5 (1970)
284. Kershaw, J. R.: S. Afr. J. Chem. *31*, 15 (1978)
285. Wilhelm, A., Hedden, K.: Erdöl und Kohle *36*, 269 (1983)
286. Wilhelm, A., Hedden, K.: p. 357 in Supercritical Fluid Technology, eds. Penninger, J. M. L., Radosz, M., McHugh, M. A., Krukonis, V. J., Elsevier, Amsterdam 1985
287. Penninger, J. M. L.: p. 323 in High Pressure Chemical Engineering, GVC-VDI, Düsseldorf, 1984
288. Ross, D., Nguyen, G.: Fluid Phase Equil. *10*, 319 (1983)
289. Bartle, K. D., Ladner, W. R., Martin, T. G., Snape, C. E., Williams, D. F.: Fuel *58*, 413 (1979)
290. Bartle, K. D., Calimli, A., Jones, D. W., Matthews, R. S., Olcay, A., Pakdel, H., Tugrul, T.: Fuel *58*, 423 (1979)
291. Kershaw, J. R., Barrass, G., Gray, D., Jezko, J.: Fuel *59*, 413 (1980)
292. Martin, T. G., Williams, D. F.: Phil. Trans. Roy. Soc. (London) *A 300*, 183 (1981)
293. Elgin, J. C., Weinstock, J. J.: J. Chem. Eng. Data *4*, 3 (1959)
294. Snedeker, R. A.: Ph. D. Thesis, Princeton University 1955
295. Kuk, M. S., Montagna, J. C.: Paper presented at the Annual Meeting AIChE, New Orleans, November 1981

296. Kuk, M. S., Montagna, J. C.: p. 101 in Chemical Engineering at Supercritical Conditions, eds. Paulaitis, M. E., Penninger, J. M. L., Gray, R. D., Davidson, P., Ann Arbor Science 1983
297. McHugh, M. A., Mallet, M. W., Kohn, J. P.: p. 113 ibid
298. Modell, M., de Filippi, R. P., Krukonis, V.: Div. Env. Chem., Am. Chem. Soc., Miami, September 1978
299. de Filippi, R. P., Krukonis, V., Modell, M.: Environmental Protection Agency Report, EPA – 600/2-80-054, March 1980
300. Kander, R. G., Paulaitis, M. E.: p. 461 in Chemical Engineering at Supercritical Conditions, eds. Paulaitis, M. E., Penninger, J. M. L., Gray, R. D., Davidson, P., Ann Arbor Science 1983
301. Fetzer, F. C., Graham, J. A., Arrendale, R. F., Klee, M. S., Rogers, L. B.: Sep. Sci. Technol. *16*, 97 (1981)
302. Rau, G.: Thesis, Saarbrücken 1985
303. Gjavotchanoff, S.: Z. Anal. Chem. *264*, 371 (1973)
304. Vitzthum, O., Hubert, P.: German patent 21 27 642 (Studienges. Kohle mbH), filed 1971

V. Non-Extractive Applications

V.1. Pest Control

The present use of gases, such as phosphine, methyl bromide and prussic acid for pest control is drastically limited by their poisonous character and the large residual amounts in the treated material. Ethylene oxide, used on a relatively large scale until recently, has likewise been forbidden for pest control on account of its carcinogenic metabolites, e.g. ethylene chlorohydrin [1]. However, many plant products are still subject to attack by pests. Suitable pest control, as far as possible not hostile to the environment, is therefore required for hygienic and economic reasons.

Stahl, Rau and Adolphi [2] based a procedure on the fact that organisms seldom survive a powerful pressure followed by rapid decompression. The experiments were carried out on the laboratory scale using the installation shown diagrammatically in Fig. 100.

The heart of this installation is a 14-litre pressure vessel B, capable of resisting pressures up to 80 bar. It is packed tightly with the drug sample in paper bags. The vessel is then largely evacuated with the help of a vacuum pump VP. The manometer M2 controls this. Valve 3 is then closed and gaseous CO_2 flows from the supply vessel G via valve V1 and the reducing valve RV into the autoclave B until the desired pressure is attained, as shown on manometer M1. After a pre-determined exposure time,

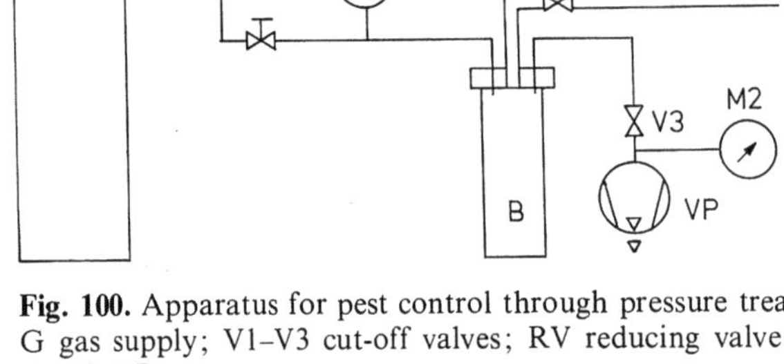

Fig. 100. Apparatus for pest control through pressure treatment (2).
G gas supply; V1–V3 cut-off valves; RV reducing valve; M1, M2 manometers; B pressure vessel; VP vacuum pump

decompression is carried out rapidly through the wide ball valve V2. The gas is led into the atmosphere through a flexible exhaust tube of large cross-section.

Numerous gases, not directly toxic, such as nitrogen, helium and carbon dioxide, were first tested and the relevant pressure ranges ascertained. Carbon dioxide, safe for the environment and of low cost, proved especially good. Its strong effect may be connected with its action as a respiratory analeptic. Through an increase in the partial pressure of the CO_2 in the haemolymph of the insects, intensified respiration is enforced, similar to the known effect with mammals.

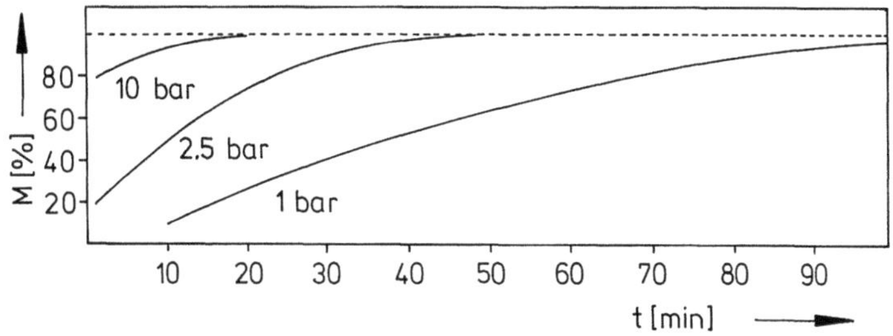

Fig. 101. Percent mortality of mature rice flour weevils at various pressures, as a function of the exposure time to carbon dioxide (2)

Fig. 101 shows the mortality percentage of mature rice flour weevils. It can be seen that increase of pressure is appreciably more effective than prolongation of the exposure time. The optimum conditions ascertained for complete destruction of the different insects and their stages of development are collected together in Table 45. These permit a sure pest control through carbon dioxide in a pressure range of 40 to 50 bar and with an exposure time of 10–20 min. As an alternative, treatment can be carried out several times at lower pressures. It would be an advantage to pack the material in an atmosphere of carbon dioxide to prevent any later contamination. Negative effects on the dry plant material thus treated have not been observed.

Table 45. Parameters for destruction by carbon dioxide of various insect pests and their stages of development

Insect	Development stage	Pressure (bar)	Time (min)
Red-brown rice flour weevil *Tribolium* sp.	beetle	10	20
Granary weevil (*Calandria granaria*)	beetle	20	10
Large rice flour weevil	larva	20	5
Yellow meal worm (*Tenebria molitor*)	larva	20	10
Tobacco beetle (*Lasioderma serricorna* F.)	larva	20	10
Tobacco beetle	eggs	40	10
Flour moth (*Ephestia kühniella*)	larva and eggs	40	5
Red-brown rice flour weevil *Tribolium* sp.	eggs	20	120
Red-brown rice flour weevil *Tribolium* sp.	eggs	30	60
Red-brown rice flour weevil *Tribolium* sp.	eggs	40	20
Red-brown rice flour weevil *Tribolium* sp.	eggs	50	10

V.2. Sterilisation

The presence of micro-organisms in medicinal plants and spices creates a special problem in the pharmaceutical domain and in foodstuff processing. Microbial contamination of these starting materials means not only a risk to health but in addition, economic disadvantages; too many micro-organisms lead to rapid decay of the product and it becomes no longer marketable. The difficulty is that the procedures hitherto used for sterilisation are only partly admissable. Thus treatment with ethylene oxide gas is permitted only for cold sterilisation of spices [1]; and treatment with γ-radiation (e.g. from ^{60}Co) is allowed in only some countries and by few organisations and then only for certain products [3]. Although this irradiation yields good results and any changes for the worse of the foodstuffs thus treated are scarcely detectable by analytical methods used today, this sort of decontamination is accepted only unwillingly by the consumer because the use of irradiation has to be declared. Other procedures of micro-organism removal include treatment with hot steam or hot alcohol [4] and mechanical effects, such as pressure extrusion [5, 6] and shock waves [7]; these have been described as alternatives but their many limitations have prevented their break-through in practice.

Stahl, Rau and Kaltwasser published the results of a study of the behaviour of various micro-organisms in gases under high pressure [8]. They evaluated the influence of dense carbon dioxide on the counts of bacteria, fungi and viruses, expressed as a function of the pressure.

It can be seen from Fig. 102 that increase in pressure leads to a reduction of the count, sometimes marked, referred to the control tests. The destruction in the pressure

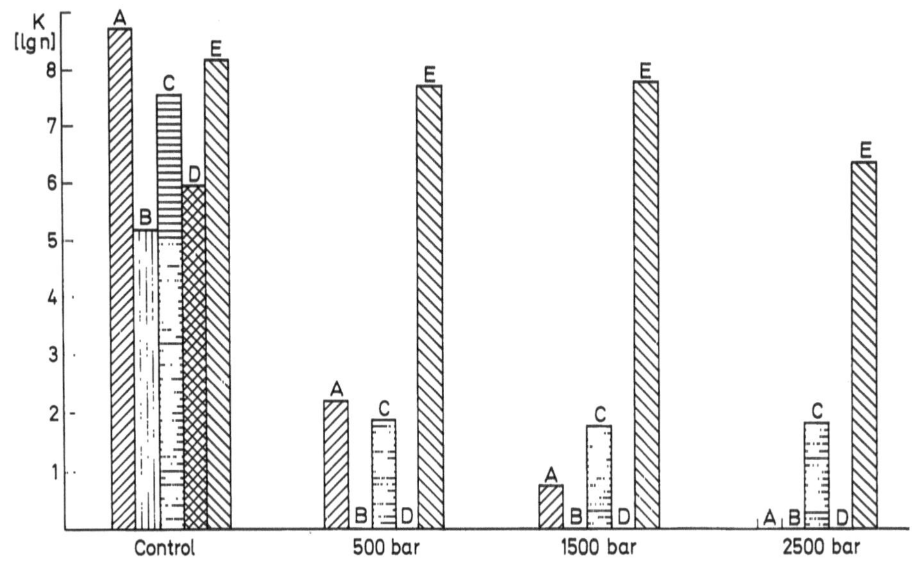

Fig 102. Influence of various pressures of carbon dioxide, acting for 20 min, on the count of bacteria, fungi and viruses.
A: *Escherichia coli*; B: *Pseudomonas aeruginosa*; C: *Bacillus subtilis*; D: *Candida albicans*; E: *Coliphage*; K: log count

region above 500 bar is especially large with bacteria, such as *Escherichia coli* (A) and *Pseudomonas aeruginosa* (B) and with the yeast *Candida albican* (D).

The influence of the CO_2-pressure on the spore-former, *Bacillus subtilis* (C), is somewhat less. The marked count reduction by over five powers of ten at 500 bar, is due to destruction of the vegetative life; further pressure increase to 2500 bar causes no additional decrease because the spores of *Bacillus subtilis* are not affected. Only a weak influence on the *Coliphages* (E) is observed in the pressure region of 2500 bar. The influence of sudden decompression after subjection to high pressures was also studied with *Escherichia coli*. This showed that fast decompression (<1 min) brought about greater inactivation than a fall in pressure over a longer period (ca. 10 min).

The reductions of count effected by pressure treatment with carbon dioxide are considerable; the D-values of 5–8 (D = log count reduction) are comparable with those accomplished by current procedures. The high pressure procedure has the disadvantage of high investment costs for the plant. Further, it must be remembered that dense carbon dioxide also dissolves lipophilic components. In some cases these may cause problems, in others, however, be an advantage when the sterilisation is combined directly with simultaneous extraction of lipophilic impurities from the starting material (e.g. pesticide removal from senna leaves, cf. Chapter IV.10) or when the extract and the carrier material are desired products (e.g. separation of essential oil from common wormwood, p. 145).

V.3. Inflation, Disruption and Comminution

The first procedures and appliances for "dilation of organic substances" by treatment with gases under pressure were described in a patent application in 1975 [9]. The use was at first limited to tobacco raw materials, with the aim of increasing their filling capacity, for example to reduce the amount of tobacco needed per cigarette [10, 11]. The use was later extended to employing CO_2 under pressure for the so-called cell cracking [12]. The effect of distension through applied pressure is combined in this method with a shock disintegration; this increases the effectiveness of the procedure and can lead in addition to a fast and thermally non-aggressive comminution of certain materials (spices) [13]. Equivalent procedures can be employed in pharmacy and biotechnology for disrupting difficultly attainable constituents [14]. Success has been had also this way in releasing more efficiently the constituents of tea plants, such as *Hibisci flos* [15].

Henry's law is the basic principle of the cell cracking procedure. According to this, the saturation activity of a gas in a liquid phase is proportional to the gas pressure over the solution. Diffusion phenomena through the cell membranes are also important. When plant material is subjected to a high gas pressure, the gas becomes more concentrated in the plant cells. This gas cannot escape quickly enough when the pressure is suddenly released. The resulting excess pressure is converted to mechanical energy which destroys the cell wall and in some cases may even lead to comminution of the material.

The amount of compressed gas in the cells and the related cracking effect when pressure is suddenly lowered depend on many factors, e.g. the nature of the plant material, the type of gas, the contact time, the pressure and the temperature.

The preference for carbon dioxide in cell cracking is based on two factors. First, carbon dioxide is neither expensive nor dangerous to handle; secondly, the solubility of carbon dioxide in lipophilic substances is an order of magnitude greater than that of, for example, air or nitrogen. The factor is even 100 in aqueous solutions.

It may be noted that 3.5 to 35 bar pressure within the cell is sufficient to burst a cell wall, provided that the medium, for example CO_2, is compressible and can expand when the excess pressure is released. The time required to introduce the relevant amount of gas into the interior of the cell of the plant material in the pressure vessel depends on the rate of diffusion. A model calculation for plant cells without fissures and pores showed that CO_2 needed about 8–9 min for a diffusion length of 1 cm [16]. Fissures and pores accelerate the penetration of the gas but if the openings are too large and the diffusion paths too short, the pressure equilibrium is established without difficulty on decompression and the excess pressure briefly needed cannot be achieved. A fast release of pressure is most important to prevent this situation as far as possible.

An additional factor is that the moisture content determines the surface structure of the material and hence the rate of diffusion of the gas. The moisture content influences at the same time the solubility of the gas in the cell plasma and the elasticity of the cells, which ought not to be too high. Solubility, rate of diffusion of the gas and the elasticity of the cell membranes are also affected by the temperature. All these factors influence the procedure in a complex way but if they are adjusted to one another in the best way, a pressure treatment bringing about distension, disintegration or comminution of the plant material can be carried out with success.

Fig. 103 shows the rate of liberation of the organic acids in an infusion of hibiscus flowers, with and without pressure treatment; it is a representative example of water-soluble active principles. CO_2 was the gas used, at 60 bar and for 20 min. Pressure was reduced to atmospheric in 15 sec, which was accompanied by a marked cooling of the drug material. The acids were determined potentiometrically on 2 g drug per 100 ml water. The values obtained, expressed as mg total acid and calculated as citric acid, are plotted in Fig. 103. The active principles were noticeably more rapidly liberated from the treated material, especially when the infusion times were short.

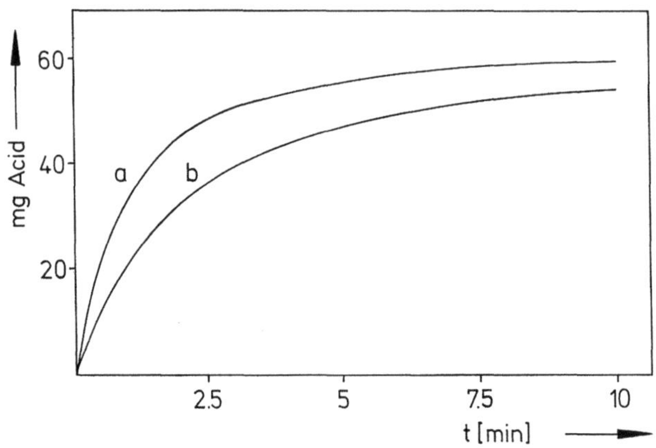

Fig. 103. Liberation of total acids from hibiscus tea as a function of the time of infusion with (a) and without (b) pressure treatment [15]

As the time of infusion was increased, the difference became less but the pressure-treated drug was better exploited.

The improvement of the imbibition value of linseed by pressure treatment is a second example. Linseed is used pharmaceutically for treating constipation on account of its good imbibition capacity. The mucus responsible for the swelling is situated in the seed husk; breaking up or gristing the seed improves the accessibility of water and yields increased swelling. However, linseed treated in this way has only a limited storage life because the polyunsaturated fats present in large amounts (42%) in the seeds become relatively rapidly rancid through atmospheric action.

An alternative to the gristing way of increasing the imbibition value, customary at present, is to burst the seeds by the action of carbon dioxide under pressure followed by rapid release of this pressure; a 50 to 60 bar pressure is enough to give good results [15]. It is to be noted, however, that when the cell membranes are destroyed, fat-splitting enzymes can be released which markedly increase the acid number of the fatty oil during storage of the treated linseed. This effect can be minimised by disrupting the membranes at a higher temperature, e.g. 80 °C, which inactivates the enzymes. The fatty oil is of high energy content, so that linseed from which the fat has been partly or wholly removed is also interesting as a diet product. It is therefore a good idea to extract the linseed fatty oil with dense carbon dioxide after the destruction of the membranes. The linseed then acquires a high imbibition value and good storage properties.

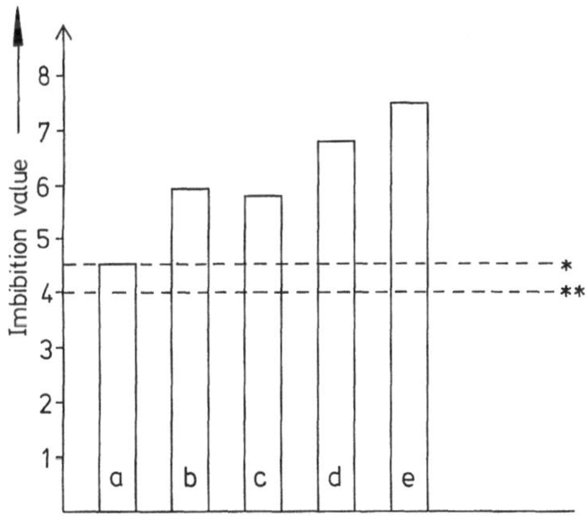

Fig. 104. Imbibition values of linseed (commercial product).
* minimum demand according to German Pharmacopoeia 8 (crushed); ** minimum demand according to German Pharmacopoeia 8 (unbroken).
a) unbroken, 42% fat; b) gristed, 42% fat; c) broken up with CO_2, 36% fat; d) broken up with CO_2, 20% fat; e) gristed, 3% fat

The imbibition values of linseed treated in various ways are compared in Fig. 104. The values of seed extracted with CO_2 (c) are quite comparable with those gristed (b); the values after partial (d) or complete (e) extraction of the fatty oil are clearly larger than those of the fat-containing materials.

V.4. High Pressure Micronising

Many crystalline lipophilic active substances have to be in a finely divided state for rapid and uniform liberation or solution. Micronised substances, especially those poorly soluble in water, are of great interest for example in pharmaceutical technological processing. The conventional micronising of crystalline compounds is carried out by crushing in grinding installations (air stream-, disc mills) or precipitating from classical solvents. Many naturally-occurring and synthetic substances are, however, sensitive to temperature increases which can occur during grinding. One disadvantage of precipitation is that undesired solvent residues may be co-precipitated.

High pressure micronising offers an alternative to these methods, yielding finely divided lipophilic substances without excessive temperatures and solvent contamination. This procedure profits from the phenomenon that solid matter dissolved in a dense gas precipitates in an extremely fine-grained form when the solvent capacity of the gas is suddenly reduced by decompression. A further advantage of the high pressure micronising is that the powder obtained has a very uniform particle size spectrum, so that no sifting is necessary, as in the grinding procedures. The solitary requirement of the substances worked on is that they are sufficiently soluble in the dense gas. Potential examples of application of the high pressure procedure are above all high grade or thermally labile natural products, fine chemicals, colouring and pharmaceutical substances which have to be made available in the finest and most uniform particle size possible.

Carotinoids, such as β-carotene, zeaxanthin, citranaxanthin, cryptoxanthin, canthaxanthin and some carotenals are of notable importance as dyes or colouring materials for foods, pharmaceutical and cosmetic preparations, and (as additives) animal feeds. Their employment is hindered by their insolubility in water and poor solubility in many of the usual solvents. Best et al. described a procedure for preparing very finely divided carotinoids using supercritical gases [17]. Carbon dioxide, ethylene and ethane were used; certain organic entrainers, such as ethanol or acetone, were added to increase solubility in the gas. The commercially available, crystalline carotinoids were dissolved in the mixture of supercritical gas and entrainer. This solution was dispersed in an aqueous colloidal matrix (e.g. gelatine-sugar mixture), the dispersion then decompressed and converted into a water-dispersable powder mixture, using a conventional technique, such as spray drying. Pressures of 80 to 400 bar and temperatures of 25 to 50 °C were used in the examples given. Dispersions were obtained of less than 1 μm crystal size.

An example from the realm of fine chemicals was given by Worthy [18]. The organo-metallic compound, ferrocene (dicyclopentadiene-iron), was dissolved in supercritical carbon dioxide at 276 bar and 55 °C and this solution expanded to atmospheric pressure. The grain size of the product separating from the dense gas was compared with that of the starting material (microscopic measurements) and found to be far less than the ca. 100 μm size of the latter.

Efforts are being made in the pharmaceutical field to increase the rate of solution and hence the resorption, of difficultly soluble medicaments by using these in as finely divided a form as possible. The examples below show that medicaments micronised by the high pressure procedure have very good properties and are superior to those micronised using classical methods.

Crystalline phenacetin (4-ethoxyacetanilide) was dissolved in supercritical carbon dioxide or trifluoromethane at pressures up to 600 bar and temperatures between 40 and 80 °C, and then micronised by releasing the pressure to 60 bar. The powder obtained in this way, and phenacetin powder prepared in a jet mill, were compared from pharmaceutical-technological points of view [19]. The powder samples had identical melting points and densities and also similar differential-calorimetric behaviours, so that polymorphic modifications could be excluded. Large differences were found, however, in particle size and shape and in the specific surface areas, obtained by permeability measurements. The specific surface areas of the phenacetin from the high pressure procedure were 23,800 cm^{-1} (from CO_2) and 34,100 cm^{-1} (from CHF_3), both values being markedly larger than the 14,500 cm^{-1} for the milled product. These considerable differences were due to differing particle shapes, as can be seen in Fig. 105. The crystals from the high pressure micronising procedure were skeletal crystals of microporous, three-dimensional net structure; they result from the high crystallisation speed imposed by the sudden decompression. The phenacetin crystals from the grinding procedure are, on the other hand, compact and almost spherical.

Despite the noteworthy difference in specific surface, the dissolving coefficients of the three micronised phenacetin samples, which gives a measure of the rate of release, are of the same order of magnitude (see Table 46).

The explanation for this is that phenacetin tends to agglomerate and is not easily wetted. This agglomeration is reduced and wetting improved by adding a hydrophilic solubilising agent. Thus the addition of 1% of Aerosil R 972 increases noticeably

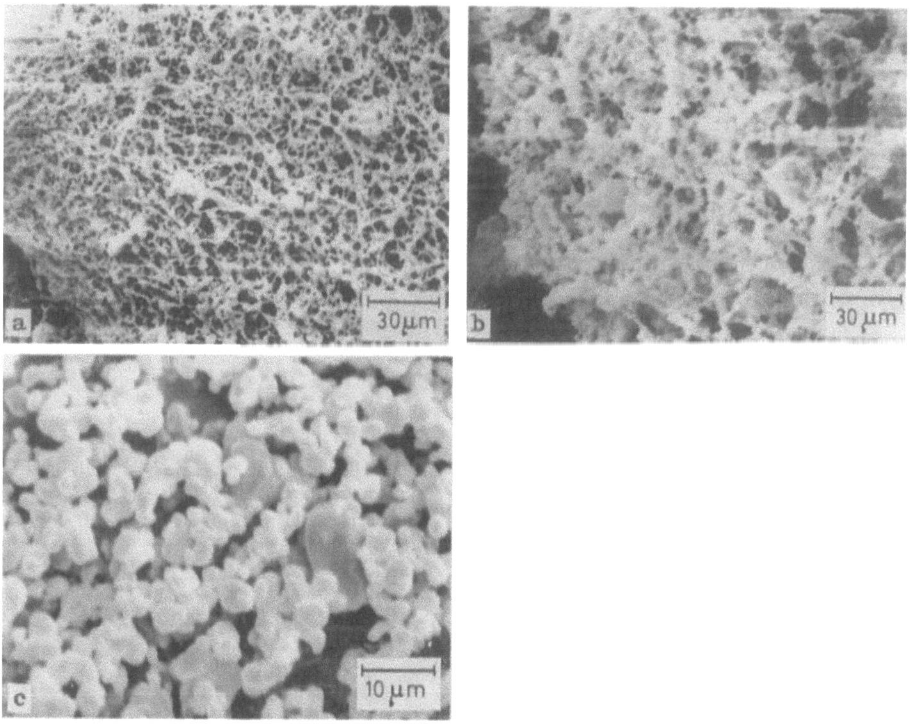

Fig. 105. Raster electron micrographs of high pressure micronised (a with CO_2; b with CHF_3) and ground (c) phenacetin

Table 46. Dissolving coefficients of micronised phenacetin preparations

Phenacetin preparation	Dissolving coefficient $k_w = \frac{1}{t'}$ (min^{-1}) of phenacetin micronised by		
	a jet mill	expansion of supercritical	
		CO_2	CHF_3
pure	0.089	0.080	0.091
mixed with:			
Aerosil R 972 (99 + 1)	0.27	0.31	0.45
mannitol (1 + 1)	2.2	2.1	2.8

t' = time in minutes in which 63.2% of the medicament dissolves

the dissolving coefficient, an effect which is much more marked with the phenacetin samples micronised by the high pressure procedure; the product from trifluoromethane dissolves fastest, as expected from its larger specific surface area.

Micronised nifedipin, a medicament for cardiac treatment, is another example from the pharmaceutical field [20]. High pressure micronising was carried out similarly to that with phenacetin, using supercritical carbon dioxide at 600 bar and 70 °C. A powder of very uniform particle size from 1 to 3 μm was obtained through decompression of the supercritical solution. An ordinary commercial, conventionally obtained (through wet grinding in a disc mill) nifedipin powder was studied for com-

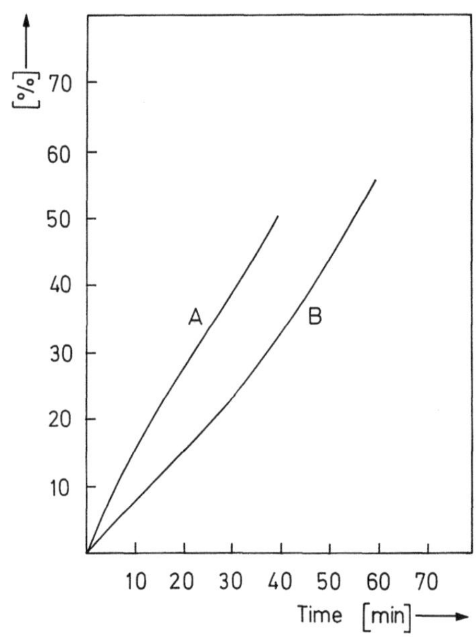

Fig. 106. Release of nifedipin tablets in pepsin-free artificial gastric juice.*
A micronised under high pressure; B micronised conventionally.
Nifedipin content of the tablets = 20%, tablet base of inert carrier, disintegrating rapidly (<5 sec)

* We thank Dr. A. Glatz, c/o the firm of Heumann Pharma GmbH, Nuremberg, GFR. For the measurements of the release of nifedipin tablets.

parison. The particle size of this powder was spread over a wider range, namely from 1 to 20 µm. 2% being still larger than 20 µm.

The nifedipin micronised in the high pressure procedure is released more rapidly than the commercial product, as a result of its smaller and more uniform particle size (Fig. 106). The kinetic dissolving data were obtained for the nifedipin tablets using the apparatus in the American Pharmacopoeia USP XXI (p. 1245). In this case also, the poor wettability of the nifedipin and its formation of agglomerates weaken the effect which would be expected as a result of the finer structure of the medicament micronised by the high pressure procedure. Addition of a suitable solubilising agent should augment the difference between the sample micronised in the classical way and that from the high pressure procedure.

Studies of the extraction of steroids with supercritical gases (see Chapter IV.4) revealed that these separated in various ways on pressure reduction. Compounds such as testosterone, androsterone, sitosterol and lanosterol, for example, precipitated always as powdery masses. Under the microscope these particles create the impression of being porous sponges of irregular structure. Gentle shearing force breaks them into very fine, practically amorphous particles of less than 1 µm size. Recrystallisation is possible, so that, if required, embedding in a film-forming colloidal solution can be carried out.

In contrast, other steroids separate always in microcrystalline form; examples are ethinyloestradiol, oestradiol, oestrone, oestriol, cortisone and tetrahydrocortisone. The crystal forms vary but the sizes are near together, in the 1 to 5 µm region. It may be mentioned in passing that the appearance of polymorphic modifications has been demonstrated [21]. The steroids were best micronised using supercritical ethane at 400 bar and 80 °C.

V.5. Fluid Chromatography

Chromatography using supercritical mobile phases has the same advantages as those of the preparative procedures of extraction with dense gases, described above. This analytical method, in accordance with the chromatographic separation principle, gives highly efficient separations. Common to both cases is the fact that supercritical gases have densities and dissolving capacities similar to those of liquids, but lower viscosities and better diffusion properties. Accordingly, dense gases used as mobile phases in chromatography should act both as substance carriers like the mobile phase in gas chromatography and also dissolve these substances like the solvents in liquid chromatography. This chromatographic variant is known as Supercritical Fluid Chromatography (SFC) and, increasingly, as Dense Gas Chromatography (DGC).

Klesper, Corwin and Turner are considered to be the discoverers of the method; they described in 1962 the separation of thermolabile porphyrin derivatives using supercritical chlorofluoromethanes at pressures up to 140 bar and temperatures from 150–170 °C [22]. The method was further developed both theoretically and experimentally later in the sixties. The publications of Giddings et al. [23–25] and of Sie et al. [26, 27] merit special mention. This development was, however, not comparable with the tempestuous growth of HPLC at about the same time. Only in recent years has SFC gained some significance as a routine procedure complementing GC and HPLC; this is shown by the large number of relevant publications.

Various reviewers have described the state of progress in particular periods, among which the following may be mentioned: Vigdergauz and Semkin [28]; Gouw and Jentoft [29–31]; Klesper [32]; Snyder and Saunders 33]; van Wasen, Swaid and Schneider [34]; Randall [35]; and Peaden and Lee [36].

Supercritical fluid chromatography can be classified as an analytical method in the domain between GC and HPLC. Its advantage over GC is that it enables also mixtures of high-molecular, poorly volatile substances and thermolabile compounds to be separated. Its advantage over HPLC is in the shorter times of separation, brought about by the higher linear flow rates of the mobile phase with a concurrent smaller pressure drop in the separating column. As seen in the newest development, the separation performance of SFC, as with GC, can be improved by using open tubular columns [37–39]. This is possible only to a limited extent in HPLC because the lower diffusion of the dissolved substances makes very thin capillaries necessary (e.g. of 10 µm internal diameter); as the mobile phase is of high viscosity, sufficiently long capillaries of this sort cannot be used. There are also problems of detection of such small amounts of substance [40].

The partition coefficient K, and hence the capacity ratio k and the retention time t_R, which determine the chromatographic behaviour, can be changed in SFC either by alteration of temperature, as in GC, or by altering the type of solvent and by adding a modifier (an entraining agent) as in HPLC [41, 42]. The pressure and the density of the mobile can also have a pronounced influence on the separation. Special separation effects and the analysis of heterogeneous samples can be accomplished accordingly in SFC through use of gradients of temperature, pressure or entrainer, alone or in combination. Fig. 107 shows the separation of Triton X-100, an alkylphenyl-polyethyleneglycol fraction (Rohm and Haas); this was done A) on a capillary column, using CO_2 as mobile phase, at 110 °C and a pressure gradient from 185 to 340 bar; or B) on an HPLC-column, using CO_2/methanol as mobile phase, at 60 °C, 307 bar and an entrainer gradient of 1–5 % methanol. A UV-detector was used for A), a flame ionisation detector for B). This example illustrates the manifold possibilities of SFC for achieving a particular separation.

The demands of apparatus for SFC are comparable with those for HPLC. Commercial HPLC equipments can be adapted to SFC with little trouble. An SFC-apparatus has been marketed by the firm of Hewlett-Packard since 1983/4. Ordinary HPLC columns of 4–5 mm internal diameter and 10–20 cm length are used as separation columns. Fine-grained (10–30 µm) adsorption agents, molecular sieves or bound phases come into consideration as stationary phases. Depending on the phase, a distinction can be made between fluid-solid and fluid-liquid chromatography. Capillary columns for SFC of 0.05 and 0.1 mm inner diameter and 10 or 20 m length with a fixed stationary phase of different film thicknesses (0.05–0.4 µm) are now available commercially. Carbon dioxide and dinitrogen monoxide, if desired with an entrainer, are the usual mobile phases considered, especially for thermolabile substances; they possess favourable critical data and additional advantages discussed already above. To separate substances of high molecular weight (over 2000) stable organic solvents, liquid under normal conditions, can also be used at appropriately high temperatures. The experience and solubility data gathered during extraction with dense gases can give useful information about choice of mobile phase and conditions of separation. In SFC, UV- [43, 44], FID-detectors [45, 46] and differential

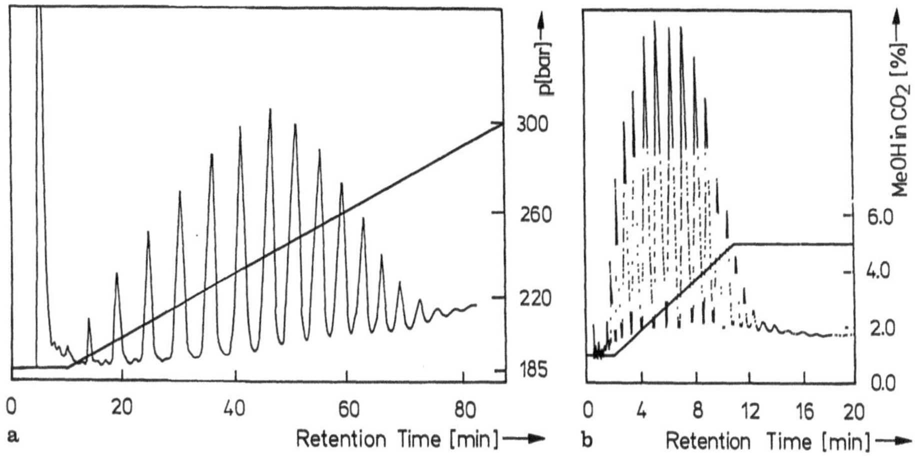

Fig. 107. SFC-separation of Triton-X-100 (Rohm and Haas).
a) Column: DB-5 (J + W Scientific), film thickness 0.25 μm, 100 μm × 10 m; mobile phase: CO_2, 110 °C, 185 bar (10 min); programmed rise of 1.5 bar/min to 340 bar; FID detector; (from a publication of the ict-Handelsgesellschaft m.b.H.)
b) Column: 10 μm PRP-1.15 cm × 4.1 mm; mobile phase: CO_2 with 1–5% methanol, 60 °C, 307 bar; UV detector at 225 nm (from a publication by Hewlett-Packard)

refractometers [47] can be considered for detection. Scintillation [43] and heat conductivity detectors [48] and also, recently, coupling with mass spectrometers [39, 49, 50] have been described, yielding sensitive and selective detection.

In the meantime, SFC has found a niche in routine analysis for solving the most varied separation problems [51]. The versatility of the method is demonstrated by the separations described in the literature. Successful analysis has been possible of, for example, alkanes [45, 46], polyaromatics [30, 45, 52, 53], siloxanes [41], plasticisers [54] and car exhaust gases [43]. Another field of application has been of petroleum products [31, 33, 55, 56], especially high-boiling fractions and distillation residues. The efficiency of SFC has also been demonstrated for separating polymers and oligomers, notably of polystyrene [32, 42, 57] and of substances of biochemical interest [25].

However, supercritical fluid chromatography is interesting not only for analytical purposes but also to obtain physico-chemical and thermodynamical values, for example, virial coefficients [26, 28, 58, 59], partition coefficients, partial molar volumes [60], activity coefficients, excess values, adsorption constants and diffusion coefficients [44, 61–64]; these values can in turn furnish valuable information in connection with the development, calculation and optimisation of extraction procedures with dense gases. This fundamental research is still at its beginnings, however, and extensive systematic investigations are still required.

References

1. Gerhard, U., Ladd Effio, J. C.: Fleischwirtsch. *62*, 1 (1982)
2. Stahl, E., Rau, G., Adolphi, H.: Pharm. Ind. *47*, 528 (1985)
3. Zehnder, H. J.: Alimenta *22*, 65 (1983)

4. Neumayr, L.: Thesis, Kulmbach (1983)
5. Gremli, H., Mor, J. R.: Europ. pat. appl. 0012813 (L. Givaudan & Cie Soc. An). filed 1979
6. Baynsik, M. J., Chen, P. H.: US Patent 4210678 (The R. T. French Co); 1980
7. Printed by Firm Extraktor AG, Bad Homburg
8. Stahl, E., Rau, G., Kaltwasser, H.: Naturwissenschaften *72*, 144 (1985)
9. Germ. pat. appl. 2503636 (Aicro. Inc. Montvale, N. J., USA) filed 1975
10. Germ. pat. appl. 2143388 (Philip Morris Inc., New York, USA), filed 1971
11. Germ. pat. appl. 2709651 (R. J. Reynolds Tobacco Co., Winston-Salem, N. C. USA), filed 1977
12. Fiedler, J., Hack, K.: Verfahren zur Zerkleinerung pflanzlicher Rohstoffe, Forschungsbericht 03 VM 197, Bundesministerium für Forschung und Technologie, 1981
13. Soma-Gewürze GmbH, Köln: Neue Fleischer Zeitung *31*, 19 (1985)
14. Loewus, M. W., Loewus, F.: Cell Disruption Bombs, printed by Parr Instrument Company, Moline, Ill., USA
15. Stahl, E., Rau, G., Carius, W.: Z. Lebens. Unters. Forsch., *182*, 33 (1986)
16. Kohlensäurewerk Deutschland GmbH, Bad Hönningen, CO_2-Technische Tabellen (1981)
17. Best, W., Müller, F. J., Schmieder, K., Frank, R., Paust, J.: Germ. pat. appl. 2943267 (BASF AG); filed 1979
18. Worthy, W.: Chem. Eng. News *59* (31), 16 (1981)
19. Loth, H., Hemgesberg, E.: Int. J. Pharmaceutics, *32*, 265 (1986)
20. Gerard, D., Quirin, K.-W.: Private communication of FLAVEX Naturextrakte GmbH
21. Glatz, A.: Thesis, Saarbrücken 1985
22. Klesper, E., Corwin, A. H., Turner, D. A.: J. Org. Chem. *27*, 700 (1962)
23. Myers, M. N., Giddings, J. C.: Sep. Sci. *1*, 761 (1966)
24. McLaren, L., Myers, M. N., Giddings, J. C.: Science *159*, 197 (1968)
25. Giddings, J. C., Myers, M. N., King, J. W.: J. Chromatogr. Sci. *7*, 276 (1969)
26. Sie, S. T., van Beersum, W., Rijnders, G. W. A.: Sep. Sci. *1*, 459 (1966)
27. Sie, S. T., Rijnders, G. W. A.: Sep. Sci. *2*, 699 and 729 and 755 (1967)
28. Vigdergauz, M. S., Semkin, V. I.: Russ. Chem. Rev. *40*, 533 (1971)
29. Gouw, T. H., Jentoft, R. E.: J. Chromatogr. *68*, 303 (1972)
30. Gouw, T. H., Jentoft, R. E.: Adv. Chromatogr. *13*, 1 (1975)
31. Gouw, T. H., Jentoft, R. E.: Chromatogr. Sci. *11*, 313 (1979)
32. Klesper, E.: Angew. Chem. *90*, 785 (1978)
33. Snyder, R. L., Saunders, D. L.: Chromatogr. Sci. *11*, 215 (1979)
34. van Wasen, U., Swaid, I., Schneider, G. M.: Angew. Chem. *92*, 585 (1980)
35. Randall, L. G.: Sep. Sci. Technol. *17*. 1 (1982)
36. Peaden, P., Lee, M.: J. Liq. Chromatogr. *5*, 179 (1982)
37. Novotny, M., Springston, S. R., Peaden, P. A., Fjeldsted, J. C., Lee, M. L.: Anal Chem. *53*, 407 A (1981)
38. Peaden, P. A. et al.: Anal. Chem. *54*, 1090 (1982)
39. Fjeldsted, J. C., Kong, R. C., Lee, M. L.: J. Chromatogr. *279*, 339 (1983)
40. Knox, J. H., Gilbert, M. T.: J. Chromatogr. *186*, 405 (1979)
41. Niemann, J. A., Rogers, L. B.: Sep. Sci. *10*, 517 (1975)
42. Klesper, E., Hartmann, W.: J. Polym. Sci. Polym. Lett. Ed. *15*, 707 (1977)
43. Jentoft, R. E., Gouw, T. H.: Anal. Chem. *48*, 2195 (1976)
44. Swaid, I., Schneider, G. M.: Ber. Bunsenges. Phys. Chem. *83*, 969 (1979)
45. Bartmann, D.: Ber. Bunsenges. Phys. Chem. *76*, 336 (1972)
46. Ecknig, W., Polster, H. J.: Chem. Tech. (Leipzig) *31*, 89 (1979)
47. Asche, W.: Chromatographia *11*, 411 (1978)
48. Bartmann, D.: Thesis, Bochum 1972
49. Randall, L. G., Wahrhaftig, A. L.: Anal. Chem. *50*, 1703 (1978)
50. Smith, R. D., Felix, W. D., Fjeldsted, J. C., Lee, M. L.: Anal. Chem. *54*, 1883 (1982)
51. Rawdon, M. G., Norris, T. A.: Int. Lab. *1984* (6), 12
52. Simonian, B. P., Rogers, L. B.: J. Chromatogr. Sci. *16*, 49 (1978)
53. Schmitz, F. P., Leyendecker, D., Klesper, E.: Ber. Bunsenges. Phys. Chem. *88*, 912 (1984)
54. Sie, S. T., Rijnders, G. W. A.: Anal. Chim. Acta *38*, 31 (1967)

55. Altgelt, K. H.: Chromatogr. Sci. *11*, 287 (1979)
56. Amos, R.: Chromatogr. Sci. *11*, 329 (1979)
57. Klesper, E., Hartmann, W.: Eur. Polym. J. *14*, 77 (1978)
58. Spertell, R. B., Chang, G. T.: J. Chromatogr. Sci. *10*, 60 (1972)
59. Ecknig, W., Polster, H. J.: Chem. Tech. (Leipzig) *31*, 245 (1979)
60. van Wasen, U., Schneider, G. M.: J. Phys. Chem. *84*, 229 (1980)
61. Balenovic, Z., Myers, M. N., Giddings, J. C.: J. Chem. Phys. *52*, 915 (1970)
62. van Wasen, U., Swaid, I., Schneider, G. M.: Ber. Bunsenges. Phys. Chem. *83*, 1130 (1979)
63. Feist, R., Schneider, G. M.: Sep. Sci. Technol. *17*, 261 (1982)
64. Wilsch, A., Feist, R., Schneider, G. M.: Fluid Phase Equil. *10*, 299 (1983)

Subject Index

MIX
Papier aus verantwortungsvollen Quellen
Paper from responsible sources
FSC® C105338

If you have any concerns about our products,
you can contact us on
ProductSafety@springernature.com

In case Publisher is established outside the EU,
the EU authorized representative is:
Springer Nature Customer Service Center GmbH
Europaplatz 3, 69115 Heidelberg, Germany

Printed by Libri Plureos GmbH
in Hamburg, Germany